SELF-ASSEMBLED SUPRAMOLECULAR ARCHITECTURES

Wiley Series on
Surface and Interfacial Chemistry

Series Editors:
Ponisseril Somasundaran
Nissim Garti

SELF-ASSEMBLED SUPRAMOLECULAR ARCHITECTURES

LYOTROPIC LIQUID CRYSTALS

Edited by

Nissim Garti
Ponisseril Somasundaran
Raffaele Mezzenga

A JOHN WILEY & SONS, INC., PUBLICATION

Published by John Wiley & Sons, Inc., Hoboken, New Jersey
Published simultaneously in Canada

For general information on our other products and services or for technical support, please contact our Customer Care Department within the United States at (800) 762-2974, outside the United States at (317) 572-3993 or fax (317) 572-4002.

Wiley also publishes its books in a variety of electronic formats. Some content that appears in print may not be available in electronic formats. For more information about Wiley products, visit our web site at www.wiley.com.

Library of Congress Cataloging-in-Publication Data:

Self-assembled supramolecular architectures : lyotropic liquid crystals / edited by Nissim Garti, Ponisseril Somasundaran, [and] Raffaele Mezzenga.
       p.   cm.
  Includes index.
  ISBN 978-0-470-28175-8 (cloth)
  1. Supramolecular chemistry.   2. Liquid crystals.   I. Garti, Nissim, 1945–
II. Somasundaran, P., 1939–   III. Mezzenga, Raffaele.
  QD878.S45   2012
  530.4'29–dc23

                                                    2012011090

Printed in the United States of America

10 9 8 7 6 5 4 3 2 1

■ CONTENTS

Lyotropic liquid crystals make up one of the most outstanding examples of complex fluids based on components as simple as water, lipids, surfactants, platelets, and rodlike particles. These also are some rare cases in which complex fluids are spontaneously found at the thermodynamic equilibrium. What is then the driving force behind this unique self-organization? And how can the specific features of these resulting materials be exploited in the various disciplines of applied and fundamental sciences? This is a question of primary importance since the relevance of lyotropic liquid crystals spans today fields ranging from biology to nanotechnology, from cosmetics to food technologies. Despite their apparent simplicity, these materials remain challenging systems that need to be understood at the molecular level. Water, which is the most frequently used solvent in lyotropic liquid crystals, adds up dramatically to the complexity of the systems, as the hydrogen bonds among water molecules and the resulting clusters of water molecules, already provide this solvent most of the features common to structured fluids. When polar lipids are mixed with water, the selectivity in the partition coefficient of water with the hydrophilic and hydrophobic parts of the lipids leads to structured fluids of complex topologies and intricate architectures organized in the few nanometer-length scales. These heterogeneities and their organization determine to a very great extent the final properties of lyotropic liquid crystals and can then be exploited for encapsulation, templating, diffusion, release, reactions, or confinement in a multitude of possible applications.

Recently, researchers are devoting much of their efforts to better understand how to manipulate the lyotropic liquid crystals and to be able to increase the lattice parameter values, to elongate and to enlarge the aqueous channels in order to use these systems as solubilization reservoirs for food additives, supplements, bioactives, and drugs.

The modified lyotropic liquid crystals (LLCs) show very significant potential to serve as microreactors, template systems for interfacial crystallization, and more.

It is our believe that the recent progress made in the last decade will bring us to a new area in which LLCs will be used in the pharmaceutical, cosmetic, and food industries as major carriers of bioactives.

This book tackles the topic of LLCs from both the fundamental and applied perspectives. The first part of the book discusses our current fundamental understanding of lyotropic liquid crystals, with emphasis on lipid-based systems and rodlike solvent dispersions, whereas the second part of the book tackles the applications landscape, with special emphasis on drug delivery applications.

In Chapter 1, Mezzenga describes the physics of self-assembly of lyotropic liquid crystals from a thermodynamic point of view, introducing recent concepts such as self-consistent field theory, to understand the structural and physical properties of these systems; the chapter also touches on the current understanding of transient states associated with order–order transitions.

Herrera and Rey, in Chapter 2, present a comprehensive review of the rheology and properties of nematic phases, including calamitic and discotic micellar solutions and wormlike micelles. Their review examines verifiable rheological liquid crystal models for lyotropic nematics highlighting the mechanisms that control orientation behavior under shear, anisotropic viscoelasticity, and non-Newtonian behavior.

In Chapter 3, Kolev, Aserin, and Garti focus on the topological properties of the columnar hexagonal phase and are critically looking at—and summarizing—the available theoretical and mathematical aspects and the most fundamental aspects of the columnar reverse hexagonal mesophase.

Chapter 4 by Hartley and Shen provides a comprehensive review of the available experimental characterization methods of lyotropic liquid crystalline systems, spanning small-angle neutrons and X-ray scattering, nuclear magnetic resonance, positron annihilation lifetime spectroscopy, electron and atomic force microscopy, and neutron reflectivity.

In Chapter 5, Yaghmur and Glatter discuss the phase diagram, emulsification procedures, structural and physical properties of LLCs confined in small nanoparticles dispersed in water. They focus on monoglyceride-based systems and also touch on potential applications of these formulations in the area of food colloids and dispersions.

Kulkarni and Glatter, in Chapter 6, review the hierarchical organization of lyotropic liquid crystals from the lipid length scale up to the macroscopic organization of emulsified liquid crystalline systems, with special emphasis on their end-use applications in dispersed systems in the form of oil-in-water emulsions, high internal phase emulsions, and gels.

Chapter 7, by Amar-Yuli, Aserin, and Garti, is a review of the current strategies available for the use of LLCs as templates for the synthesis and alignment of nanostructured materials. The chapter reviews carefully the templating effects in both inorganic and organic materials throughout the main classes of available lyotropic mesophases.

Libster, Aserin, and Garti, in Chapter 8, open the discussion on the use of LLCs as drug delivery vehicles. The different intrinsic diffusion of guest molecules as a function of the mesophase structure are discussed in details, and different means available to tune this diffusion are discussed, considering both external stimuli and "doping" strategies, such as the use of membrane piercing agents.

Chapter 9, by Boyd and Fong, is a discussion of the different types of external stimuli to induce a release of drugs on demand. Recent developments in the field are carefully reviewed and external stimuli such as temperature, pH, and light are discussed in details.

Chang and Nylander (Chapter 10) discuss the problem of dealing with nonlamellar lyotropic liquid crystalline nanoparticles, mostly hexasomes and cubosomes, at interfaces and surfaces. The chapter provides a state of the art on the techniques available to detect the adsorption and structures of these nanoparticles at interfaces, their formation, and implications for biological interfaces and drug delivery.

Chapter 11, by Géral and colleagues, concludes the book by providing a practical direct example of the relevance of lyotropic liquid crystalline nanoparticles, by discussing their use as nanocarriers for targeting of cells expressing brain receptors. The chapter screens self-assembled lipid systems suitable for the encapsulation of neurotrophic peptides and demonstrate their potential in targeting neuronal therapetic applications.

To summarize, the present book gathers an impressive set of contributions by leading scientists in the area, overarching the entire framework in which lyotropic liquid crystals are today investigated worldwide; this is done by moving from a very fundamental perspective into the possible practical applications area of these materials. Such a contribution was missing from available literature, and it is in the scope of this book to timely fill this gap. It is to be hoped that this outstanding set of reports will not only serve scientists working in the field but will also become an inspiring source for younger scientist generations and also serve students in their education process in the area of nanosctructured self-associating soft materials.

RAFFAELE MEZZENGA
NISSIM GARTI
PONISSERIL SOMASUNDARAN

CONTRIBUTORS

Idit Amar-Yuli, Casali Institute of Applied Chemistry, The Institute of Chemistry, The Hebrew University of Jerusalem, Jerusalem Israel

Borislav Angelov, Institute of Macromolecular Chemistry, Academy of Sciences of the Czech Republic, Prague, Czech Republic

Angelina Angelova, CNRS, University of Paris Sud, Châtenay-Malabry, France

Abraham Aserin, Casali Institute of Applied Chemistry, The Institute of Chemistry, The Hebrew University of Jerusalem, Jerusalem Israel

Ben J. Boyd, Drug Delivery, Disposition and Dynamics, Monash Institute of Pharmaceutical Sciences, Monash University, Parkville, Victoria, Australia

Debby P. Chang, Physical Chemistry, Lund University, Lund, Sweden

Wye-Khay Fong, Drug Delivery, Disposition and Dynamics, Monash Institute of Pharmaceutical Sciences, Monash University, Parkville, Victoria, Australia

Nissim Garti, Casali Institute of Applied Chemistry, The Institute of Chemistry, The Hebrew University of Jerusalem, Jerusalem Israel

Claire Géral, CNRS, University of Paris Sud, Châtenay-Malabry, France

Otto Glatter, Department of Chemistry, University of Graz, Graz, Austria

Patrick G. Hartley, CSIRO Materials Science and Engineering, Clayton, Victoria, Australia

E. E. Herrera-Valencia, Department of Chemical Engineering, McGill University, Montréal, Quebec, Canada

Vesselin Kolev, Casali Institute of Applied Chemistry, The Institute of Chemistry, The Hebrew University of Jerusalem, Jerusalem Israel; Department of Chemical Engineering, Faculty of Chemistry, Sofia University, Sofia, Bulgaria

Chandrashekhar V. Kulkarni, Department of Chemistry, University of Graz, Graz, Austria

Sylviane Lesieur, CNRS, University of Paris Sud, Châtenay-Malabry, France

Dima Libster, Casali Institute of Applied Chemistry, The Institute of Chemistry, The Hebrew University of Jerusalem, Jerusalem, Israel

Raffaele Mezzenga, ETH Zurich, Food & Soft Materials Science, Institute of Food, Nutrition & Health, Zürich, Switzerland

Valérie Nicolas, Cell Imaging Platform, University of Paris Sud, Châtenay-Malabry, France

Tommy Nylander, Physical Chemistry, Lund University, Lund, Sweden

Alejandro D. Rey, Department of Chemical Engineering, McGill University, Montréal, Quebec, Canada

Hsin-Hui Shen, CSIRO Materials Science and Engineering, Clayton, Victoria, Australia

Anan Yaghmur, Department of Chemistry, University of Graz, Graz, Austria; Department of Pharmacy, School of Pharmaceutical Sciences, Faculty of Health and Medical Sciences, University of Copenhagen, Copenhagen, Denmark

■■■■■ CHAPTER 1

Physics of Self-Assembly of Lyotropic Liquid Crystals

RAFFAELE MEZZENGA

ETH Zurich, Food & Soft Materials Science, Institute of Food, Nutrition & Health, Zürich, Switzerland

Abstract

We review recent advances in the understanding of self-assembly principles in lipid-based lyotropic lipid crystals, from the original rationalization achieved using the critical packing parameter up to recent, more sophisticated thermodynamics approaches, such as the self-consistent field theory, which can be efficiently used to minimize the total free energy of a lipid–water system and identify stable mesophases. We highlight the importance of reversible hydrogen bonding as one of the key parameters ruling the self-assembly in these systems and examine the implications this may have also in real applications. We finally discuss the current understanding on the dynamics of phase transitions and review the status of the art on current atomistic approaches to investigate the relaxation dynamics in these systems.

Self-Assembled Supramolecular Architectures: Lyotropic Liquid Crystals, First Edition.
Edited by Nissim Garti, Ponisseril Somasundaran, and Raffaele Mezzenga.
© 2012 John Wiley & Sons, Inc. Published 2012 by John Wiley & Sons, Inc.

1.1 INTRODUCTION

The term *lyotropic liquid crystals* generally refers to systems in which any form of liquid crystallinity is induced or affected by the presence of a solvent. The simplest form of lyotropic liquid crystal is the nematic phase, which is given by the nonisotropic orientation of rigid particles or molecules, called mesogens. In what follows we shall focus on lyotropic liquid crystals of superior order and, in particular, in those observed in surfactants (surface-active agents) in the presence of water, with special emphasis on lipid–water systems. Disregarding other forms of aggregation of lipids in water to consider only those characterized by a periodic order, lipids most frequently found to form lyotropic liquid crystals are neutral lipids such as monoglycerides or phospholipids (Krog, 1990). These two systems have in common one or two fatty acid tails and a polar head with total neutral charge (in the case of phospholipids most often the positive and negative charges are present in stoichiometric ratio; one talks then about zwitterionic lipids). These structured fluids are periodically organized on the nanometer length scale and have been known for half a century since the pioneering work of Luzzati and Husson (1962). One particular feature that is breathing new life into these systems is the fact that, when redispersed in excess water, some specific liquid crystalline structures can be maintained. This makes them particularly suitable for food, pharmaceutics, and cosmetic applications (Fong et al., 2009; Mezzenga et al., 2005a; Mohammady et al., 2009; Yaghmur and Glatter, 2009). Additionally, in their bulk form, they have been used as nanoreactors to run and control regioselective reactions (Garti et al., 2005; Vauthey et al., 2000) due to the environment constituted by a very large lipid–water interfacial area. The most frequently found structures in lipid–water lyotropic liquid crystals are the isotropic fluid (L_{II}), the lamellar phases (L_α or L_C depending on whether the alkyl tail is amorphous or has crystallized), inverted columnar hexagonal cylinders (H_{II}), and bicontinuous double gyroid (Ia3d), double diamond (Pn3m), and primitive (Im3m) cubic phases (de Campo et al., 2004; Mezzenga et al., 2005b; Qiu and Caffrey, 2000). The presence of charges on the lipid–water interface tends to disrupt the interface, and, thus, lyotropic liquid crystals, with a great nonzero total charge, are rarely found. Nonetheless, cationic and anionic surfactants have been used as doping agents for these systems, to engineer other phases or to displace boundaries in the phase diagram (Borne et al., 2001). Finally, a new class of lyotropic liquid crystals has recently emerged, that is, the macromolecular amphiphilic systems: the most common, Pluronics, is formed by a diblock copolymer with one hydrophilic (generally polyethyleneoxide, PEO) and one hydrophobic (generally polypropyleneoxide, PPO) blocks, so that in the presence of water a strong partitioning effect leading to self-assembly and microphase segregation of the two blocks is found.

Figure 1.1 shows a typical example of a phase diagram for a commercial form of monoglyceride, monolinolein, in the presence of water. Remarkably, several transitions are found within a few degrees of temperature or small

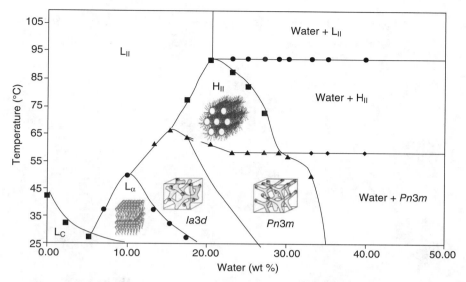

Figure 1.1 Phase diagram of a monolinolein–water system, with a schematic sketch of the different liquid crystalline phases found in the various regions of the phase diagram. [Reproduced with permission from Mezzenga et al. (2005a).]

percentage changes in the composition. Below, we will briefly review the state of the art on the driving forces for the order–order transitions among different mesophases and the order–disorder transitions of a mesophase into the isotropic micellar fluid.

1.2 CRITICAL PACKING PARAMETER

Without any doubt, the simplest, the easiest, and the most diffuse approach to the rationalization of the different structures found in the lyotropic liquid crystals is based on the concept of the critical packing parameter (CPP), a criterion developed originally by Israelachvili and colleagues (Israelachvili, 1991; Israelachvili et al., 1976). The CPP is a geometrical value consisting of the ratio between the volume of the hydrophobic lipid tail, v, and the product of the cross-sectional lipid head area, A, and the lipid chain length, l. Following the changes of the CCP, one can approximately predict order–order transitions associated with the change in the curvature of the water–lipid interface. Figure 1.2 gives a schematic overview of the correspondence between various mesophases and their corresponding CCP.

The CCP predicts essentially two classes of morphologies. For a value of CCP $v/(Al) < 1$ "oil-in-water" morphologies are expected, which correspond to the so-called direct liquid crystalline phases in which the polar heads are

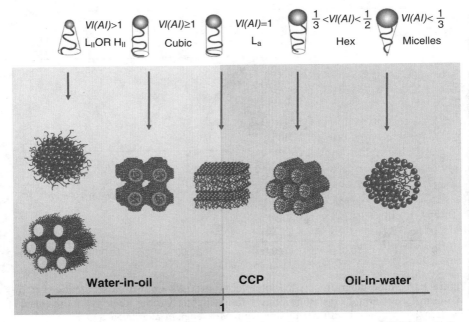

Figure 1.2 Relationship between the critical packing parameter (CCP) and the expected morphology in lyotropic liquid crystals.

forming a convex interface against water. On the other hand, for $v/(Al) > 1$, a phase inversion occurs and "water-in-oil" morphologies are found, with concave lipid head surfaces against the water. The flat interfaces, corresponding to the L_α lamellar phase are found for $v/(Al) = 1$. More in details, inverted micelles/inverted hexagonal, inverted cubic phases, lamellar, hexagonal phases, and direct micelles are expected when the CPP has a value of $v/(Al) > 1$, $v/(Al) \geq 1$, $v/(Al) \approx 1$; $\frac{1}{3} < v/(Al) < \frac{1}{2}$, and $v/(Al) < \frac{1}{3}$, respectively (Israelachvili et al., 1991; Jonsson et al., 2001).

The CPP has been widely employed to predict and rationalize differences observed among liquid crystalline phases and can capture some of the physical changes. For example, changes occurring on CCP with temperature and composition can be understood to some extent. Increases in temperature leads to partial breaking of hydrogen bonds, with the number of water molecules hydrating the polar heads of the lipids, and this leads to an increase of the CPP because it decreases the effective head area. This can well explain the cubic-to-hexagonal transition, for example (Qiu and Caffrey, 2000). Similar arguments can be used to explain the L_α to cubic transition induced by temperature raises. However, other transitional changes, such as the concentration-induced lamellar $\rightarrow$ cubic transition remains unexplained by the application of the CCP concepts. The concept of the CPP is limited to a qualitative interpretation of the phase diagrams and cannot be used to bring insight into the structural complexity of these mesophases, nor on the physical mechanisms regulating their self-assembly behavior.

1.3 ANALOGIES AND DIFFERENCES BETWEEN LIPIDS AND BLOCK COPOLYMERS

Alternative approaches can be taken in order to develop a physical under-standing of lyotropic liquid crystals. A good starting point is to draw analogies and differences with another important organic self-assembling system, that is, block copolymers.

Block copolymers bear many similarities to surfactants. As for lipids, they are also formed by two blocks attached into the same amphiphilic molecule. If the two blocks differ enough in chemical composition, they will induce a microphase segregation into nanostructured morphologies spanning a few nanometers in periodicity (Bates and Fredrickson, 1999). The types of struc-tures found are also very similar: lamellar, columnar hexagonal, and bicontinu-ous cubic (Ia3d) phases are present in both classes of materials. The Pn3m bicontinuous cubic phase, which is missing for a pure diblock copolymer phase diagram, can be recovered when additional components are added. Indeed, when comparing the morphologies from the two classes of materials, one must realize that since the water–lipid system is a two-component system, for direct comparison with block copolymers one should take the case of a block copo-lymer plus a second component compatible with one of the two constitutive blocks of the copolymer. This second component, be it a solvent or a homo-polymer, has the role to mimic water in the lipidic mesophases. Figure 1.3 shows the phase diagram of a hypothetical diblock copolymer, with a graphic of the morphologies found.

Beside this encouraging start, some important differences should be men-tioned. Let us take a classical diblock copolymer system: Onset of microphase separation occurs when the segregation power, χN, where χ is the Flory–Huggins parameter and N the polymerization degree of the entire block copolymer, is of the order of ≈ 10 or more (Leibler, 1980). By adding the third block mimicking the role of water, small perturbations on this threshold cri-terion can be expected, without, however, changing it significantly. Because typical polymer pairs typically have an unfavorable χ of ≈ 0.1 or less at room temperature, this indicates that in order to observe microphase separation at standard temperatures the polymerization degree must be 100 or more.

In the case of lipids, the scenario is very different. The Flory–Huggins parameter of the polar versus unipolar species can be estimated by measur-ing the activity of alkanes partitioning in water at room temperature, which indicates values of the order of 3 for χ (Mezzenga et al., 2005a). This represents such a high unfavorable mixing enthalpy that microphase segregation between the polar head (plus water) and the hydrophobic tail occurs at much lower polymerization degrees.

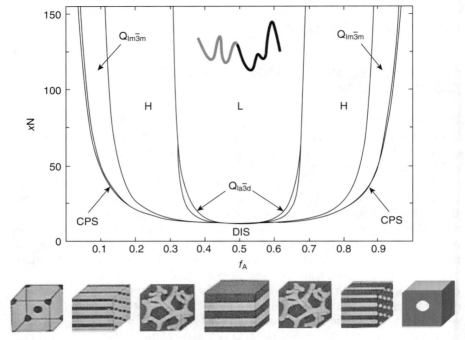

Figure 1.3 Theoretical phase diagram for a diblock copolymer system [Readapted with permission from Matsen and Bates (1996). Copyright 1996, American Chemical Society.]

This explains why microphase segregation in lipid–water mixtures is nearly always observed. Therefore, the loss of entropy in packing together the different lipid molecules is always small compared to the enthalpic gain driving microphase segregation, and self-assembly is driven nearly entirely by enthalpy gain. These concepts are summarized in Figure 1.4.

Another important difference between block copolymers and lipids is the sequence of phases found, for example, on a isothermal, concentration-titrating experiment. Matsen has formulated a theoretical phase diagram for an A–B block copolymer plus an A homopolymer pair (Matsen, 1995a,b), which is exactly the system needed to draw direct comparisons with the lipid–water system of interest.

Consider the arrow in Figure 1.5 in which the phase diagram of such a system, an A–B diblock copolymer, is reported as a function of the block copolymer volume fraction of block A, f, versus the volume fraction ϕ of the additional homopolymer A. Upon adding the homopolymer A, the system goes through spheres, hexagonal, gyroid, and lamellar, which is also referred to as a "normal" sequence since the curvature is progressively decreasing as a consequence of the increasing overall A fraction. The analog experiment

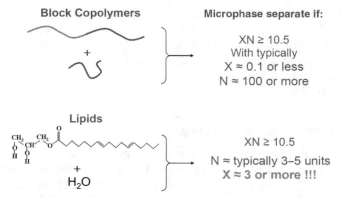

Figure 1.4 Differences in the enthalpic driving force to microphase separation in lipids and block copolymers.

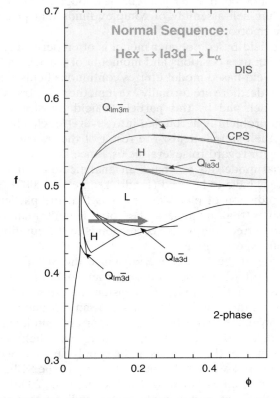

Figure 1.5 Theoretical phase diagram of an A–B + B diblock copolymer–homopolymer blend as a function of the homopolymer volume fraction ϕ and block asymmetry f (low f is the monomer fraction of species A in diblock A). [Redrawn with permission from Matsen (1995a). Copyright 1995, American Chemical Society.]

carried out in the lipid–water system, however, would lead to a transition from lamellar to Ia3d, and thus, with an increasing curvature (see Fig. 1.1). This sequence, also refered to as "anomalous," is a signature of complexity of the lipid–water system and illustrates well why attempts at modeling the phase diagram in these complex fluids has been very unsuccessful compared to other systems such as block copolymers.

1.3.1 Self-Consistent Field Theory

Self-consistent field theory (SCFT) is a coarse-grained field-theoretic model solved within the mean-field approximation (De Gennes 1969; Edwards, 1965). This method has been successfully applied to many different examples of polymer systems (Fredrickson et al., 2002; Matsen and Bates, 1996; Matsen and Schick, 1994). In the case of diblock copolymers, the agreement found between theory (Matsen and Bates, 1994) and experiments (Khandpur et al., 1995) is remarkable. This shows that SCFT is a reliable enough framework for describing the equilibrium self-assembly of complex fluids that possess structured lyotropic and thermotropic phases.

A mesoscopic field-based polymer model is often derived from a particle-based model, which uses positions and momenta of particles as the degree of freedom while a field-based model utilizes continuous fields. Once a particle-based model is made, there are formally exact methods for transforming it into a field-based model, and by this particle-to-field transformation, a many-interacting-chain problem is decoupled into several single-chain problems in the presence of the external fields. The result of the transformation is functional integrals with regard to several fields. Since these functional integrals for any nontrivial model do not have an analytic solution in closed form, numerical methods have to be used (Fredrickson et al., 2002). The main idea behind SCFT prediction of phase diagrams is that the particular profile of segmental densities that minimizes the total Hamiltonian of the system expressing the total free energy also corresponds to the equilibrium structure that should occur experimentally.

For the purpose of illustration, we introduce the SCFT formalism in the context of a blend of AB diblock copolymer and A homopolymer using the continuous Gaussian chain model. This system has some common features with lipid–water mixtures since a lipid has the same structure and role as an AB diblock copolymer, notwithstanding the shorter chain length, presence of unsaturated bonds, and rigid head group. Similarly, an A homopolymer can be viewed as a "solvent" for the A blocks of the copolymer, analogous to the role that water plays in lipid–water systems. A summary of the SCFT equations for this system is given in an earlier work (Mezzenga et al., 2006).

The results of SCFT on such a model system indicate that all the phases— lamellar, hexagonal, and bicontinuous—can be recovered by the theory and found to correspond to energy minima (e.g., equilibrium morphologies). Unfortunately, however, their locations in the phase diagram do not correspond to the real sequences of phases, and neither the thermotropic nor the lyotropic behavior is interpreted meaningfully. This indicates that, while SCFT is a robust method that should capture the structural complexity of this phases, the physical model interpreting lipid–water as a block copolymer homopolymer pair is much too simplistic to generate quantitative or even semiquantitative agreement with experiments.

1.3.2 The SCFT of Water and Hydrogen-Bonded Lipids

The technical challenge of implementing SCFT for the latter types of systems does not lie in either the theoretical framework or the numerical methods but rather in accurate parameterization of the more complicated *interactions* and *short-scale molecular details* present in lipid–water mixtures. Specific complicating factors in these systems are the presence of hydrogen bonding among hydrophilic heads and water molecules, unsaturated bonds in the hydrocarbon tails, and short tail length. Muller and Schick (1998) studied the equilibrium self-assembly of a simple model of glycerolmonoolein in water using SCFT techniques. The rotational isomeric state (RIS) model for representing a tail structure was applied, and the head of glycerolmonoolein was treated as a rigid rod. To obtain the single lipid chain partition function at fixed field Monte Carlo simulations were applied. This stochastic method is computationally very expensive (in comparison with the deterministic method of computing partition functions described above), which made determining phase boundaries difficult. Although the RIS model of Muller and Schick described the lipid tail conformations quite realistically, including the presence of unsaturated bonds, their results found limited agreement with the experimental phase diagram (Muller and Schick, 1998; Qiu and Caffrey, 2000). It seems likely that the neglect of specific interactions in the model related to head-group hydration and hydrogen bonding is responsible for this discrepancy. Nevertheless, the Muller–Schick work is an impressive first step toward extending SCFT methods to food-grade systems.

While keeping the Gaussian chain approximation for the lipid tail, a step toward a successful SCFT description of the self-assembly process in lipid–water mixtures has been the recent work of Lee and co-workers, who have been the first to introduce the concept of reversible hydrogen bonds in the energetic description of the lipid–water system (Lee et al., 2007, 2008). In that work, a simplistic scenario is considered: The lipid and water can possibly occupy two states: lipid and water unbound (state OFF) and lipid and water bound (state ON). The two states differ in the sense that the volume of the polar head, its enthalpy with the surrounding water (hydrophobic effect), and that with the alkyl tail, all change depending on which of the two states is occupied. The simultaneous statistical occurrence of ON and OFF states is simply weighted by a Boltzmann term, with energy corresponding to the energetic gain of one bound molecule of water. This physical approach has the merit to describe in a very realistic way the self-assembly process but has also

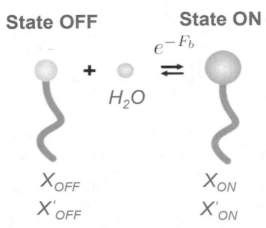

Figure 1.6   Schematics of the fundamental concepts used in the H-bonding SCFT treatment. [Reproduced with permission from Lee et al. (2007).]

the drawback to introduce a number of unknowns, which complicates the numerical simulations (e.g., the binging energy F_b, the two head-to-tail Flory–Huggins parameters χ_{ON} and χ_{OFF}, and the two head-to-solvent χ'_{ON} and χ'_{OFF}). Nonetheless, this approach clearly allows an unprecedented physical description of the self-assembly process. Figure 1.6 summarizes the main physical ingredients used for the hydrogen-bonding SCFT treatment.

An insight into the model results show that some remarkable predictions can be made by implementing hydrogen bonding into SCFT. Figure 1.7a shows the predicted phase diagram and compares it to the experimental one established by Qiu and Caffrey (2000) shown in Figure 1.7b. An effective temperature is used in the phase diagram, expressed by the inverse of the energy gain for hydrogen bonding $(-1/F_b)$: When the temperature increases, hydrogen bonds are progressively broken, with a progressive reduction of the energy gain (e.g., $-1/F_b$ increases).

The anomalous transition sequence—isotropic fluid $\rightarrow$ Ia3d $\rightarrow$ L$_\alpha$—is now correctly predicted, which represents a breakthrough in the lyotropic behavior interpretation. The theory is even capable of predicting thermodynamic coexistence regions such as the L$_\alpha$ + Ia3d. Additionally, the thermotropic sequence, L$_\alpha$ $\rightarrow$ Ia3d $\rightarrow$ hexagonal $\rightarrow$ isotropic fluid, is also well predicted. The exact location of the phase boundaries depends on the amount of complexity added to the model. For example, implementing multiple water molecule–polar head binding sites for hydrogen bonds and handling the multiple hydrogen bonds in a procedure similar to that discussed by the groups of Tanaka and Pincus (Bekiranov et al., 1997; Matsyama and Tanaka, 1990) is anticipated to lead to a finer agreement between the experiments and the theory. Therefore, SCFT implementing reversible H bonds clearly settles a new direction for the physical description of phase diagrams in lipid–water systems.

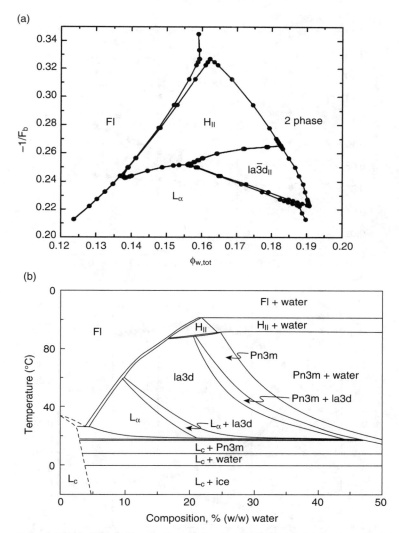

Figure 1.7 (a) Predicted phase diagram using SCFT implementing H bonding. (b) Experimental phase diagram for monoolein–water mixtures. [Reproduced with permission from Lee et al. (2007).]

1.3.3 Validation of SCFT Predictions

Being a statistical thermodynamic model, the approach described above also gives access to a number of unique features of the system. For example, it can predict the fraction of water molecules bound to the lipid polar head and those that are free: Both classes of molecules can be predicted in a statistical way as a function of temperature and composition. Figures 1.8a and 1.8b show the

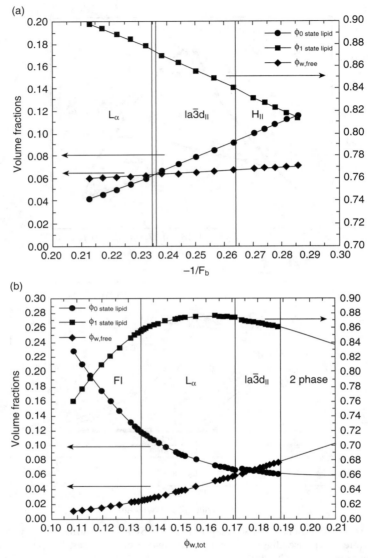

Figure 1.8 Prediction of the volume fraction of free water, H-bonded lipid, and H-bond free lipids, as a function of (a) temperature and (b) composition. [Reproduced with permission from Lee et al. (2007).]

predictions on free water, H-bonded lipid, and H-bond free lipids, as a function of temperature and composition, respectively.

As expected intuitively, the amount of free water increases progressively with increasing temperature or total water content. Particularly interesting is the case of dependence of free water on the total amount of added water (Fig. 1.8b): More than 50% of the water is bound at any composition. These trends have been already assessed experimentally by Garti and colleagues by measuring the subzero (°C) water behavior in nonionic surfactants via differential scanning calorimetry (Garti et al., 1996). These authors found that free water can only be measured at high water volume fractions, but that at typical compositions used for microemulsions and lyotropic liquid crystals a high amount of water (>50%) is typically bound to the lipid polar heads.

These findings have not only fundamental relevance but might also assist the design of lyotropic liquid crystalline phases for real applications. For example, by varying the amount of free water contained within the water channels, one can also tune the water activity, and ultimately also the state of molecules encapsulated within the hydrophilic regions of the mesophases. One clear example would be the control of activity of enzymes encapsulated within the mesophases by varying their activity via the amount of free and bound water.

Another very successful achievement of SCFT implementing H-bond features has been the prediction of the exact size of the water channels in the inverted lyotropic liquid crystals, as a function of physical parameters such as concentration and temperature. Figure 1.9 shows the density profiles predicted by SCFT for the hexagonal, Ia3d, and Pn3m phases (Lee et al., 2008). The water channels are compared directly to the size (black circles) predicted by either simple geometrical consideration, such as in the case of the hexagonal, or by triply periodical minimal surfaces (TPMS).

Again, this has a direct practical relevance in view of lyotropic liquid crystals as encapsulating agents, as it is know that the capacity of these mesophases to encapsulate macromolecular compounds depends on the ratio between the drug diameter and the size of the hosting water channels (Mezzenga et al., 2005c).

1.4 DYNAMICS OF ORDER–ORDER TRANSITIONS AND ATOMISTIC SIMULATIONS

All the approaches discussed above only describe the status of lipids and water at the final thermodynamic equilibrium. Of particular interest, however, is also the dynamic of structural transitions from an ordered state to another ordered one, since this has direct relevance also in the field of biology in which lipid membrane fusion has a vital role in phenomena such as viral infections, cell adhesion, penetration, and the like. This is a complex phenomenon that involves several intermediate steps occurring within very short times and, as

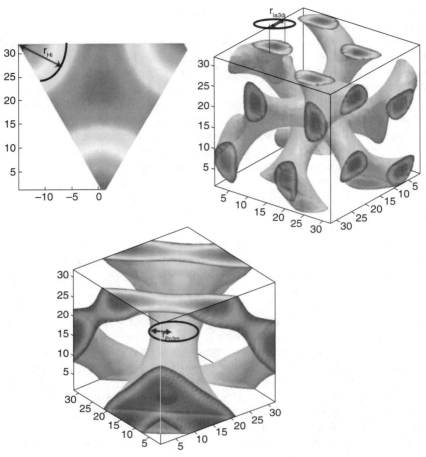

Figure 1.9 SCFT predictions of density profiles (colored) in (a) H_{II}, (b) Ia3d, and (c) Pn3m and their matching to size predicted by geometrical considerations (black circles). [Reproduced with permission from Lee et al. (2008).]

such, is difficult to assess experimentally. As a consequence, most of the work carried out in this field is either theoretical or numerical. Pioneering work from Siegel and co-workers based on cryogenic transmission electron microscopy (Siegel and Epand, 1997; Siegel et al., 1994) has demonstrated the presence of transient structures observed during transit through order–order transitions. Furthermore, in recent times, an experimental description of intermediate steps involving long-living intermediate structures when moving from a lyotropic liquid-crystalline phase to another one has been possible via synchrotron X-ray diffraction (Angelov et al., 2009). Very recently, work from

Mulet et al. (2009) has again demonstrated the presence of intermediate transient states by using both X-ray diffraction and cryogenic transmission electron microscopy. These works have settled important milestones toward the experimental assessment of transient structures accompanying order–order transitions.

A suitable theoretical description of transient states accompanying order–order transitions such as the evolution of lamellar phases into inverted hexagonal or inverted cubic phases, has been given by Siegel in the so-called modified stalk theory (Siegel, 1999). In this approach, the stability of intermediate states is tested by calculating their corresponding curvature elastic free energy [following the method proposed by Helfrich (1973)] and by inferring that the preferred structural evolution pathway is that which maintains the energy at its lower limit.

The Helfrich free energy consists of two terms: a term accounting for the bending displacement from an equilibrium curvature c_0, and another term associated with the bending following an imposed Gaussian curvature:

$$F = \int \left[\frac{1}{2} k_c \left(\frac{1}{R_1} + \frac{1}{R_2} - 2c_0 \right)^2 + k_g \frac{1}{R_1 R_2} \right] dS \tag{1.1}$$

where R_1 and R_2 are the curvatures of the membrane and k_c and k_g are elastic constants referred to as bending and saddle-splay moduli, respectively. The term $((1/R_1) + (1/R_2))$ is called the mean curvature, whereas the term $1/(R_1 R_2)$ represents the Gaussian curvature.

Siegel was able to determine the energy of intermediate states referred to as stalks, transmonolayer contacts (TMCs), and interlamellar attachments (ILAs). The various steps of the structural evolutions predicted by the modified stalk theory are shown in Figure 1.10. In short, two lipid bilayers are predicted to touch each other (e.g., by fluctuation-induced contacts) and lead to the formation of a first critical step, which is a stalk. A stalk can grow in size but at the expense of energy; a second critical step then that allows reducing energy is the transformation of a stalk into a TMC. The evolution of TMCs into pores follows two different energetic scenarios: In the case of lamellar → cubic transition, the third critical structural change involves the transformation of a TMC into an ILA, which can then grow spontaneously in size. In the case of lamellar → hexagonal transition, the mechanism involves the attraction of different TMCs, which then finally evolve into a hexagonal columnar lattice.

The modified stalk theory has been successful in explaining a number of experimental evidences. Nonetheless, considering the short time scales involved in the structural changes (seconds or less) simulations are obviously to be called up for help.

With the increasing computational power and speed, in recent years, some specific features of lipid-based lyotropic liquid crystals have been correctly captured by coarse-grained molecular dynamic simulations. One of the first

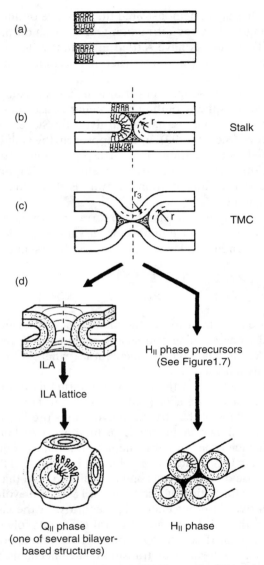

Figure 1.10   Main steps of the modified Stalk theory describing intermediate steps in order–order transitions in lyotropic liquid crystals. [Reproduced with permission from Siegel (1999).]

works in this respect is that of Marrink and Tieleman (2001), which has successfully modeled the exact structure of a Pn3m monoolein-inverted cubic phase by considering both lipid and water molecules in an explicit manner. Then, by imposing periodic boundary conditions, infinitely large, bulk systems could be predicted. Other contributions have followed on the atomistic simulations and dynamics of lipid-based lyotropic liquid crystals (Horta et al., 2010; Marrink and Tieleman, 2002), and it is to be expected that increasing computational speed will further trigger these types of simulation in the near future. Clearly, this approach has the power to identify and locate each individual molecule and to follow the dynamic of each of those at very short time scales (nanoseconds). The main limitation, however, is that presently only the lyotropic liquid crystal group spaces with the smallest lattice parameters such as the columnar hexagonal or cubic Pn3m can be studied, and systems such as the bicontinuous Ia3d cubic or the micellar Fd3m cubic still constitute a challenge.

1.5 OUTLOOK AND CONCLUSIONS

Although self-assembly of lyotropic liquid crystals has been known since the early 1960s, these systems are experiencing new interest today. Only in recent years has the complexity of the thermodynamics and dynamics of self-assembly in these intriguing systems been fully realized, which has catalyzed new studies and developments. Recent progress in the field has been carried out in all directions: New applications have flourished, new methods have been developed and assessed in order to shed light on the structure of these systems, and when experiments have proved challenging to be carried out, simulations have filled some of the gaps left opened. Last, but not least, new theoretical approaches have been put in place that have elucidated some of the unexplained features of these unique systems. It is to be hoped and expected that this continuous learning progress will continue and increase our understanding of these systems, remarkable for their compositional simplicity, but challenging from the point view of self-assembly complexity.

REFERENCES

Angelov, B., Angelova, A., Vainio, U., Garamus, V. M., Lesieur, S., Willumeit, R., and Couvreur, P. (2009). Long-living intermediates during a lamellar to a diamond-cubic lipid phase transition: A small-angle X-ray scattering investigation. *Langmuir*, *25*, 3734–3742.

Bates, F. S., and Fredrickson, G. H. (1999). Block copolymers—Designer soft materials. *Physics Today*, *52*, 32–38.

Bekiranov, S., Bruinsma, R., and Pincus, P. (1997). Solution behavior of polyethylene oxide in water as a function of temperature and pressure. *Physical Review E*, *55*, 577–585.

Borne, J., Nylander, T., and Khan, A. (2001). Phase behavior and aggregate formation for the aqueous monoolein system mixed with sodium oleate and oleic acid. *Langmuir*, *17*, 7742–7751.

de Campo, L., Yaghmur, A., Sagalowicz, L., Leser, M. E., Watzke, H., and Glatter, O. (2004). Reversible phase transitions in emulsified nanostructured lipid systems. *Langmuir*, *20*, 5254–5261.

De Gennes, P. G. (1969). Some conformation problems for long macromolecules. *Reports on Progress in Physics*, *32*, 187–206.

Edwards, S. F. (1965). Statistical mechanics of polymers with excluded volume. *Proceedings of the Physical Society*, *85*, 613–624.

Fredrickson, G. H., Ganesan, V., and Drolet, F. (2002). Field-theoretic computer simulation methods for polymers and complex fluids. *Macromolecules*, *35*, 16–39.

Fong, W. K., Hanley, T., and Boyd, B. J. (2009). Stimuli responsive liquid crystals provide "on-demand" drug delivery in vitro and in vivo. *Journal of Control Release*, *135*, 218–226.

Garti, N., Aserin, A., Ezrahi, S., Tiunova, I., and Berkovic, G. (1996). Water behavior in nonionic surfactant systems. 1. Subzero temperature behavior of water in nonionic microemulsions studied by DSC. *Journal of Colloid and Interface Science*, *178*, 60–68.

Garti, N., Spernath, A., Aserin, A., and Lutz, R. (2005). Nano-sized self-assemblies of nonionic surfactants as solubilization reservoirs and microreactors for food systems. *Soft Matter*, *1*, 206–218.

Helfrich, W. (1973). Elastic properties of lipid bilayers—theory and possible experiments. *Journal of Biosciences*, *C28*, 693–703.

Horta, B. A. C., de Vries, A. H., and Hunenberger, P. H. (2010). Simulating the transition between gel and liquid-crystal phases of lipid bilayers: Dependence of the transition temperature on the hydration level. *Journal of Chemical Theory and Computation*, *6*, 2488–2500.

Israelachvili, J. N. (1991). *Intermolecular and Surfaces Forces*, 2nd ed., Academic, New York.

Israelachvili, J. N., Mitchell, D. J., and Ninham, B. W. (1976). Theory of self-assembly of hydrocarbon amphiphiles into micelles and bilayers. *Journal of the Chemical Society Farady Transactions*, *72*, 1525–1568.

Jonsson, B., Lindman, B., Holmberg, K., and Kronberg, B. (2001). *Surfactants and Polymers in Aqueous Solutions*, Wiley, Chichester, England.

Khandpur, A. K., Forster, S., Bates, F. S., Hamley, I. W., Ryan, A. J., Bras, W., Almdal, K., and Mortensen, K. (1995). Polyisoprene-polystyrene diblock copolymer phase diagram near the order-disorder transition. *Macromolecules*, *28*, 8796–8806.

Krog, N. J. (1990). Food emulsifiers and their chemical and physical properties, In S. E. Friberg and K. Larsson (Eds.), *Food Emulsions*, Marcel Dekker, New York, p. 141.

Lee, W. B., Mezzenga, R., and Fredrickson, G. H. (2007). Anomalous phase sequences in lyotropic liquid crystals. *Physical Review Letters*, *99*, 187901–187904.

Lee, W. B., Mezzenga, R., and Fredrickson, G. H. (2008). Self-consistent field theory for lipid-based liquid crystals: Hydrogen bonding effect. *Journal of Chemical Physics*, *128*, 74504.

Leibler, L. (1980). Theory of microphase sepration in block copolymers. *Macromolecules*, *13*, 1602–1617.

Luzzati, V., and Husson, F. (1962). Structure of liquid crystalline phase of lipid-water systems. *Journal of Cell Biology*, *12*, 207–219.

Marrink, S. J., and Tieleman, D. P. (2001). Molecular dynamics simulation of a lipid diamond cubic phase. *Journal of American Chemical Society*, *123*, 12383–12391.

Marrink, S. J., and Tieleman, D. P. (2002). Molecular dynamics simulation of spontaneous membrane fusion during a cubic-hexagonal phase transition. *Biophysical Journal*, *83*, 2386–2392.

Matsen, M. W. (1995a). Phase behaviour of block copolymer homopolymer blends. *Macromolecules*, *28*, 5765–5773.

Matsen, M. W. (1995b). Stabilizing new morphologies by blending homopolymer with block copolymer. *Physical Review Letters*, *74*, 4225–4228.

Matsen, M. W., and Bates, F. S. (1996). Unifying weak- and strong-segregation block copolymer theories. *Macromolecules*, *29*, 1091–1098.

Matsen, M. W., and Schick, M. (1994). Stable and unstable phases of a diblock copolymer melt. *Physical Review Letters*, *72*, 2660–2663.

Matsuyama, A., and Tanaka, F. (1990). Theory of solvation-induced reentrant phase separation in polymer solutions. *Physical Review Letters*, *65*, 341–344.

Mezzenga, R., Schurtenberger, P., Burbidge, A., and Michel, M. (2005a). Understanding foods as soft materials. *Nature Materials*, *4*, 729–740.

Mezzenga, R., Meyer, C., Servais, C., Romoscanu, A. I., Sagalowicz, L., and Hayward, R. C. (2005b). Shear rheology of lyotropic liquid crystals: A case study. *Langmuir*, *21*, 3322–3333.

Mezzenga, R., Grigorov, M., Zhang, Z., Servais, C., Sagalowicz, L., Romoscanu, A. I., Khanna, V., and Meyer, C. (2005c). Polysaccharide-induced order-to-order transitions in lyotropic liquid crystals. *Langmuir*, *21*, 6165–6169.

Mezzenga, R., Lee, W. B., and Fredrickson, G. H. (2006). Design of liquid-crystalline foods via field theoretic simulations. *Trends in Food Science Technology*, *17*, 220–226.

Mohammady, S. Z., Pouzot, M., and Mezzenga, R. (2009). Oleoylethanolamide-based lyotropic liquid crystals as vehicles for delivery of amino acids in aqueous environment. *Biophysical Journal*, *96*, 1537–1546.

Mulet, X., Gong, X., Waddington, L. J., and Drummond, C. J. (2009). Observing self-assembled lipid nanoparticles building order and complexity through low-energy transformation processes. *ACS Nano*, *3*, 2789–2797.

Muller, M., and Schick, M. (1998). Calculation of the phase behavior of lipids. *Physical Review E*, *57*, 6973–6978.

Qiu, H., and Caffrey, M. (2000). The phase diagram of the monoolein/water system: Metastability and equilibrium aspects. *Biomaterials*, *21*, 223–234.

Siegel, D. P. (1999). The modified stalk mechanism of lamellar/inverted phase transitions and its implications for membrane fusion. *Biophysical Journal*, *76*, 291–313.

Siegel, D. P., and Epand, R. M. (1997). The mechanism of lamellar-to-inverted hexagonal phase transitions in phosphatidylethanolamine: Implications for membrane fusion mechanisms. *Biophysical Journal*, *73*, 3089–3111.

Siegel, D. P., Green, W. J., and Talmon, Y. (1994). The mechanism of lamellar to inverted hexagonal phase transition: A study using temperature-jump cryoelectron microscopy. *Biophysical Journal, 66*, 402–444.

Yaghmur, A., and Glatter, O. (2009). Characterization and potential applications of nanostructured aqueous dispersions. *Advances in Collord and Interface Science, 147*, 333–342.

Vauthey, S., Milo, C., Frossard, P., Garti, N., Leser, M. E., and Watzke, H. J. (2000). Structured fluids as microreactors for flavor formation by the Maillard reaction. *Journal of Agricultural and Food Chemistry, 48*, 4808–4816.

■■■■■ CHAPTER 2

Rheological Theory and Simulation of Surfactant Nematic Liquid Crystals

ALEJANDRO D. REY and E. E. HERRERA-VALENCIA

Department of Chemical Engineering, McGill University, Montréal, Quebec, Canada

Abstract

This chapter presents a comprehensive review of rheological theory, modeling, and simulation of surfactant nematic liquid crystalline phases, including calamitic and discotic micellar solutions and wormlike micelles. A review of verifiable rheological liquid crystal models for lyotropic nematics highlighting the mechanisms that control orientation behavior under shear, anisotropic viscoelasticity, and non-Newtonian behavior. Since defects and textures are essential characteristics of these materials that affect the flow properties, an in-depth review of physical and rheophysical defects is presented, including defect nucleation and coarsening processes. The theory for micellar nematics is applied to textures, flow birefringence, phase transition under shear, orientation fluctuations, and flow alignment, and the predictions are compared with experimental data. The theory is finally applied to transient shear flows of wormlike micellar nematic solutions, and the predicted banded textures and transient stress responses are compared to rheological experiments. The predictions provide a new way to extract additional information from experimental rheological data and allow to distinguish the role of liquid crystalline properties such as viscoelastic anisotropy, flow alignment, coupling between orientation kinematics, and flow kinematics. The rheological predictions show a strong similarity with other nematic materials, including low-molar-mass thermotropes and lyotropic nematic polymers.

Self-Assembled Supramolecular Architectures: Lyotropic Liquid Crystals, First Edition.
Edited by Nissim Garti, Ponisseril Somasundaran, and Raffaele Mezzenga.
© 2012 John Wiley & Sons, Inc. Published 2012 by John Wiley & Sons, Inc.

2.1   INTRODUCTION

Liquid crystals (LCs) are anisotropic viscoelastic mesophases with partial degrees of orientational and positional order. When the symmetry-breaking transitions that occur in going from isotropic liquids to orientational ordered nematics, to layered smectics, and to hexagonal phases is driven by lowering temperature, they are known as thermotroes, and, if the transitions are triggered by increasing concentration, they are lyotropes. Liquid crystals are part of the soft-matter-materials family since their elastic moduli is in the range of piconewtons, and they also belong to the complex fluids group since they exhibit non-Newtonian and complex viscoelastic behavior. In addition, these soft mesophases exhibit defects and textures as hard anisotropic materials, with disclinations, dislocations, and walls, arising often from incompatible orienting effects. The characteristics associated with anisotropic soft matter, complex fluids, and textured materials provide the richness associated with mesophases as functional and structural materials. Surfactant lyotropic liquid crystals are composed of self-assembled aggregates that display the multifunctionality associated with mesophases and find industrial applications in drug delivery, lubrications, detergency, viscosifiers, and turbulence suppression (Rey, 2010). Given the practical importance of flow and rheology, this chapter focuses on rheological modeling of surfactant nematic solutions emphasizing the connection between orientational order, flow kinematics, and order kinematics, reviewing recent works on surfactant liquid crystal (Alves et al., 2009; Erni et al., 2009; Fernandes and Figueiredo Neto, 1994; Fernandes et al., 2006; Kuzma et al., 1989; Lee, 2008; Palangana et al., 1990; Richtering, 2001; Sampaio et al., 2005; Simoes et al., 2000, 2005; Sonin 1987).

2.1.1   Lyotropic Liquid Crystals

Synthetic and biological lyotropic liquid crystalline phases arise in semiflexible or rigid rodlike macromolecules in suitable solvents at sufficiently high concentration (Rey, 2010). Typically nematic or chiral nematic phases arise when the packing factor $\varphi L_e/D_e \approx 4$, where φ is the volume fraction and L_e/D_e the effective molecular length-to-diameter ratio due to minimization of excluded volume. Examples of lyotropic biopolymer solutions include tobacco mosaic virus, poly-γ-benzyl-L-glutamate, silk protein solutions, and deoxyribonucleic acid (DNA). The pitch of chiral nematics is on the order of microns and is a function of temperature, pH, and external fields such as shear rate. Synthetic lyotropic polymers include polyamides, which are precursors to high-performance fibers. The anisotropic elasticity in these materials arises from orientation gradients, whose basic splay, bend, and twist distortion modes are associated with elastic moduli in the range of piconewtons. The flow behavior of lyotropic nematic polymers is non-Newtonian, anisotropic, with anomalous shear responses such as sign transitions in the first normal stress differences, shear thinning, damped stress oscillations under shear startup and flow reversal, flow birefringence in the dilute isotropic phase, and banded texture formation under weak shear. The interaction between orientation elasticity, couplings between flow and orientation kinematics, defects, and anisotropy has been shown to be behind the complex rheology of nematic lyotropes.

In Table 2.1, some surfactant systems and their respective phases as a function of the sample concentration are shown.

2.1.2 Micellar Nematic Liquid Crystals

Lyotropic liquid crystalline phases are based on amphiphilic molecules. Surfactant liquid crystals include: (i) nematic, (ii) cholesteric, (iii) smectic A, (iv) hexagonal, (v) lamellar, and (vi) cubic phases, according to ratios of head to tail molecular lengths, single or multiple tails or heads, presence of co-surfactants, presence of chiral dopants, and salt concentration; see Figure 2.1.

Lyonematics are found in mixtures of alcohol and aqueous solutions of ionic surfactants with longer chains. Binary water solutions of short-chain fluorocarbon derivatives and certain nonionic surfactants also show nematic ordering. Depending on the shape of the micellar aggregates, the nematic phases can be discotic nematic N_d, calamitic nematic N_c, and biaxial nematic N_b, where the latter is only found with co-surfactants. Phases changes with temperature and concentration correspond to micellar shape changes. Analysis of fluctuations reveals that the viscoelatic coupling between reorientation and flow is consistent with LC physics of thermotropic low-molar-mass nematics. Standard shear rheometry shows that the N_c and N_d phases align as other low-molar-mass thermotropic nematics. The addition of chiral molecules, either amphiphilic or nonamphiphilic, to a nematic phase gives rise to chiral nematic, which again can be discotic, calamitic, or biaxial. Rheological data appears to be consistent with helix reorganization processes present in thermotropic cholesterics.

TABLE 2.1 Examples of Nematic and Wormlike Surfactant Liquid Crystals

System	Length/Diameter	Application	Concentration/Solvent	LC Organization	Reference
Potassium laurate (KL) and sodium decylsulfate (SDS) Temperature = 30°C	$L_e \sim 200$ Å and $D_e \sim 20$ Å	Biological sensors and mechanical vibrations	27.5/6.24/66.26 KL/decanol/water 37/7/58 wt % SDS/decanol/water	Rodlike (R_L) Nematic phase Disk-like (N_D) Nematic phase	Larson (1999)
Cetylpyridinium chloride-hexanol (CPCL-Hex) Temperature = 30°C	$L_e \sim 150$ Å and $D_e \sim 15$ Å	Biological sensors and mechanical vibrations and oil recovery operations	33.7/76.3 wt % CPCL-hex/water 35.4/64.6 wt % CPCL-hex/water 38.8/51.2 wt % 40/60 wt %	Biphasic L_1-N_c Nematic biphasic Hexagonal phase	Larson (1999); Cates and Candau (1990); Berni et al. (2002); Lu et al. (2008)
Penta(ethylene glycol) monododecyl ether $C_{12}E_5$ Temperature = 66–71°C	$L_e \sim 200$ Å and $D_e \sim 20$ Å	Biological sensors and mechanical vibrations and oil recovery operations	<30 wt %and Sal < 12 wt % 30% < wt % < 35% ➤ 40 wt % Solvent: water	Isotropic L_1-nematic-calamitic (N_L) Hexagonal	Lu et al. (2008)
Hexadecyltrimethylammonium p-toluenesulfonate CTAB Temperature = 30°C	$L_e \sim 150$ Å and $D_e \sim 20$ Å	Biological sensors and mechanical vibrations and oil recovery operations	<33 wt %and Sal < 14 wt % 30% < wt % < 35% ➤ 40 wt % Solvent: hexanol/water	Isotropic L_1-nematic-calamitic (N_L) Hexagonal phase	Lu et al. (2008)
Sodium docecyl sulfate (SDS) Temperature = 30°C	$L_e \sim 200$ Å and $D_e \sim 15$ Å	Biological sensors and mechanical vibrations and oil recovery operations	<31 wt %and Sal < 13 wt % 30% < wt % < 35% ➤ 38 wt % Solvent: pentanol-dodecadene	Isotropic L_1-nematic-calamitic (N_L) Hexagonal phase	Lu et al. (2008)

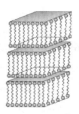

Figure 2.1 Schematics of discotic nematics, calamitic nematics, lamellar, and hexagonal phases. The symmetry breaking between nematic and lamellar involves positional order.

2.1.3 Wormlike Nematic Solutions

Wormlike micelles are semiflexible long rodlike aggregates that form nematic phases at high concentration. They arise in cationic surfactants at ambient temperature with aliphatic chains of approximately 16 carbons (Richtering, 2001). The addition of simple salt to ionic surfactant facilitates growth by screening electrostatic repulsion between head groups in such a way that the average micellar length $\bar{L}$, according to mean-field estimates (Cates and Candau, 1990), increases exponentially:

$$\bar{L} = \varphi^{1/2} \exp\left(\frac{E_c}{kT}\right) \tag{2.1}$$

where φ is the surfactant volume fraction, and E_c is the micelle endcap energy. At surfactant concentrations above 20–30 wt %, wormlike micellar solutions have received considerable attention during the past decades because of their remarkable structural and complex viscoelastic rheology behavior (Auffret et al., 2009; Berni et al., 2002; Burghardt and Fuller, 1991; Cates, 1987; Chen et al., 1977; Clawson et al., 2006; De Melo filho et al., 2007; Doi and Ohta, 1991; Hendrix et al., 1993; Itri and Amaral, 1990, 1993; Khokhlov and Semenov, 1981; Larson and Doi, 1991; Lawson and Flautt, 1967; Leadbetter and Wrighton, 1979; Lu et al., 2008; Lukascheck et al., 1995; Moldenaers et al.,1991; Muller et al., 1997; Onsager, 1949; Ostwald and Pieransky, 2000; Quist et al., 1992; Ramos, 2001; Richtering et al.,1994; Savage et al., 2010; Schmidt, 2006; Schmidt et al., 1988; Semenov and Khokhlov, 1988; Siebert et al., 1997; Spenley and Cates, 1994; Spenley et al., 1993, 1996; Srinivasarao, 1992; Tschierske, 2007; Wang et al., 2005; Yada et al., 2003; Yang, 2002).

Nematic phases exhibit long-range orientational and translational order with no positional order, whereas hexagonal phases show both orientational and translational long-range order of the centers of mass of the micelles (Semenov and Khokhlov, 1988); see Figure 2.1.

At concentrations below the isotropic/nematic transition, viscoelastic concentrated surfactants are characterized by an entangled network of large wormlike micelle structures. These structures break and reform during flow, exhibiting variable and rich rheological behavior (Cates, 1987; Cates and Candau 1990; Larson, 1999; Larson and Doi, 1991; Spenley and Cates, 1994; Spenley et al., 1993, 1996; Yang, 2002). These systems exhibit Maxwell-type behavior in small-amplitude oscillatory shear flow and saturation of the shear stress in steady simple shear, which leads to shear banding flow (Cates, 1987; Cates and Candau 1990; Larson, 1999; Larson and Doi, 1991; Spenley and Cates, 1994; Spenley et al., 1993, 1996; Yang, 2002). In the nonlinear viscoelastic regime, elongated micellar solutions also exhibit remarkable features, such as the presence of a stress plateau in steady shear flow past a critical shear rate accompanied by slow transients to reach steady state (Cates, 1987; Cates and Candau 1990; Larson, 1999; Larson and Doi, 1991; Spenley and Cates, 1994; Spenley et al., 1993, 1996; Yang, 2002). Prediction of the flow behavior of viscoelastic surfactants by constitutive equations has been a challenging issue (Acierno et al., 1976; Calderas et al., 2009; De Kee and Fong, 1994; Giesekus, 1966, 1982, 1984, 1985; Herrera et al., 2009, 2010; Marrucci, 1985). In the nematic phase, the rheology of wormlike micelles has similar features as lyotropic nematic polymers and other complex systems (Calderas et al., 2009). The rheological classification of uniaxial nematic LCs is to a large extent based on a single parameter known as the "reactive" or "tumbling" parameter, which describes the ability of a mesogen to align in a simple shear flow. The tumbling parameter λ is a ratio of the aligning effect of strain to the rotational effect of vorticity. For rodlike nematics, $\lambda > 0$, whereas for disklike nematic, $\lambda < 0$. The name *reactive* refers to the fact that the entropy production is independent of this parameter and hence its sign is undetermined. Hence we find the following cases (Larson, 1999; Rey, 2007, 2010):

1. Rodlike nematics: flow aligning for $\lambda > 1$, nonaligning for $0 < \lambda < 1$.
2. Disklike nematics: flow aligning for $\lambda < -1$, nonaligning for $-1 < \lambda < 0$.

When $|\lambda| > 1$, strain overcomes vorticity effects, and the average orientation is close to the velocity for rods and the velocity gradient for disks, respectively, while for $|\lambda| < 1$ complex periodic and steady three-dimensional (3D) orientation modes arise (Tsuji and Rey, 2000). A general expression for the tumbling parameter is given by the product of a shape function $\mathbb{R}(p)$ and a thermorheological function $g(T, \varphi, S(T, \varphi, \dot{\gamma}))$ (Rey, 2010):

$$\lambda = \frac{\text{aligning effect of strain}}{\text{rotation effect of vorticity}} = \beta(T, \varphi) g(S(T, \varphi, \dot{\gamma})) \qquad (2.2)$$

where β is a function of the effective aspect ratio, T is the temperature, φ is the mesogen volume fraction, $\dot{\gamma}$ is the shear rate, and S is the molecular alignment along the average orientation. For monomeric rodlike nematics with a lower temperature smectic A phase, a sufficient decrease in T affects $\beta(T, \varphi)$ and triggers a rheological aligning-to-nonaligning transition, as in the case of 8CB. Likewise in monomeric discotic nematics with a lower columnar phase, decreasing T disrupts flow alignment. For lyotropic nematic polymers, flow alignment is dominated by $g(S(T, \varphi, \dot{\gamma}))$, such that in the isotropic phase $\lambda > 1$, while in the nematic phase at low shear $\lambda < 1$ and at high shear $\lambda > 1$. The actual orientation field is a balance between space-dependent elastic and flow torques, and banded patterns often arise at shear rates less than 1 s^{-1}. Thus the important variables to control flow alignment of monomeric nematics is T, while for nematic polymers it is $\dot{\gamma}$.

2.1.4 Diamagnetic Anisotropy

This section is based on the review of diamagnetic susceptibility of lyotropic liquid crystals by Sonin (1987). Previous reports showed that the nematic lyophase of composition NaDS–36%, DeOH–7%. D$_2$O–50%. Na$_2$SO$_2$–7% has a positive diamagnetic anisotropy $\chi_a = \chi_{\|} - \chi_{\perp} > 0$ (Lawson and Flautt, 1967; Sonin, 1987). Here $\chi_{\|}$ and $\chi_{\perp}$ are the diamagnetic susceptibilities in the direction of the director and perpendicular to it, respectively.

Some author established the existence of nematic mesophase having a negative diamagnetic anisotropy ($\chi_a < 0$) (Radley and Reeves, 1975; Radley et al., 1976). They are formed in the same NaDS–DeOH–H$_2$O system, either upon a small change in the concentrations of the starting components or upon adding the counterions K*, Li*, or Cs*. Systems having $\chi_a > 0$ are called type I and systems having $\chi_a < 0$ type II. The hypothesis was advanced (Charvolin et al., 1979) that the diamagnetic anisotropy in the NaDS–DeOH–H$_2$O system arises from the shape of the micelles: Calamitics have $\chi_a > 0$, and discotic $\chi_a < 0$. However, the nematic phase in the cesium pentadecafluorooctanoate (CsPFO)–H$_2$O system with micelles of discotic form has $\chi_a > 0$ Boden et al. (1979). This contradiction stimulated the study of other systems having polymethylene and perfluoropolymethylene chains (Boden et al., 1981). The former have negative diamagnetic anisotropy ($\chi_a < 0$), and the latter, positive ($\chi_a > 0$). Consequently, it proved possible to "construct" nematic discotics and calamitics, both with $\chi_a > 0$ and $\chi_a < 0$. Examples of such systems are presented in Table 2.2. Thus it was shown that the sign of the diamagnetic susceptibility is not determined by the type of micelles (Boden et al., 1981). This viewpoint

TABLE 2.2 Lyotropic Systems Having Different Diamagnetic Anisotropies at a Temperature of 20°C

χ_a	Type of Micelles	Symbol	MTAB$_r$[a]	MTAΦSO$_3$[b]	H$_2$O	DeOH	NH$_4$B$_r$
>0	Cylinder	N$_c^+$	36,0	—	64.0	—	—
<0	Disk	N$_d^-$	26,6	—	63.3	3.8	6.3
<0	Cylinder	N$_c^-$	—	38.0	62.0	—	—
>0	Disk	N$_d^+$	—	14.1	59.2	3.7	7.4

Source: Adapted from Sonin (1987).

[a]MTAB$_r$ is myristyltrimethylammonium bromide.

[b]MTAΦSO$_3$ is myristyltrimethyl-ammonium toluensulfonate.

was confirmed when discotics were found (Boden et al., 1979) with $\chi_a > 0$ and calamitics with $\chi_a < 0$ in the $NaH_xB-Na_2SO_4-DeOH-H_2O$ system.

The cause of the diamagnetic anisotropy of micelles has been examined in detail (Amaral, 1983; Boden et al., 1979, 1981, 1985; Charvolin et al., 1979; Radley and Reeves, 1975; Radley et al., 1976; Vedenov and Levchenko, 1983). The orientation of a diamagnetic ellipsoid having the mean susceptibility χ placed in a medium having the susceptibility χ_0 does not depend on the sign of $\overline{\chi} - \chi_0$ and is determined unambiguously—the principal axis of the ellipsoid is established parallel to the field. Therefore cylindrical micelles will orient with their long axes and discotic micelles with their short axes along the field.

If the difference in energies corresponding to configurations in which the principal axis of the ellipse is parallel or perpendicular to the magnetic field is ΔW_{shape}, then we have

$$\Delta W_{shape} \cong -\frac{V}{2}H^2(\overline{\chi}-\chi_0)(D_c - D_a)\overline{\chi} \qquad (2.3)$$

where V is the volume of the ellipsoid, and D_c and D_a are the diamagnetic factors in the directions of the least and greatest axes of the ellipsoid, respectively.

The greatest value of ΔW_{shape} will occur in the case of asymmetric ellipsoids ($a \approx 10c$). In this case we have $D_a = 0$, while $D_c = 4\pi$ for an oblate and 2π for a prolate spheroid. If we assign typical values of χ and χ_0 (see below), then we find from the condition $\Delta W_{shape} > kT$ that $H^2V \sim 1G^2cm^3$. Hence, for the usually employed values of the magnetic fields (1 kG), this implies that $V \sim 10^{-6}$ cm^3. This means that only macroscopic diamagnetic particles can be subjected to orientation due to shape.

As is known (Gelbart et al. 1994; de Gennes, 1993; Kuzma and Saupe, 1997), orientation owing to the diamagnetic anisotropy of molecules depends on the sign of χ_0. The difference between the energies of micelles corresponding to configurations along the magnetic field and perpendicular to it involves the number N of molecules on the micelle, its volume V_0, and the order parameter S_{mol}, which characterizes the ordering of the molecules in the micelle:

$$\Delta W_{mol} = W - W_\wedge = -\frac{H^2NV_0}{2}S_{mol}\chi_a \qquad (2.4)$$

Yet, if the ordering of the principal axes of the micelles in a nematic is described by the order parameter S_{mic}, then the difference the energies indicated above for a system of M micelles will be

$$\Delta W_{mol} = -\frac{H^2MNV_0}{2}S_{mol}S_{mic}\chi_a \qquad (2.5)$$

One can define the quantity $S_{mol}S_{mic} = S$ as the macroscopic parameter. We shall examine these quantities in greater detail below in Section 2.2. By using

Eqs. (2.3) and (2.5), we can easily find the ratio of the energy differences caused by the shape and the molecular anisotropies:

$$\frac{\Delta W_{mol}}{\Delta W_{shape}} = -\frac{S_{\chi a}}{(\bar{\chi} - \chi_a)(D_c - D_a)\bar{\chi}} \tag{2.6}$$

The calculation of this ratio is impeded by the lack of experimental data of the diamagnetic susceptibilities and anisotropy. However, one can estimate the latter by using the additive scheme of atomic susceptibilities. Thus, for $\bar{\chi} = (\chi_{||} + 2\chi_{\perp})/3$ the following values were obtained [in centimeter–gram–second (CGS) units/mole]: for NaDS, 165×10^{-6}; for KI, 155×10^{-6}; and for DeOH, 124×10^{-6}. By using the known compositions (see above), one can adopt the typical values $\bar{\chi} = -6 \times 10^{-7}$ CGS units/cm^3, and $\chi_0 = -7 \times 10^{-7}$ CGS unit/cm^3. Estimates of $\chi_0 \sqrt{\chi}$ have been obtained under the assumption that the diamagnetic anisotropy arises only from the anisotropy of the methylene groups. Typical values for all the systems are equal to about 0.1. Upon using these values of the diamagnetic constant as well as the typical values of S equal to 0.5–0.6 we find that

$$\frac{\Delta W_{mol}}{\Delta W_{shape}} \approx 10^4 \tag{2.7}$$

Hence it is clear that the orientation of a lyotropic nematic is caused by the molecular diamagnetic susceptibility (Boden et al., 1985; Sonin, 1987).

2.2 FLOW MODELING OF NEMATIC LIQUID CRYSTALS

2.2.1 Quadrupolar Order Parameter

The Landau–de Gennes theory of liquid crystals describes the viscoelastic behavior of nematic liquid crystals by means of the second moment of the orientation distribution function, known as the tensor order parameter **Q** (de Gennes and Prost, 1993; Rey, 2007, 2009, 2010):

$$\mathbf{Q} = \int \left(\mathbf{uu} - \frac{\mathbf{I}}{3} \right) f(\mathbf{u}) d^2\mathbf{u} \tag{2.8}$$

where **u** is the molecular unit vector, and **I** is unit tensor. The tensor order parameter **Q** is expressed in terms of the orthonormal director triad (**n, m, l**) and the scalar order parameters (S,P):

$$\mathbf{Q} = S\left(\mathbf{nn} - \tfrac{1}{3}\mathbf{I}\right) + \tfrac{1}{3}P\left(\mathbf{mm} - \tfrac{1}{3}\mathbf{ll}\right) \tag{2.9}$$

where the following restrictions apply: $\mathbf{Q} = \mathbf{Q}^T$; $\mathbf{Q}:\mathbf{I} = 0$; $-\leq 2S \leq 2$; $-3 \leq 2P \leq 3$. The uniaxial director **n** corresponds to the maximum eigenvalue $\mu_n = 2S/3$,

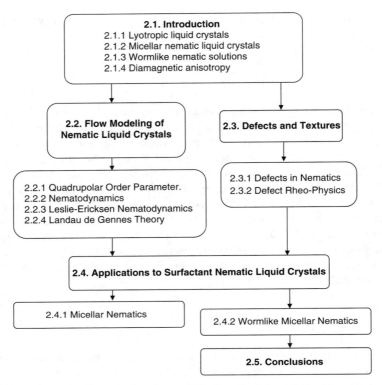

Figure 2.2 Organization of the chapter. Basic principles of liquid crystal physics, organized in terms of experimental and theoretical rheology are used to describe the complex behavior of micellar nematic liquid crystals.

the biaxial director $\mathbf{m}$ corresponds the second largest eigenvalue $\mu_m = -(S - P)/3$, and the second biaxial director $\mathbf{l} = \mathbf{n} \times \mathbf{m}$ corresponds to the smallest eigenvalue $\mu_l = -(S + P)/3$. The orientation is defined completely by the orthogonal director triad $(\mathbf{n}, \mathbf{m}, \mathbf{l})$. The magnitude of the uniaxial scalar order parameter S is a measure of the molecular alignment along the uniaxial director $\mathbf{n}$ and is given as $S = 3(\mathbf{n} \cdot \mathbf{Q} \cdot \mathbf{n})/2$. The magnitude of the biaxial scalar order parameter P is a measure of the molecular alignment in a plane perpendicular to the direction of uniaxial director $\mathbf{n}$, and is given as $P = 3(\mathbf{m} \cdot \mathbf{Q} \cdot \mathbf{m} - \mathbf{1} \cdot \mathbf{Q} \cdot \mathbf{1})/2$ (de Gennes and Prost, 1993; Rey, 2007, 2009, 2010).

The ordering (P, S) triangle shows the possible states of rods for a fixed set of the orthogonal director triad $(\mathbf{n}, \mathbf{m}, \mathbf{l})$. Considering the upper triangle (regions B, D with $P \geq 0, S \geq 0$) we find isotropic $(P = S = 0)$, maximal positive uniaxial (parallel rods at end of α line), and maximal negative uniaxial (plane at end of γ line, upper right).

Figure 2.3 shows that imposing a strong uniaxial extension flow along $\mathbf{x}$ (left) in the nematic disklike phase reorientes the director $\mathbf{n}$ toward the com-

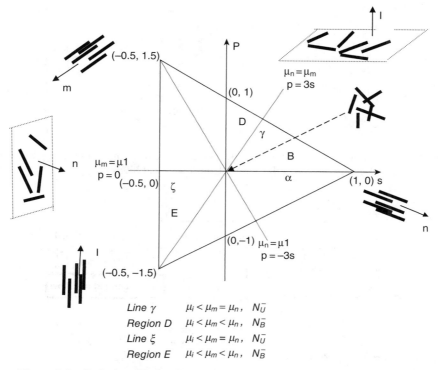

Figure 2.3 Ordering (P, S) triangle showing the different rod configurations.

pression (y, z) plane (middle and right panel). The three lines from each vertex to the midpoint of the opposite side represent uniaxial states, and they meet at the isotropic state, while the rest of the triangle represents the biaxial states. By using symmetry considerations, equivalent states are found in the other sectors. For example, at the corners of the triangle there is maximal positive unixiality along each of the three directors. The relevant eigenvalue inequalities and the superscript sign denotes the sign of the eigenvalue corresponding to the unique eigenvector or for biaxial cases to prolate (+) or oblate (−) shapes. A thermodynamic or flow process is a trajectory in the ordering triangle.

Another case is when temperature changes and provokes a modification in the micellar shape from uniaxial prolate to biaxial to uniaxial oblate. This is represented as a trajectory from a point in the uniaxial α line through the B region, which ends on a point in the γ ($P = 3S$) uniaxial line. In general, processes involving (**n, m, l**) flow-induced orientation changes are represented as trajectories in the orientation unit sphere (Boden et al., 1981; Rey, 2009). For example, Figure 2.4 shows that imposing a strong uniaxial extension flow along

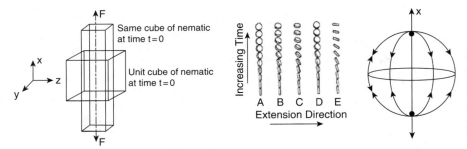

Figure 2.4 Different orientations of the nematic disklike phase upon imposing a strong uniaxial extensional flow. [Adapted from Lu et al. (2008).]

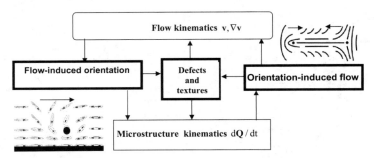

Figure 2.5 Schematic of velocity microstrusture kinematics defect and texture couplings that build the nematodynamic cycle. The effect of flow on **Q** includes flow-induced orientation of the N phase and flow birefringence of the I phase. The effect of reorientation (d**Q**/dt) on velocity includes hydrodynamic interactions. The bottom left schematic represents flow-induced orientation with defect nucleation (dark circle) close to a surface due to director tumbling and the top right shows backflow (arrows) generated by $\pm\frac{1}{2}$ annihilating defects.

x (left) in the nematic disklike phase reorients the director **n** toward the compression (y, z) plane (middle and right panel).

2.2.2 Nematodynamics

Flow processes of LCs can be considered in terms of a microstructural reactor, where orientation, domains, defects, and textures are manipulated by shear and extension deformation rates and where the flow patterns, secondary flows, and hydrodynamic interactions are tightly coupled to liquid crystalline order (Rey, 2009, 2010).

This LC flow analysis paradigm shown in Figure 2.5 is a closed loop that couples the flow velocity and order parameter through flow-induced orientation (lower left schematic) and its converse, orientation-induced flow (lower right schematic), and with both processes feeding into defects and textures. While flow-induced orientation is well known and characterized in flexible and rigid-rod polymers, the full range of effects arising from orientation-induced flow, such as backflow, transverse flow, and hydrodynamic interactions during defect–defect annihilation, is less characterized. Furthermore, flow-induced textural transformations can only be understood using defect physics and rheophysics. The closed loop shown in Figure 2.5 can be achieved at a macroscopic level using the Leslie–Ericksen (LE) $\mathbf{n}$-vector description (de Andrade Lima and Rey, 2003a–c, 2004a–e, 2005, 2006a–e; Larson, 1999; Rey 2007, 2009, 2010) or at the mesoscopic level using the $\mathbf{Q}$-tensor Landau–de Gennes (LdG) model (Murugesan and Rey, 2010; Rey, 2007, 2010; Tsuji and Rey, 1998). Only the latter captures important features such as singular defect nucleation, singular defect–defect reactions, and singular defect–flow interactions. In order to avoid repetitions (Rey, 2007, 2009, 2010) and concentrate on objectives, we provide a qualitative brief presentation of nematodynamics; for reviews and other accounts on this active field see, for example, Larson (1999) and Rey (2007, 2009, 2010), and references therein.

The Landau–de Gennes model has an external length scale ℓ_e and an internal length scale ℓ_i as follows (Rey, 2009, 2010; Rey and Denn, 2002; Rey and Tsuji, 1998):

$$\ell_e = H \qquad \ell_i = \sqrt{\frac{L}{3ckT^*}} \qquad \mathbb{R} = \left(\frac{\ell_e}{\ell_i}\right)^2 = \frac{3ckT^*}{L/H^2} \gg 1 \qquad (2.10)$$

where H is the system size, L (energy/length) is a characteristic orientation elasticity constant associated with gradients in the directors $(\mathbf{n}, \mathbf{m}, \mathbf{l})$ and ckT^* is the energy per unit volume associated with molecular elasticity (S, P); c is the concentration per unit volume, k is the Boltzmann constant, and T^* is the isotropic-nematic transition temperature. The external scale is associated with micron-scale changes in $(\mathbf{n}, \mathbf{m}, \mathbf{l})$ and the internal length scale ℓ_i is associated with nanoscale changes in (S, P), as in the disclination core shown in Figure 2.5. The ratio of molecular ordering energy to orientation elasticity R, or ratio of square length scales, is of the order of 10^6–10^9. In the LE model, R is assumed to be infinity, and hence the scalar order parameters (S, P) are not taken into account. The external τ_e and internal τ_i time scales of the LdG model are ordered as follows (Grecov and Rey, 2003a,b; Rey 2007, 2009, 2010):

$$\tau_e = \frac{\eta H^2}{3L} \qquad \tau_i = \frac{1}{D_r} \qquad \tau_e \gg \tau_i \qquad (2.11)$$

where D_r is the bare rotational diffusivity and $\eta = ckT^*/D_r$. The external time scale describes slow orientation variations, and the internal length scale describes fast order parameter variations. In the LE model $\tau_i = 0$ and no molecular dynamics are taken into account. Finally, the presence of shear flow of rate $\dot{\gamma}$ introduces a flow time scale τ_f and a flow length scale ℓ_f:

$$\tau_f = \frac{1}{\dot{\gamma}} \qquad \ell_f = \sqrt{\frac{D_{orient}}{\dot{\gamma}}} \qquad D_{orient} = \frac{LH^2}{3\tau_e} = \frac{L}{\eta} \qquad (2.12)$$

where D_{orient} is the characteristic orientation diffusivity (length^2/time) in the system. With regard to the values of Deborah numbers we have two processes: (a) Orientation process (De $\ll 1$): The time scale ordering is $\tau_i < \tau_f < \tau_e$, the orientation processes dominate the rheology, and the scalar order parameter is close to its equilibrium value. In this regime the flow affects the eigenvectors of $\mathbf{Q}$ but does not affect the eigenvalues of $\mathbf{Q}$. Since LCs are anisotropic, shear thinning, nonmonotonic stress growth, and first normal stress differences are possible. (b) Molecular process (De > 1): The time scales ordering is $\tau_f < \tau_i < \tau_e$, and the flow affects the eigenvectors and eigenvalues of $\mathbf{Q}$. The dimensionless Ericksen number Er and Deborah number De are given by (Larson and Doi, 1991; Rey, 2007, 2009, 2010; Rey and Denn, 2002; Tsuji and Rey, 1998):

$$\text{Er} = \left(\frac{\ell_e}{\ell_f}\right)^2 = \frac{3\tau_e}{\tau_f} = \frac{\dot{\gamma}H^2\eta}{L} \qquad \text{De} = \frac{\text{Er}}{R} = \left(\frac{\ell_i}{\ell_f}\right)^2 = \frac{\tau_i}{6\tau_f} = \frac{\dot{\gamma}}{6D_r} \qquad (2.13)$$

To characterize the degree of ordering in the LC phase, a dimensionless concentration $U = 3c/c^*$ is used. Hence the most general parametric space for LdG nematodynamics is given by $(1/U, \mathbb{R}, \text{Er})$, while for the LE nematodynamics it is Er.

2.2.3 Leslie–Ericksen Nematodynamics

2.2.3.1 Bulk and Interfacial Equations
The LE equations consist of the linear momentum balance and director torque balance with additional constitutive equations for stress tensor $\mathbf{T}$, elastic torque $\mathbf{\Gamma}^e$, and viscous torque $\mathbf{\Gamma}^v$ (Larson and Doi, 1991; Murugesan and Rey, 2010; Rey, 2001a,b, 2007, 2009, 2010).

The mass and linear momentum balance equations are

$$\nabla \cdot \mathbf{v} = 0 \qquad \rho\dot{\mathbf{v}} = \mathbf{f} + \nabla \cdot \mathbf{T} \qquad (2.14)$$

where $\mathbf{f}$ is the body force per unit volume, $\mathbf{v}$ is the linear velocity, and a superposed dot represents the material time derivative of the velocity. The constitutive equation for the total stress tensor $\mathbf{T}$ is given by

$$\mathbf{T} = -p\mathbf{I} - \frac{\partial F_g}{(\partial \nabla \mathbf{n})^T} \cdot \nabla \mathbf{n} + \alpha_1(\mathbf{nn} : \mathbf{A})\mathbf{nn} + \alpha_2\mathbf{nN}$$
$$+ \alpha_3\mathbf{Nn} + \alpha_4\mathbf{A} + \alpha_5\mathbf{nn} \cdot \mathbf{A} + \alpha_6\mathbf{A} \cdot \mathbf{nn} \qquad (2.15)$$

$$2\mathbf{A} = (\nabla\mathbf{v} + (\nabla\mathbf{v})^T) \qquad \mathbf{N} = \dot{\mathbf{n}} - \mathbf{W} \cdot \mathbf{n} \qquad 2\mathbf{W} = (\nabla\mathbf{v} - (\nabla\mathbf{v})^T) \qquad (2.16)$$

where p is the pressure, $\mathbf{I}$ is the unit tensor; α_i, $i = 1, \ldots, 6$ are the six Leslie viscosity coefficients; $\mathbf{A}$ is the symmetric rate of deformation tensor; $\mathbf{N}$ is the Zaremba–Jaumann time derivative of the director; and $\mathbf{W}$ is the vorticity tensor. The Frank elastic energy density f_g is given by

$$2f_g = K_{11}(\nabla \cdot \mathbf{n})^2 + K_{22}(\mathbf{n} \cdot \nabla \times \mathbf{n})^2 + K_{33}(\mathbf{n} \times \nabla \times \mathbf{n})^2 \qquad (2.17)$$

where $\{K_{ii}; ii = 11, 22, 33\}$ are the temperature-dependent three elastic constants for splay, twist, and bend, respectively. Anisotropies and thermal dependence of the elastic constants are discussed in the literature (de Gennes and Prost, 1993; Larson and Doi, 1991; Rey 2007, 2009, 2010; Rey and Denn, 1987, 1998a–c, 2002). The director torque balance equation is given by the sum of the viscous $\mathbf{\Gamma}^v$ and the elastic $\mathbf{\Gamma}^e$ torque:

$$\mathbf{\Gamma}^v + \mathbf{\Gamma}^e = 0 \qquad \mathbf{\Gamma}^v = \mathbf{n} \times \mathbf{h}^v \equiv -\mathbf{n} \times (\gamma_1 \mathbf{N} + \gamma_2 \mathbf{A} \cdot \mathbf{n}) \mathbf{\Gamma}^e$$

$$= \mathbf{n} \times \mathbf{h}^e \equiv -\mathbf{n} \times \left(\frac{\partial f_g}{\partial \mathbf{n}} - \nabla \cdot \frac{\partial f_g}{\partial (\nabla \mathbf{n})^T} \right) \qquad (2.18)$$

where $\mathbf{h}^v$ is the viscous molecular field, $\mathbf{h}^e$ is the elastic molecular field, $\gamma_1 = \alpha_3 - \alpha_2$ is the rotational viscosity, and $\gamma_2 = \alpha_6 - \alpha_3 = \alpha_3 + \alpha_2$ is the irrotational torque coefficient. For thermotropic LCs, the rheological behavior is controlled by the temperature-dependent reactive parameter λ, given by

$$\lambda = -\frac{\gamma_2}{\gamma_1} = -\frac{\alpha_2 + \alpha_3}{\alpha_3 - \alpha_2} \qquad (2.19)$$

2.2.3.2 *Rheological Functions*   The LE nematodynamics predicts the following rheological functions (de Gennes and Prost, 1993; Larson and Doi, 1991; Rey, 2007, 2009, 2010; Rey and Denn, 2002):

Shear Flow Alignment   At sufficiently large Er, when $\lambda > 1$ (rods) or $\lambda < -1$ (disks), the stable shear flow alignment angle θ_{al} is given by (de Gennes and Prost, 1993)

$$2\theta_{al} = \cos^{-1}\left(\frac{1}{\lambda} \right) \qquad (2.20)$$

where θ_{al} is the angle between $\mathbf{n}$ and the velocity $\mathbf{v}$ in the shear plane ($\mathbf{v} - \nabla \mathbf{v}$ plane).

Shear Viscosities   The three Miesowicz shear viscosities (η_1, η_2, and η_3) that characterize viscous anisotropy are measured in a steady simple shear flow between parallel plates with fixed director orientations along three characteristic orthogonal directions: η_1 when the director is parallel to the velocity direction, η_2 when it is parallel to the velocity gradient, and η_3 when it is parallel to the vorticity axis, given by

$$\eta_1 = \frac{\alpha_3 + \alpha_4 + \alpha_6}{2} \qquad \eta_2 = \frac{-\alpha_2 + \alpha_4 + \alpha_5}{2} \qquad \eta_3 = \frac{\alpha_4}{2} \qquad (2.21)$$

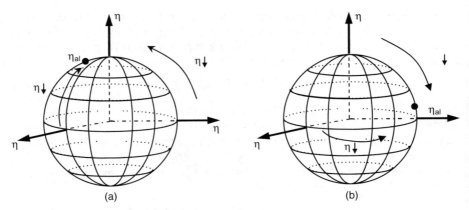

Figure 2.6 Orientation reduction mechanism predicted by the LE nematodyanmics for (a) disks and (b) rods.

For disks (rods), it is found that $\eta_1 > \eta_3 > \eta_2$ $(\eta_1 < \eta_3 < \eta_2)$ (de Gennes and Prost, 1993; Grecov and Rey, 2003a–c, 2004, 2005, 2006; Rey, 2007, 2009, 2010; Rey and Denn, 2002). The shear viscosity under flow alignment, defined by η_{al}, is

$$\eta_{al} = \frac{1}{2}(\eta_1 + \eta_2 - \gamma_1) + \frac{1}{4}\alpha_1\left(1 - \left(\frac{1}{\lambda}\right)^2\right) \qquad (2.22)$$

and it is slightly larger than η_2 for disks and η_1 for rods. Hence when $|\lambda| > 1$, as the shear rate increases, the LE nematodynamics describes an orientation viscosity reduction mechanism (OVR), as shown in Figure 2.6 (Grecov and Rey, 2003a–c, 2004, 2005, 2006; Rey, 2007, 2009, 2010). For example, if a CM sample is sheared with a random director orientation distribution, the increasing effect of shear is to narrow the distribution with a peak that is parallel to the flow alignment angle (close to the shear gradient direction), and hence the apparent viscosity will decrease with increasing shear since the flow alignment angle is close to the minimum possible viscosity, which in the case for disks is η_2.

Steady shear flow simulations indicate that for any arbitrary Er, the LE nematodynamics can be fitted with the Carreau–Yasuda LC model (Grecov and Rey 2003b; Rey, 2009):

$$\eta_s = \frac{\eta - \eta_{al}}{\eta_0 - \eta_{al}} = \left[1 + (\tau\,\text{Er})^a\right]^{(n-1)/a} \qquad (2.23)$$

where η_s is the scaled shear viscosity, n is the "power-law exponent," a is a dimensionless parameter that describes the transition region between the zero shear rate region and the power law region, τ is a dimensionless time constant, and η_0 is the zero shear rate viscosity. The first transition region in the Carreau–

Yasuda LC model is defined by $Er = 1/a$, which is the value of the Ericksen number at which flow significantly affects the orientation. The second transition between the power law shear thinning regime and the flow alignment regime is

$$Er_{ST\text{-}FA} = \frac{1}{\tau}\left(c^{a/(n-1)} - 1\right)^{1/a} \tag{2.24}$$

where c is of the order of 10^{-3}.

Backflow This process is the opposite to orientation-driven flow (Rey, 2010). Except for pure homogeneous twist reorientation, changes in the director orientation **n** create flow. The reorientation viscosities associated with splay, twist, and bend deformations (shown in Fig. 2.10) are defined by Rey (2007, 2009, 2010) and de Andrade Lima and Rey (2003c):

$$\eta_{\text{twist}} = \gamma_1 \qquad \eta_{\text{splay}} = \gamma_1 - \frac{\alpha_3^2}{\eta_1} \qquad \eta_{\text{bend}} = \gamma_1 - \frac{\alpha_2^2}{\eta_2} \tag{2.25}$$

These transient reorientation viscosities are given by the rotational viscosity (γ_1) decreased by a factor introduced by the backflow effect. The general expression for the reorientation viscosities can be rewritten in a more revealing general form (de Andrade Lima and Rey, 2003c; Rey, 2000a–c, 2007, 2009, 2010):

$$\eta_\alpha = \gamma_1 - \frac{(TC_i)^2}{\eta_i} \tag{2.26}$$

where η_i denotes the corresponding Miesowicz viscosity and TC_i the corresponding torque coefficient. Since twist is the only mode that creates no backflow (Rey, 2010), then $\eta_{\text{twist}} = \gamma_1$. For a bend distortion the backflow is normal to **n**, and hence the torque coefficient is α_2, and the Miesowicz viscosity is η_2. On the other hand for a splay distortion, the backflow is parallel to **n** and hence the torque coefficient is α_3, and the Miesowicz viscosity is η_1. The ordering in the reorientation viscosities is

$$\eta_{\text{twist}} > \eta_{\text{splay}} \qquad \eta_{\text{twist}} > \eta_{\text{bend}}$$

Secondary Flows Whenever the director deviates from the shear plane, a transverse flow will be generated since the viscosity tensor C_{ijkl} in the extra stress tensor [see Eq. (2.27)] is a function of the director (Rey, 2007, 2009, 2010; Stewart, 2004):

$$T_{ij}^{\text{extra}} = C_{ijkl}A_{lk} + D_{ijk}N_k \tag{2.27}$$

Hence, under shear flow in the x direction and three-dimensional (3D) orientation $\mathbf{n} = (n_x, n_y, n_z)$, a velocity of the form $\mathbf{v} = (v_x, 0, v_z)$ must be considered at a minimum.

First Normal Stress Difference N_1   For nematic liquid crystals N_1 is a strong function of orientation and can have positive or negative values. Expressions for N_1 in terms of the director components n_x and n_y are (Grecov and Rey, 2003a,b; Rey, 2007, 2009, 2010)

$$N_1 = t_{xx} - t_{yy} = \dot{\gamma} n_x n_y \left(\gamma_2 + \alpha_1 \left(n_y^2 - n_x^2 \right) \right) \tag{2.28}$$

As the director circles the shear plane, the total number N_T of sign changes in N_1 is (de Andrade Lima and Rey, 2003b,c, 2004d)

$$N_T = N_{SC}(n_x) N_{SC}(n_y) N_{SC}\left(\left(\gamma_2 + \alpha_1 \left(n_y^2 - n_x^2 \right) \right) \right) = 4 N_{SC}\left(\left(\gamma_2 + \alpha_1 \left(n_y^2 - n_x^2 \right) \right) \right) \tag{2.29}$$

where N_{SC} denotes the number of sign changes. The number of sign changes in n_x is 2, and similarly for n_y. The two following material property-dependent outcomes are found:

$$|\gamma_2| < |\alpha_1|: N_T = 4 \qquad |\gamma_2| > |\alpha_1|: N_T = 4 \times 2 = 8 \tag{2.30}$$

The nonlinearity orientation introduced by α_1 can increase the frequency of sign changes from four to eight. Equation (2.30) embodies the orientation-driven first normal stress sign change mechanism (ONSC). In the flow alignment regime, Eq. (2.28) becomes

$$\lim_{E > E_{IST\text{-}FA}} N_1 = N_{1al} = \dot{\gamma} \left(\gamma_2 - \frac{\alpha_1}{\lambda} \right) \frac{\sqrt{\lambda^2 - 1}}{2\lambda} \tag{2.31}$$

and is proportional to the shear rate. Shearing an LC, with a heterogeneous director field and sufficiently high material nonlinearity (i.e., large $|\alpha_1|$), at increasing rates, it will narrow and shift the orientation distribution function toward the Leslie angle, an orientation process that causes N_1 to change sign (Grecov and Rey, 2003a,b):

$$N_1(\mathbf{n}_1(x), E_1) < 0 \rightarrow N_1(\mathbf{n}_2(x), E_2 > E_2) > 0 \tag{2.32}$$

2.2.4   Landau de Gennes Nematodynamics

2.2.4.1   Bulk and Interfacial Equations   The governing equations for LC flows follow from the dissipation function Δ (de Andrade Lima et al., 2006; Farhoudi and Rey, 1993a–c; Murugesan and Rey, 2010; Rey 2007, 2009, 2010; Soule et al., 2009):

$$\Delta = \mathbf{t}^s : \mathbf{A} + ckT\mathbf{H} : \hat{\mathbf{Q}} \tag{2.33}$$

where $\mathbf{t}^s$ is the viscoelastic stress tensor, $\mathbf{H}$ is the dimensionless molecular field, and $\hat{\mathbf{Q}}$ is the Jaumann derivative of the tensor order parameter. The molecular field $\mathbf{H}$ is the negative of the variational derivative of the free energy density f:

$$\frac{f}{ckT} = \frac{1}{2}\left(1-\frac{1}{3}U\right)\mathbf{Q}:\mathbf{Q}-\frac{1}{3}U\mathbf{Q}:(\mathbf{Q}\cdot\mathbf{Q})+\frac{1}{4}U(\mathbf{Q}:\mathbf{Q})^2$$
$$+\frac{L_1}{2ckT}\nabla\mathbf{Q}:(\nabla\mathbf{Q})^T+\frac{L_2}{2ckT}(\nabla\cdot\mathbf{Q})\cdot(\nabla\cdot\mathbf{Q}) \tag{2.34}$$

where the first line is the homogeneous (f_h) and the second is the gradient (f_g) contribution; L_1 and L_2 are the Landau coefficients. In this format, comparing Eqs. (2.12) and (2.29) gives $L_1 = K_{22}/2S^2$, $L_2 = K - K_{22}/S^2$, and $K = K_{11} = K_{33}$. The presence of the homogeneous energy allows the resolution of defect cores and the prediction of defect nucleation and coarsening. Expanding the forces $(\mathbf{t}^s, \hat{\mathbf{Q}})$ in terms of fluxes $(\mathbf{A}, ckT\mathbf{H})$ in Eq. (2.33), and taking into account thermodynamic restrictions and the symmetry and tracelessness of the forces and fluxes, the equations for $\mathbf{t}^s$ and $\hat{\mathbf{Q}}$ can be obtained. The dynamics of the tensor order parameter is given by

$$\mathrm{Er}\,\hat{\mathbf{Q}}^* = \mathrm{Er}\left[\frac{2}{3}\beta\mathbf{A}^* + \beta\left[\mathbf{A}^*\cdot\mathbf{Q}+\mathbf{Q}\cdot\mathbf{A}^*-\frac{2}{3}(\mathbf{A}^*:\mathbf{Q})\mathbf{I}\right]\right.$$
$$-\frac{1}{2}\beta[(\mathbf{A}^*:\mathbf{Q})\mathbf{Q}+\mathbf{A}^*\cdot\mathbf{Q}\cdot\mathbf{Q}+\mathbf{Q}\cdot\mathbf{A}^*\cdot\mathbf{Q}$$
$$+\mathbf{Q}\cdot\mathbf{Q}\cdot\mathbf{A}^*-\{(\mathbf{Q}\cdot\mathbf{Q}):\mathbf{A}^*\}\mathbf{I}]\Big]$$
$$-\frac{3}{U}\cdot\frac{\mathbb{R}}{(1-\frac{3}{2}\mathbf{Q}:\mathbf{Q})^2}\left[\left(1-\frac{1}{3}U\right)\mathbf{Q}-U\mathbf{Q}\cdot\mathbf{Q}+U\left\{(\mathbf{Q}:\mathbf{Q})\mathbf{Q}+\frac{1}{3}(\mathbf{Q}:\mathbf{Q})\mathbf{I}\right\}\right]$$
$$+\frac{3}{(1-\frac{3}{2}\mathbf{Q}:\mathbf{Q})^2}\left[\nabla^{*2}\mathbf{Q}+\frac{1}{2}L_2^*\left[\nabla^*(\nabla^*\cdot\mathbf{Q})\right.\right.$$
$$+\{\nabla^*(\nabla^*\cdot\mathbf{Q})\}^T-\frac{2}{3}\mathrm{tr}\{\nabla^*(\nabla^*\cdot\mathbf{Q})\}\mathbf{I}\Big]\Big] \tag{2.35}$$

where $t^* = \dot{\gamma}t$, $\mathbf{A}^* = \mathbf{A}/\dot{\gamma}$, $\mathbf{W}^* = \mathbf{W}/\dot{\gamma}$, $\nabla^* = H\nabla$, $L_2^* = L_2/L_1$, $\dot{\gamma}$ is a characteristic shear rate, and β is a shape parameter. The first brackets denote flow-induced orientation, the second phase ordering, and the third gradient elasticity. The total extra stress tensor $\mathbf{t}^t$ for liquid crystalline materials is given by the sum of symmetric viscoelastic stress tensor $\mathbf{t}^s$, antisymmetric stress tensor, and Ericksen stress tensor $\mathbf{t}^{\mathrm{Er}}$ (Grecov and Rey, 2004; Rey, 2007, 2009, 2010):

$$\mathbf{t}^t = \mathbf{t}^s + \mathbf{t}^a + \mathbf{t}^{\mathrm{Er}} \tag{2.36}$$

Summing up all the contributions and nondimensionalizing the following we found:

$$
\tilde{\mathbf{t}}^t = \frac{\mathrm{Er}}{\mathbb{R}}\left(v_1^*\mathbf{A}^* + v_2^*\left\{ \mathbf{Q}\cdot\mathbf{A}^* + \mathbf{A}^*\cdot\mathbf{Q} - \frac{2}{3}(\mathbf{Q}:\mathbf{A}^*)\mathbf{I} \right\} \right)
$$
$$
+ \frac{\mathrm{Er}}{\mathbb{R}}\left(v_4^*\left[(\mathbf{A}^*:\mathbf{Q})\mathbf{Q} + \mathbf{A}^*\cdot\mathbf{Q}\cdot\mathbf{Q} + \mathbf{Q}\cdot\mathbf{A}^*\cdot\mathbf{Q} \right.\right.
$$
$$
+ \mathbf{Q}\cdot\mathbf{Q}\cdot\mathbf{A}^* - \left\{(\mathbf{Q}\cdot\mathbf{Q}):\mathbf{A}^*\right\}\mathbf{I}])
$$
$$
+ \frac{3}{U}\left[-\frac{2}{3}\beta\,\mathbf{H} - \beta\left\{ \mathbf{H}\cdot\mathbf{Q} + \mathbf{Q}\cdot\mathbf{H} - \frac{2}{3}(\mathbf{H}:\mathbf{Q})\mathbf{I} \right\} \right]
$$
$$
+ \frac{3}{2U}\beta[(\mathbf{H}:\mathbf{Q})\mathbf{Q} + \mathbf{H}\cdot\mathbf{Q}\cdot\mathbf{Q} + \mathbf{Q}\cdot\mathbf{H}\cdot\mathbf{Q} + \mathbf{Q}\cdot\mathbf{Q}\cdot\mathbf{H} - \{(\mathbf{Q}\cdot\mathbf{Q}):\mathbf{H}\}\mathbf{I}]
$$
$$
+ \frac{3}{U}(\mathbf{H}\cdot\mathbf{Q} - \mathbf{Q}\cdot\mathbf{H}) + \frac{3}{R}\left[-\nabla^*\mathbf{Q}:(\nabla^*\mathbf{Q})^T - \frac{L_2}{L_1}(\nabla^*\cdot\mathbf{Q})\cdot(\nabla^*\mathbf{Q})^T \right] \tag{2.37}
$$

where

$$
\tilde{\mathbf{t}}^t = \frac{t^t}{ckT^*} \qquad v_1^* = \frac{v_1\,6D_r}{ckT^*} \qquad v_2^* = \frac{v_2\,6D_r}{ckT^*} \qquad v_4^* = \frac{v_4\,6D_r}{ckT^*} \tag{2.38}
$$

The total dimensionless extra stress tensor [Eq. (2.38)] is neither symmetric nor traceless. In this model there are three viscosity coefficients. The term introduced by β indicates backflow. The term $\mathbf{H}\cdot\mathbf{Q} + \mathbf{Q}\cdot\mathbf{H}$ is the asymmetric stress, and the last is the purely elastic Ericksen stress.

2.2.4.2 Rheological Functions

Projecting the LdG into the LE model, we find the six Leslie coefficients (Grecov and Rey, 2003b, 2004; Rey, 2009):

$$
\alpha_1 = \bar{\eta}\left(2v_4^*S^2 - \beta^2 S^2\left(\frac{8}{9} - \frac{8}{9}S + \frac{S^2}{12} \right) \right)
$$
$$
\alpha_2 = \bar{\eta}\left(-S^2 - \frac{1}{3}\beta S\left(2 + S - \frac{S^2}{2} \right) \right)
$$
$$
\alpha_3 = \bar{\eta}\left(S^2 - \frac{1}{3}\beta S\left(2 + S - \frac{S^2}{2} \right) \right)
$$
$$
\alpha_4 = \bar{\eta}\left(v_1^* - \frac{2}{3}v_2^*S + \frac{1}{3}v_4^*S^2 + \frac{4}{9}\beta^2\left(1 - S - \frac{S^2}{4} \right) \right) \tag{2.39}
$$
$$
\alpha_5 = \bar{\eta}\left(v_2^*S + \frac{1}{3}\beta S\left(2 + S - \frac{1}{2}S^2 \right) + \frac{1}{3}\beta^2 S(4 - S - S^2) \right)
$$
$$
\alpha_6 = \bar{\eta}\left(v_2^*S - \frac{1}{3}\beta S\left(2 + S - \frac{1}{2}S^2 \right) + \frac{1}{3}\beta^2 S(4 - S - S^2) \right)
$$

where $\bar{\eta} = ckT^*/6D_r$. According to experimental data on rodlike nematics, the Miesowicz viscosities are connected as follows (Rey, 2009, 2010; Simoes and Domiciano, 2003):

$$\eta_1 + \eta_2 + 8\eta_3 = C_1 + C_2(\eta_1 - \eta_2) \tag{2.40}$$

where C_1 is a constant and C_2 is bounded by $2.77 < C_2 < 3.84$. In the present model, if we only retain linear terms in S in Eq. (2.40) we find that

$$C_2 = \frac{8\beta^2 + 16v_2^*}{v_2^* + 4\beta} \tag{2.41}$$

For aligning rods ($\beta > \frac{6}{5}$) the linearized model is consistent with experiments if $v_2^* > 0.57$. Expressions (2.39) allow to express the reactive parameter λ, the shear viscosities (η_1, η_2, η_3), and the normal stress difference N_1 in terms of the scalar order parameter S. For example, the reactive parameter is (Grecov and Rey, 2003b)

$$\lambda = \frac{\beta(4 + 2S - S^2)}{6S} \tag{2.42}$$

Rods will always align if $\beta > \frac{6}{5}$ and disks if $\beta < -\frac{6}{5}$. In this model β is interpreted in terms of the geometry of the rheological flowing unit. For De < 1, the predictions are obtained by replacing S by its equilibrium value S_{eq}:

$$S_{eq} = \frac{1}{4} + \frac{3}{4}\sqrt{1 - \frac{8}{3U}} \tag{2.43}$$

For De $\gg$ 1 numerical solutions are required. In this regime the Carreau–Yasuda model becomes (Grecov and Rey, 2003b)

$$\eta_s = \frac{\eta - \eta_\infty}{\eta_{al} - \eta_\infty} = \left[1 + (\tau \, \mathrm{Er})^a\right]^{(n-1)/a} \tag{2.44}$$

where η_∞ is the plateau viscosity when S is close to 1, and the parameters τ, a, n refer to the De $\gg$ 1 regime. Hence the LdG model predicts a viscosity curve with three plateaus and two shear thinning regions (Grecov and Rey, 2003b). As shown in Lhuillier and Rey (2004a,b), the LdG model emerges from the Doi–Hess molecular model based on the extended Maier–Saupe potential. For further discussions of the Doi–Hess molecular model, related nematodynamic models, and rheological applications see Larson and Doi (1991), Larson (1999), Rey (2007, 2009, 2010), and Rey and Denn (2002).

2.3 DEFECTS AND TEXTURES

2.3.1 Defects in Nematics

Defects in nematic liquid crystals (NLCs) are classified as singular and non-singular (Larson, 1999; Larson and Doi, 1991; Rey 2009, 2010). Singular defects include point and edge and twist disclination lines; the quantized strength of a disclination line M ($\pm\frac{1}{2}, \pm1, \pm\frac{3}{2}, \ldots$) describes the amount of director rotation when encircling the defect. Singular disclination lines either form loops or end at other defects or bounding surfaces. Since the elastic energy of a defect scales with M^2, the most abundant ones are the $M = \pm\frac{1}{2}$. In the LdG model, the cores of singular disclination lines correspond to the unstable saddles of Eq. (2.45) and singular points correspond to unstable nodes (de Luca and Rey, 2004; Rey, 2009, 2010). Figure 2.7 shows a schematic representation of the stable (sink) uniaxial nematic root ($S = S_{eq}$, $P = 0$), the unstable (source) defect point ($S = P = 0$), and the unstable (saddle) disclination line, which are the roots of

$$\left(1 - \tfrac{1}{3}U\right)\mathbf{Q} - U\mathbf{Q} \cdot \mathbf{Q} + U\left\{(\mathbf{Q}:\mathbf{Q})\mathbf{Q} + \tfrac{1}{3}(\mathbf{Q}:\mathbf{Q})\mathbf{I}\right\} = \mathbf{0} \qquad (2.45)$$

In addition to singular defects, nonsingular defects arise in the form of lines and inversion walls (de Luca and Rey, 2003, 2004, 2006a–c; Farhoudi and Rey 1993a–c; Grecov and Rey, 2003a,b; Hwang and Rey, 2006a,b; Larson

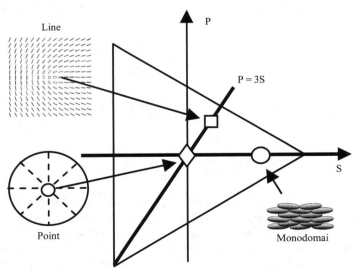

Figure 2.7 Schematic representation of stable and unstable roots of the free energy [Eq. (2.45)]. Defect free monodomain (circle) is a stable root, the point defect (diamond) is an unstable node, and the disclination line is an unstable saddle.

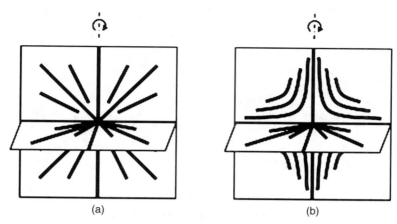

Figure 2.8 Schematic representation of (a) a radial +1 and (b) a hyperbolic +1 point defect.

and Doi, 1991; Lhuillier and Rey, 2004a,b; Rey, 2000a–c, 2001a,b, 2007, 2009, 2010; Soule et al., 2009; Stewart, 2004; Tsuji and Rey, 1988, 1997, 2000; Yan and Rey, 2002; Zhou et al., 2006).

Nonsingular defects are captured by the LE model since they do not involve changes in the order parameters. Nonsingular disclination lines appear to minimize the energy through out-of-plane director escape, thus avoiding orientation singularities. Inversion walls appear in the presence of external fields, when the field-induced orientation has a degeneracy such that clockwise and anticlockwise rotations are equally possible.

Nonsingular orientation wall defects are two-dimensional (2D) defects that may arise under the presence of external fields, such as flow fields and electromagnetic fields (de Gennes and Prost, 1993; Rey, 2009).

2.3.1.1 Defects Point defects are singular solutions to the static LE equations in spherical coordinates (de Gennes and Prost, 1993; Rey, 2009). Figure 2.8 shows schematics of radial and hyperbolic hedgehog point defects of strength $M = +1$. The charge M_p of a point defect is defined by de Andrade Lima et al. (2006a–d) and de Andrade Lima et al. (2006):

$$M_p = \left| \frac{1}{8\pi} \oiint d\mathbf{S} \cdot \boldsymbol{\varepsilon} : (\nabla \mathbf{n} \times \nabla \mathbf{n}) \cdot \mathbf{n} \right| \qquad (2.46)$$

which indicates that radial and hyperbolic point defects carry the same charge and can be transformed into each other continuously. In nematics the combined charge of two point defects is either $M_p = |M_{p1} - M_{p2}|$ or $M_p = |M_{p1} + M_{p2}|$, and hence different outcomes are possible from the reaction of two points. The director energies associated with these bulk point defects are

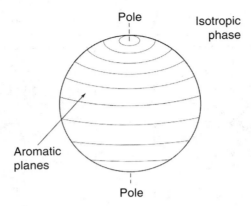

Figure 2.9   Schematic representation of the Brooks and Taylor nematic spherulite. At the poles there is a +1 defect point and the surface orientation is edge-on.

$E_{\text{radial}} = 8\pi KR < E_{\text{hyper}} = 8\pi KR/3$. Point defects appear in the formation of CMs through nucleation, growth, and impingement of nematic spherulites. For planar director orientation point defects appear on the surface of the spherulite. According to differential geometry, the Euler characteristic of a surface χ is the average Gaussian curvature K $[K = (1/R_1)(1/R_2); R_{1,2}$ is the radius of curvature] and for a sphere of radius R the Euler characteristic is (Rey et al., 2005)

$$\chi = \frac{1}{2\pi} \oiint \frac{1}{R^2} \, dS = 2 \tag{2.47}$$

According to the Poincaré–Brouwer theorem, the total charge (i.e., strength of the singularities) of a vector field on a surface is equal to the Euler characteristic of the surface (Rey, 2009). This theorem can then be used to predict surface defects on NLCs. For tangential surface orientation, the theorem predicts that in a nematic spherulite the total charge of surface defects is +2. For carbonaceous mesophase spherulites, it was found that in the micron range two defect points of charge +1 located each at the poles are observed, as indicated in Figure 2.9 and in agreement with the prediction of the Poincaré–Brouwer theorem (Rey, 2009). Using the static LE model with elastic isotropy, it is shown that the elastic energy associated with the director field of a spherulite of radius R is $E = 5\pi KR$, which is less than the spherulite with a +1 defect at the center of the sphere with $E = 8\pi KR$ (Rey, 2009).

Point defects also arise under capillary confinement, as shown in Figure 2.10 (de Luca and Rey, 2006a–c, 2007). For rodlike nematics in capillaries and homeotropic anchoring, a periodic array of alternating +1 and −1 point defects along the axis are usually observed; the defect–defect separation distance is of the order of the capillary radius. This texture forms when the escape mechanism is selected, but since clockwise and anticlockwise escape is possible,

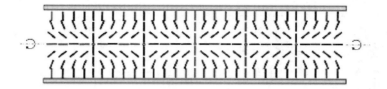

Figure 2.10 Schematic representation of an NLC confined in a capillary, displaying a periodic array of +1 and −1 point defects along its axis. This texture is known as escaped radial with point defects, ERPD.

defect point texture arises. This metastable texture is known as an escape radial with point defects (ERPD). Hence we would expect that for micellar nematics under capillary confinement a similar periodic point defect texture may arise.

In general, texturing due to confinement requires that the characteristic system of size H be greater than the extrapolation length ℓ_{extra}:

$$\ell_{extra} = \frac{K}{\gamma_2} \ll H \qquad (2.48)$$

and in principle for a given LC this length can be changed by modifying the interface.

Point defects have another aspect of interest to the understanding of texturing processes. Figure 2.11a shows the connection between point defects and disclination loops, given in terms of director trajectories. The figure shows that a radial defect point is equivalent to a $-\frac{1}{2}$ disclination loop and that a hyperbolic defect point is equivalent to a $\frac{1}{2}$ disclination loop. Figure 2.11b shows the corresponding LdG computations of a defect loop at the center of a sphere; this figure and associated computations demonstrate that radial +1 points are disclination loops and that the radius of the loop is approximately four to five times the molecular length scale.

2.3.1.2 Disclination Lines Wedge and twist disclinations are possible according to the rotation axis: Wedge (twist) lines are parallel (perpendicular) to the rotation axis (de Gennes and Prost, 1993; Rey, 2009). Wedge disclination lines are associated with planar orientation given by $\mathbf{n} = (\cos\varphi, \sin\varphi, 0)$, $\varphi = M\phi + \phi_0$, where φ is the director angle with respect to the x direction, ϕ is the polar angle of a cylindrical coordinate system, M is the strength of the disclination, and ϕ_0 is a constant that determines the overall orientation of the defect. Figure 2.12 shows director fields around typical singular wedge disclination lines. Since defects have a core and a distortion energy associated, short- and long-range energy calculations are used to establish their stability (Rey, 2009). It turns out that the defect energy is proportional to the disclination

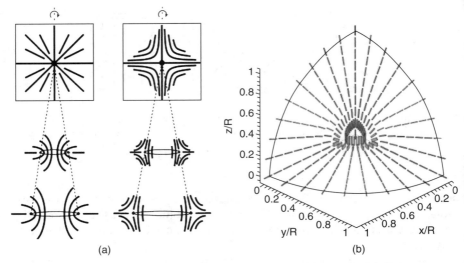

Figure 2.11   (a) Radial (hyperbolic) point defects are equivalent to $\frac{1}{2}(-\frac{1}{2})$ disclination loops. (b) Computed **Q** tensor visualization at the center of a sphere: Point defects are disclination loops. [Adapted from Tsuji and Rey (1998).]

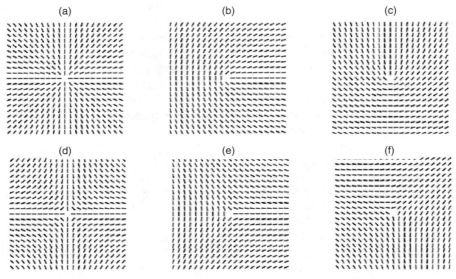

Figure 2.12   Director fields around singular wedge disclinations. (a) $M = +1$, (b) $M = +\frac{1}{2}$, (c) $M = +\frac{1}{2}$ and $\phi_0 = \pi/4$, (d) $M = -1$, (e) $M = -\frac{1}{2}$, and (f) $M = -\frac{1}{2}$ and $\phi_0 = \pi/3$.

strength squared and to the three Frank elastic constants (K_{11}, K_{22}, K_{33}) involved in the deformation associated with the disclination. It has been shown that elastic anisotropy controls the stability of the different classes of disclinations, and thus the relative abundance of certain types of defects (Rey, 2009, 2010). For discotic nematics it is known that $K_{22} > K_{11}$, $K_{22} > K_{33}$. Furthermore it is expected that for low-molar-mass discotics $K_{11} > K_{33}$, and for larger molecular weight $K_{33} > K_{11}$ (Rey, 2009; Wang and Rey, 1997).

The effects of elastic anisotropy on the stability of $S = \pm\frac{1}{2}$ lines are: (a) wedge disclinations are favored when $K_{22} > (K_{11} + K_{33})/2$ and are stable against out-of-plane perturbations; and (b) twist disclinations are favored when $K_{22} < (K_{11} + K_{33})/2$ and are unstable against out-of-plane perturbations. As a consequence, the predictions are that discotic mesophases wedge disclinations of $S = \pm\frac{1}{2}$ should be more abundant than twist disclinations of the same strength. Zimmer and White (1982) report the existence of both. For $S = 1$ it is found that the only stable wedge lines against out-of-plane perturbations are those with purely radial or azimuthal director orientation (Rey, 2009; Zimmer and White, 1982). Whether these lines have singular or nonsingular cores will depend on the degree of anisotropy, the nature and size of the core, and the confinement. For relatively weak elastic anisotropy, it is expected that nonsingular cores will prevail. The nature of the core in disclinations of unit strength has been characterized by Zimmer and White (Rey, 2009; Zimmer and White, 1982), and they show that $S = \pm1$ lines have nonsingular cores in the bulk but discontinuous near free surfaces. As shown above for C/C composites made of millimeter size fibers with a fiber arrangement of $N = 4, -1$ nonsingular lines are observed. Nevertheless we can expect that for submicron fibers two $-\frac{1}{2}$ singular disclinations have less energy than a single -1 nonsingular line.

2.3.2   Defect Rheophysics

In this section we discuss some fundamental concepts necessary to characterize defects and textures in sheared flow-aligning LCs and CMs.

2.3.2.1   Defect Nucleation Processes   Flow-induced defect nucleation in nonaligning lyotropic crystal polymer (LCP) solutions is associated with the lack of steady flow alignment and the presence of spatial gradients of rotational director kinetics (Tsuji and Rey, 1998). Two neighboring regions whose average molecular orientation rotate at different speeds will create interfacial gradients that will be compatibilized by defect nucleation. Flow-induced defect nucleation in flow-aligning LCs is not associated with tumbling processes as in sheared nonaligning LCs since the latter tend to align within the shear plane and close to the shear flow direction at an angle known as the Leslie angle θ_L [see Eq. (2.18)]. On the other hand, other shear-induced processes can clearly lead to defect nucleation in flow-aligning LCs, including (Grecov and Rey, 2003a,b): (1) defect loop emission by the Frank–Reed surface mechanism

(Gupta and Rey, 2005), (2) stretching and pinching of existing loops in the bulk, and (3) heterogeneous reorientation upon flow start up. Direct numerical simulation of all these three coexisting nucleation processes for flow-aligning LCs is beyond existing computational power since the ratio of a defect core to typical shear cell sizes [the energy ratio $\mathbb{R}$ in Eq. (2.10)] is at least five or more orders of magnitude. This critical limitation fuels the motivation of using simplified models and theoretical frameworks that provide insights to texture transformations. In a previous work, the loop emission process and its impact on rheology were investigated (Rey, 1993a,b). Numerical simulations based on heterogeneous reorientation (Gupta and Rey, 2005; Yan and Rey, 2002, 2003) indicate that the defect nucleation rate $\mathbb{N}$ for $\mathbb{R} = 10^4$–10^6 is well fitted by a power law model (Grecov and Rey, 2003a):

$$\mathbb{N} \simeq 0.01 \Upsilon (\mathrm{Er} - \mathrm{Er_{ADL}}) \sqrt{\mathrm{Er} - \mathrm{Er_{ADL}}} \qquad (2.49)$$

where Υ is the Heaviside function and $\mathrm{Er_{ADL}}$ is the minimum Ericksen number for defect nucleation processes ($\mathrm{Er_{ADL}} = 9 \times 10^4$ for $R = 10^5$). Note that $\mathrm{Er_{ADL}}$ describes the transition between the antisymmetric processes and defect lattice mode. The length scale of the texture $\ell_t = H/\mathbb{N}$ is given by (Grecov and Rey, 2003b)

$$\ell_t = \frac{H}{c \Upsilon (\mathrm{Er} - \mathrm{Er_{ADL}}) \sqrt{\mathrm{Er} - \mathrm{Er_{ADL}}}} \qquad (2.50)$$

where c is a constant and H is the system size. Thus in the absence of coarsening, the texture length scale predicted by LdG decreases with a $-\frac{1}{2}$ power law (Grecov and Rey, 2003b).

2.3.2.2 Texture Coarsening Processes Defect coarsening processes occur simultaneously with defect nucleation. Texture refinement with shear indicates that the defect nucleation rate is higher that the coarsening rate. The coarsening process includes (1) defect–defect annihilation, (2) defect–boundary annihilation, and (3) wall pinching and retraction.

Numerical simulation based on one-dimensional (1D) LdG nematodynamics that take into account the three mechanisms mentioned above predict that in the presence of nucleation and coarsening the texture length scale l_t, given by the system size divided by the number of defects, follows a decreasing function of slope close to $-\frac{1}{2}$, reaches a minimum close to De = 1, and then diverges close to De ≈ 2, as shown in Figure 2.13 for $\mathbb{R} = 10^6$ (Grecov and Rey, 2003a).

According to LdG nematodynamics (Grecov and Rey, 2003a,b, 2004), defect nucleation is only a function of the Ericksen number, and hence changing the temperature U will only affect coarsening processes. The transition dimensionless temperature that indicates the boundary between polydomain and monodomain textures for $\mathbb{R} = 10^6$ is $T/T^* = 1/U \approx 0.4 - 0.18\mathrm{De}^{0.2}$. The predictions indicate that as the temperature of the LC lowers the De number needed to attain a monodomain increases.

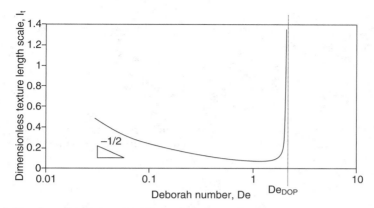

Figure 2.13 One-dimensional LdG predictions of dimensionless texture length scale l_t as a function of Deborah number, showing an initial refinement of slope $-\frac{1}{2}$, reaching a minimum scale close to De = 1 and diverging when De is close to 2. [Adapted from Grecov and Rey (2003a).]

2.4 APPLICATIONS TO SURFACTANT NEMATIC LIQUID CRYSTALS

2.4.1 Micellar Nematics

2.4.1.1 Textures Cross-polar Schlieren textures commonly observed in thermotropes are found to form and coarsen due to confinement effects; confinement in N_c and N_d phases affects some features of the textures. These structures are in the micrometer range and are interpreted in terms of spatial variations of nematic director through the sample. The LdG model of the IN transition quenching replicates the Schlieren texture through the Kibble mechanism (droplet impingement) and through interfacial defect shedding at the IN boundary. Figure 2.14 shows simulations based on the LdG model of Schlieren textures as seen between cross polars. The connection between the brushes and the disclination strength is indicated in the caption.

Figure 2.15 shows computed texture visualizations of coarsening processes through disclination–disclination annihilation. Texture coarsening is a self-similar process with a scaling law in d dimension for the defect density of $\rho(t)^{-1/d} \approx (Dt)^n$, where for 2D, $n = 0.35$–0.37, while 5CB = 0.5 was reported. The LdG-predicted exponent n is a function of the length scale ratio R, and for $R \to \infty$ the dispersive limit $n = \frac{1}{2}$ is recovered. Coarsening under strong confinement effects and other external fields modifies both the defect–defect interaction and the backflow associated with annihilation.

2.4.1.2 Flow Birefringence in the Isotropic Phase Flow birefringence in the isotropic phase is observed in all classes of liquid crystals and is a manifestation of flow-induced orientation and ordering due to viscous torques acting on the anisodiametric mesogens. Flow birefringence in the I phase of

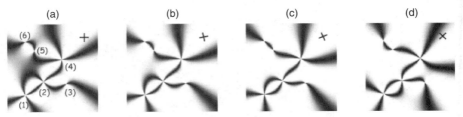

Figure 2.14   Computed Schlieren textures for different orientations of the cross polars indicated by the cross. The charge of the wedge disclinations enumerated on the left panel are: (1) −1, (2) +1, (3) −$\frac{1}{2}$, (4) −1, (5) +$\frac{1}{2}$, (6) −$\frac{1}{2}$.

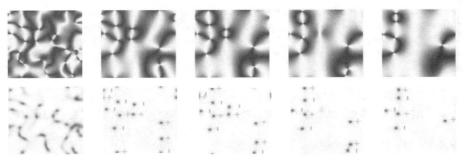

Figure 2.15   Computed visualization based on the LdG model of texture generation by the Kibble mechanism and coarsening through defect annihilation. The top is the polarized light intensity and the bottom the +$\frac{1}{2}$ and −$\frac{1}{2}$ disclinations. [Adapted from Tsuji and Rey (1998).]

potassium laurate/decanol/water has been characterized experimentally for conditions close to the I/lamellar transition. The flow birefringence due to transient shear flow was detected using optical transmittance, which is proportional to the square of the order parameter S produced by the shear flow. By fitting experimental data with the LdG, the orientational diffusivity and the intermicellar characteristic length scale can be estimated. Here, the LdG model for flow birefringence under steady flow is used to illustrate the estimation of rheological parameters from the experimental signal. Using a planar director field, simple shear, and steady uniaxial nematodynamics [Eq. (2.33)], the following relation between flow alignment θ_{al} and S is predicted (Rey, 2010):

$$\cos 2\theta_{al} = \frac{1}{\lambda} = \frac{6S}{\beta\left(4 + 2S - S^2\right)} \qquad 1 + \left(\left(\frac{(3-U)S - US^2 + 2US^3}{D_e}\right)^2 - 1\right)\lambda^2 = 0$$

$$(2.51)$$

where for simplicity the equality $18\eta = \zeta$ is used to relate translational (η) and rotational (ζ) viscosities [see, e.g., Eq. (2.30) of Rey (2010)]. Figure 2.16a shows the computed scalar order parameter S and the alignment angle as a function of De for steady shear of an isotropic solution, for $\beta = 0.95$ and several U values; the shapes of S curves are consistent with experiments (Gobeaux et al., 2007; Rey, 2010). Flow-induced birefringence is due to increasing S and decreasing θ_{al} as shear rate increases.

From Figure 2.16a it is seen that the limits of a shear rate (Deborah number) scan then are

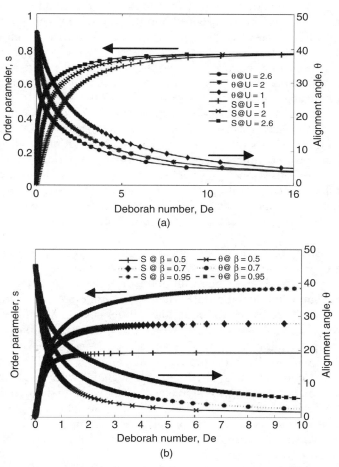

Figure 2.16 Flow-induced birefringence ($S > 0, 0 < \theta_{al} < \pi/4$) as a function of De of a sheared isotropic solution of rods: (a) different concentration of rods (U) and $\beta = 0.95$; increasing U increases the initial slope dS/dDe. (b) Same but for $U = 2$ and different β values; increasing β increases the plateau value of S. [Adapted from Rey (2010).]

$$\text{De} \ll 1 \qquad S_0 \approx 0 \qquad \theta_{al} \approx \pi/4 \qquad \text{De} \gg 1$$
$$S_\infty \approx (1 - 3/\beta) + \sqrt{(1 - 3/\beta)^2 + 4} \qquad \theta_{al} \approx 0 \tag{2.52}$$

and hence S (θ_{al}) increases (or decreases) monotonically from zero ($\pi/4$) to saturate at a β-dependent plateau S_∞ (0). Figure 2.16b shows that flow birefringence increases with increasing β (or rheological shape anisotropy); this prediction can be used to estimate β and then λ (the reactive parameter) through Eq. (2.51).

2.4.1.3 Phase Transition Phenomena under Shear Superposing thermodynamics and rheology provides information of the effect of temperature and deformation rates in the ordered and disordered phases as well as nonequilibrium phase transitions. The application of flow extends and modifies the thermodynamic branches (Golmohammadi and Rey, 2009). For lyotropic nematic polymers it has been shown that the zero shear viscosity as a function of rod concentration has a strong peak around the transition region (Larson, 1999). For lyotropic nanotube nematics a similar peak is observed at the transition (Kuzma et al., 1989). Here, similar phenomena for smetic nematic liquid crystals (SNLCs) are discussed. The thermorheology of lyotropic nematic liquid crystals (LNLCs) surfactant solutions such as sodium dodecyl sulfate (SDS)/decanol/water indicate that by increasing temperature the expected phase and transitions $N_d \rightarrow N_c \rightarrow I$ are reflected in the low-shear-rate viscosity. Using capillary linear rheology, the viscosity and the corresponding phases (Kuzma et al., 1989) show a superposition of the Arrhenius behavior of viscosity superposed to the orientation and morphology-dependent effects; see Figure 2.17.

The viscosity peaks are associated with transitional phenomena including spherulite formation of the new phase. The viscosity of the N_d phase is lower than the N_c phase, and this is the result of the orientation effect and the change in micellar shape and nematic state. The orientation effect on viscosity in the N_d phase is described by the LE equations. Under capillary confinement, face-on anchoring at the walls and no flow, the three possible onion textures are director escape $E+$ along the flow, director escape $E-$ opposite to the flow, and singular disclination $+1$ defect core D_c at the centreline; see Figure 2.18.

The escape core textures ($E+$, $E-$) are associated with low elasticity and high dissipation and the D_c with high elasticity and low dissipation. Hence the apparent viscosity is a strong function of the core structure. Using Eq. (2.53), the apparent viscosities for the three onion textures under capillary flow are shown in dimensionless format as a function of the Ericksen number (dimensionless pressure drop) in Figure 2.19.

The onion textures are shear thickening (thickening slope close to one-fifth) and the zero shear apparent viscosities are ordered as follows (de Andrade Lima and Rey, 2003a, 2004e):

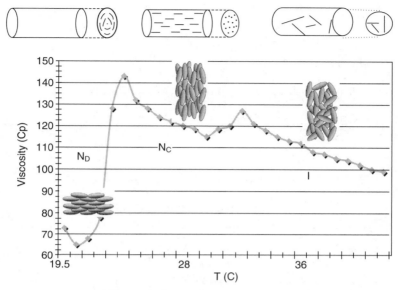

Figure 2.17 Apparent shear viscosity as a function of temperature for a 35/7/58 wt % SDS/decanol/water solution obtained using capillary rheometry. The N_d, N_c, and I regions are indicated. The peak viscosity regions are the result of transitional effects, and the viscosity decrease is due to the usual Arrhenius effect. The top schematics show corresponding side views of orientation patterns in the capillary rheometer. [Adapted from Kuzma et al. (1989).]

$$\eta_{appE+} = \eta_{appE-} = \eta_1 \left\{ 1 + 8\left(1 - \frac{\eta_2}{\eta_1}\right)\left[1 + \left(1 - 2\frac{\eta_2}{\eta_1}\right)\ln\left(4\frac{\eta_2}{\eta_1}\right)\right.\right.$$
$$\left.\left. + 2\frac{\eta_1/\eta_2\left(\frac{1}{2} - \eta_2/\eta_1\right)^2 - (1 - \eta_2/\eta_1)}{\sqrt{\eta_1/\eta_2 - 1}}\tan^{-1}\sqrt{\frac{\eta_1}{\eta_2} - 1}\right]\right\}^{-1} > \eta_{appD_c} = \eta_2$$

$$(2.53)$$

The lowest zero shear apparent viscosity is that of the singular onion textures. For typical thermotropes it was found that $\eta_{appE+} = \eta_{appE-} \approx 10\eta_{appD_c}$ (de Andrade Lima and Rey, 2003a, 2004e). Hence core restructuring can lead to viscosity decrease. Another factor that can account for the lower N_d viscosity with respect to N_c is that the viscosity of the latter is associated with a flow-aligned state at low shear if the wall anchoring is along the flow direction. In this case, the N_c viscosity is just η_1 because the director is along the flow direction and no texture appears. The LE model is consistent with experimental data if $10\eta_2(N_d, T_1) < \eta_1(N_c, T_2)$; $T_1 < T_2$. In the I phase the viscosity is $\alpha_4(I)$. For calamitic thermotropes, it is found that extrapolating $\alpha_4(I)$ to lower

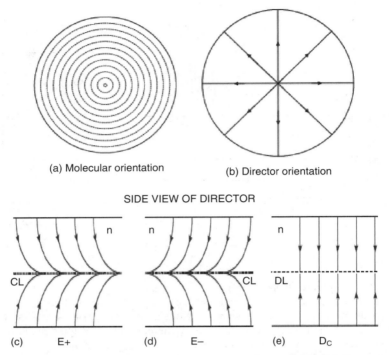

(a) Molecular orientation   (b) Director orientation

SIDE VIEW OF DIRECTOR

(c)   E+   (d)   E−   (e)   D$_C$

Figure 2.18   (a) Cross-sectional view of a discotic onion texture. (b) Corresponding radial director **n** orientation. (c–e) The right panel shows the side view of the three possible director field lines of the onion textures: positive escape $E+$, negative escape $E-$, and singular +1 disclination defect core D_c; CL denotes the centerline.

temperatures the viscosity superposes with $\eta_3(N_c)$, which is greater than $\eta_1(N_c)$ (Chandrasekhar, 1992). This trend is identical to that observed with the viscosities of the N_c and I in this SDS solution.

2.4.1.4   *Orientation Fluctuations, Backflow*

Thermally excited orientation fluctuations have a relaxation time that is proportional to the viscosity of the fluid and to the inverse of the effective Frank elastic modulus. The orientation diffusity $D_{0,i}$ for splay, twist, and bend modes is $D_{o,i}^{j} = K_{ii}/\eta_i$; i = splay(11), twist(22), bend(33); j = calamitic (C), discotic (D), measuring $D_{o,i}^{j}$ provides information on material viscoelastic anisotropies (K_{ii}/η_i) and the relative magnitude of the ratio $D_{o,i}^{j}/D_{o,k}^{j}$ for the different modes also reflects the molecular shape (disk or rod) (Santos and Amato, 1999). In addition since the transient reorientation viscosities for splay and bend contain backflow information, this important rheological process can be accessed through the orientation diffusion measurements (Santos and Amato, 1999). Extensive data with potassium

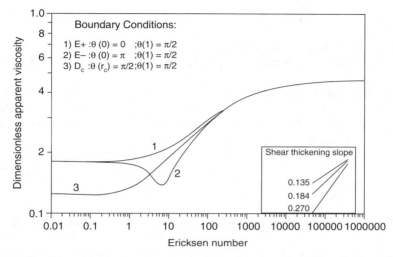

Figure 2.19 Dimensionless apparent viscosity (η) as a function of the Ericksen number. Er for the escaped (1: $E+$ and 2: $E-$) and (3: D_c) onion textures. [Adapted from de Andrade Lima and Rey (2004e).]

TABLE 2.3 **Comparison of Orientational Diffusivity Coefficients of Surfactant Nematic Calamitic (N_c) and Disklike (N_d) Nematic Phases with a Thermotropic Calamitic (MBBA)**

Orientation Diffusivity $D_{0,j}$ (m^2/s)	Calamitic N_c (LDhW) $T = 29.4°C$	Discotic N_d (LDhW) $T = 19°C$	Discotic N_d (DHCDhW) $T = 29.4°C$	Calamitic Thermotropic MBBA $T = 25°C$
$D_{0,\text{splay}}$	$\sim 0.5 \times 10^{-12}$	11.5×10^{-12}	4.5×10^{-12}	56×10^{-12}
$D_{0,\text{twist}}$	$\sim 0.42 \times 10^{-12}$	1.6×10^{-12}	0.30×10^{-12}	43×10^{-12}
$D_{0,\text{bend}}$	10.7×10^{-12}	—	0.33×10^{-12}	430×10^{-12}
$D_{0,\text{splay}}/D_{0,\text{twist}}$	~ 1	7.2	15	1.3
$D_{0,\text{bend}}/D_{0,\text{twist}}$	25	1	1.1	10

Source: Adapted from Santos and Amato (1999).

laureate/decanoe/D_2O (N_c), DHCDhW (N_d) were compared with MBBA, an N_c thermotrope; see Table 2.3.

Table 2.3 shows that the anisotropy in thermotropic and lyotropic N_c phase is similar:

$$D_{o,\text{splay}}^C \approx D_{o,\text{twist}}^C \qquad D_{o,\text{bend}}^C \gg D_{o,\text{twist}}^C \qquad (2.54)$$

which is attributed to the fact that splay backflow is weak while bend backflow is strong: $\alpha_3^2/\eta_1 \ll \alpha_2^2/\eta_2$, $\gamma_1 \gg \eta_{\text{bend}}$. On the other hand, for N_d lyotropic phases the anisotropy is reversed:

$$D_{o,splay}^{D} \gg D_{o,twist}^{D} \qquad D_{o,bend}^{D} \approx D_{o,twist}^{D} \tag{2.55}$$

which is attributed to the fact that for discotics, bend backflow is weak while splay backflow is strong: $\alpha_3^2/\eta_1 \gg \alpha_2^2/\eta_2$, $\gamma_1 \gg \eta_{splay}$ (Santos and Amato, 1999).

Backflow plays an important role in startup, cessation, reversal of flow, and magnetic reorientation. For startup shear flow, for example, backflow or reorientation-induced flow is a rotational viscosity reduction mechanism since the rotation viscosity γ_1 will be reduced for planar modes:

$$\mathbb{R} = \gamma_1 - \frac{\left(\alpha_2 \sin^2 \theta - \alpha_3 \cos^2 \theta\right)^2}{\eta_1 \cos^2 \theta + \left(\eta_2 + \alpha_1 \cos^2 \theta\right)\sin^2 \theta} \tag{2.56}$$

where θ is the director angle from the flow direction. Pattern formation under magnetic fields is observed in lyotropic rodlike nematic polymers (Berret, 1997; Berret et al., 2001; Roux et al., 1995) and also in potassium laurate/decanol/water solutions (Kuzma et al., 1989) where the viscosity reduction due to backflow produces a spatially periodic fastest growing mode; for static fields, the patterns in both systems have different symmetries. For lyotropic nematic polymers, Figure 2.20 shows the initial (left) and final stage (right) of the response to an imposed magnetic field H along x.

Banded patterns of nematic polymers in a thin film of thickness D of wavelength L arise for sufficiently large external magnetic fields H normal to an initially planar director with the growth rate s:

$$s = \frac{\chi_a (HD/\pi)^2 - \left(K_{22} + K_{33}(D/L)^2\right)}{\eta_{bend} - \eta_{twist}(L/D)^2} \tag{2.57}$$

where the numerator is the net driving force given by the field minus the resisting elastic torque, and the numerator is the operating viscosity η_{bend} due to bending of the director in the plane with a superposed twist viscosity η_{twist} due to confinement in thickness in the y direction. The pattern symmetry in

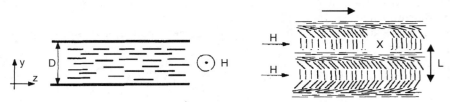

Figure 2.20 Pattern formation of a nematic LC with positive magnetic susceptibility under an external magnetic field H (left). The fastest growing mode has a periodicity L in the z direction (right).

the surfactant N_d solution is different because of the different sign in the magnetic susceptibility χ_a [see Eq. (2.57)].

2.4.1.5 Flow Alignment As mentioned earlier (Section 2.2), the ability of the director **n** to orient in a shear flow field is a dominant rheological property for oscillatory, steady, startup, and cessation of shear flow. Replication of 8CB (thermotropic N_c) experimental steady and transient shear rheology using the LE model in the aligning and tumbling modes has been presented elsewhere (Kundu et al., 2009). The N_c and N_d phases of SDS/decanol/water were investigated in a Couette flow cell for both steady flow $(0.02 \text{ s}^{-1} < \dot{\gamma} < 300 \text{ s}^{-1})$, step change $(0.1 \text{ s}^{-1} < \dot{\gamma} < 10 \text{ s}^{-1})$, and flow reversal. The study indicates that under steady shear, the alignment angles for N_d and N_c are 105° and 12° with respect to the flow direction, respectively. The wall orientation appears to be planar for N_c and face-on for N_d. In order to understand the theory explained in Section 2.2.3, we can analyze the data for the N_c and N_d phases of SDS solutions with the LdG and LE model predictions (Grecov and Rey, 2006). (i) From the reported alignment angles (N_c: 12°; N_d: 105°) and Eq. (2.51), it is found that: $\lambda^C = 1/\cos(2 \times 12) = 1.09$, and $\lambda^D = 1/\cos(2 \times 105) = -1.15$. For thermotropic N_c the following is found: $\lambda^C = 1.02$ (MBBA at 25°C), 1.01 (PAA at 122°C), 1.1 (5CB at 25°C). For thermotropic N_d such as polydispersed carbonaceous mesophases, λ^D values as low as −1.45 were estimated. (ii) The steady-state viscosity as a function of shear rate for N_c and N_d phases has a low shear rate plateau up to $\dot{\gamma} < 10 \text{ s}^{-1}$and a power law shear thinning region of $\eta \sim (\dot{\gamma})^{-m}$, $m \approx 0.55, 0.4, 0.5$, where the first two values are for two N_c phases and the latter for the N_d phase. Using the LdG model, the predicted shear thinning viscosity of a thermotropic N_d was shown to be consistent with experiments. For this case the shear thinning was consistent with texturing under shear, where the texture length scale decreases with a power law of $-\frac{1}{2}$.

Figure 2.21 shows the predictions for the shear viscosity (in the power law region), texture length scale, and visualization of the director field across the shear cell gap as functions of strain (Kundu et al., 2009). In general the Carreau–Yasuda LC model [Eq. (2.16)] includes several viscosity reduction mechanisms, such as the viscous anisotropy and texturing that arise in the LE and LdG models. These mechanisms affect the three parameters τ, a, and n so that specific flow, wall-anchoring, and material conditions need to be specified, as shown elsewhere (Grecov and Rey, 2003a–c). Under flow reversal conditions the stress exhibits oscillations followed by a long relaxation. For shear step-down the stress relaxes quickly with no oscillations. In addition to the particular stress relaxation, the signals superpose by plotting the shear stress as a function of strain scaled with the steady-state value: $\tau(\dot{\gamma}t)/\tau_{ss}(\dot{\gamma})$.

Simulations based on the LE and LdG models for flow aligning N_d and N_c phases with and without defects predict stress overshoots and undershoots as well as scaling behavior under certain orientation kinematics. Stress overshoots and undershoots under transient shear are due to a nonmonotonic

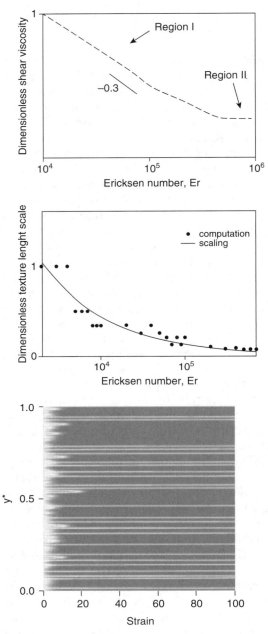

Figure 2.21   Predicted dimensionless viscosity as a function of Erciksen number (left), predicted and measured dimensionless texture length scale as a function of Ericksen number (middle), and director gray-scale visualization as a function of distance and strain for $E = 8 \times 10^5$. As the texture refines, the viscosity decreases with exponent $-\frac{1}{3}$. [Adapted from Kundu et al. (2009).]

stress surface as a function of the in-plane tilt angle and out-of-plane twist angle.

Figure 2.22 shows a gray-scale plot dimensionless shear stress surface τ as a function of twist and tilt angles. In this plot, white corresponds to the local maximum shear stress ($\tau = \tau_{max} = 2.6$) and black corresponds to the local minima ($\tau = \tau_{max} = 0.85$). A good characterization of shear stress response during shear startup is to consider the evolution from the three Miesowicz director orientations toward the steady state (Han and Rey, 1994a–c). By

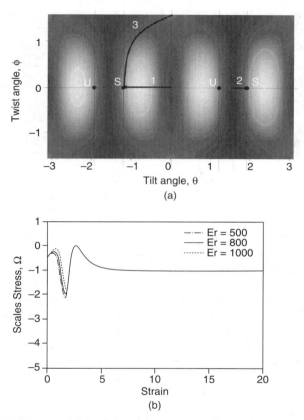

Figure 2.22 (a) Gray-scale visualization of the shear stress surface as a function of the twist and tilt angles for N_d. Light corresponds to local stress maxima and dark to minima. U and S denote the unstable and stable orientation, respectively. Path 1: evolution from velocity direction to flow alignment produces 1 overshoot; path 2: from velocity gradient to flow alignment produces monotonic change; path 3: from vorticity direction to flow alignment produces 1 undershoot. (b) Scaled dimensionless shear stress during flow-start up for three Ericksen numbers, with initial alignment at 60° to flow direction. Overshoots, undershoots, and scaling are predicted. [Adapted from Grecov and Rey (2003c).]

direct observation of Figure 2.22a we obtain the following characteristic transient rheological responses: path 1: evolution from velocity direction ($\theta = 0$, $\theta = 0$), the shear stress, will present one overshoot (path denoted by 1); path 2: evolution from velocity gradient direction ($\theta = 90$, $\theta = 0$), the shear stress will increase monotonically; path 3: evolution from vorticity direction ($\theta = 0$, $\theta = 90$); the shear stress will present an undershoot. Figure 2.22b shows that the shear stress landscape is dense with local maxima and local minima and that monotonic responses are more the exception than the rule. Anisotropic effects introduced by director stress coupling do provide an important mechanism of stress nonmonotonicity in transient shear startup flows. The viscosity maxima, minima, and their difference in the figure are:

$$\eta_{max} = \frac{\alpha_4}{2} + \frac{\alpha_1}{2} + \frac{\alpha_5 - \alpha_6}{\alpha_3 - \alpha_2}\left(\frac{\alpha_3}{2} - \frac{\alpha_2 + \alpha_3}{4}\right) + \frac{\alpha_6}{2} \tag{2.58}$$

$$\eta_{min} = \frac{\alpha_4}{2} + \frac{\alpha_5 - \alpha_6}{\alpha_3 - \alpha_2}\frac{\alpha_3}{2} + \frac{\alpha_6}{2} \tag{2.59}$$

The viscosity jump between these extreme, $\Delta\eta = \eta_{max} - \eta_{min}$, is given by

$$\Delta\eta = \frac{\alpha_1}{4} - \frac{(\alpha_5 - \alpha_6)}{\alpha_3 - \alpha_2}\frac{\alpha_2 + \alpha_3}{4} \tag{2.60}$$

Equations (2.58)–(2.60) have been used to assess the degree of the accuracy of the planar transversal isotropic fluid, and of the nonplanar Leslie–Ericksen equations, by comparing them with experimental measures (Ericksen, 1969; Han and Rey, 1994a–c; Rey, 1993a,b).

2.4.1.6 Scaling Ericksen Theory   Ericksen (1969) has shown that there are certain scaling properties of the LE theory that are useful for rheological characterization. Below we present the results of Ericksen's scaling work, which is used in the discussions of results. For a fixed material constant set, the viscosity η scaling for simple shear flow is given by

$$E_\alpha = E_\beta \quad \text{or} \quad \eta(\dot{\gamma}_\alpha h_\alpha^2) = \eta(\dot{\gamma}_\beta h_\beta^2) \quad \text{if} \quad E_\alpha = E_\beta \quad \text{or} \quad \dot{\gamma}_\alpha h_\alpha^2 = \dot{\gamma}_\beta h_\beta^2 \tag{2.61}$$

where α and β denote two different conditions. Ericksen's scaling also shows that time (t) for different plate spacing h scales with h^2, in addition to the length scaling with h. Equation (2.61) may be more general by including the time scaling, for a different combination of shear rate $\dot{\gamma}$ and plate spacing h, as follows:

$$\eta(\dot{\gamma}_\alpha h_\alpha^2, t_\alpha) = \eta(\dot{\gamma}_\beta h_\beta^2, t_\beta) \tag{2.62}$$

where the relationship between different times can be readily shown to be

$$t_\alpha = \left(\frac{h_\alpha}{h_\beta}\right)^2 t_\beta \tag{2.63}$$

In terms of shear strain γ, Eq. (2.62) can be written in the following alternative but equivalent form:

$$\eta(\dot{\gamma}_\alpha h_\alpha^2, \gamma_\alpha) = \eta(\dot{\gamma}_\beta h_\beta^2, \gamma_\beta) \tag{2.64}$$

$$\gamma_i = \dot{\gamma}_i t \qquad (i = \alpha, \beta) \tag{2.65}$$

Using Eq. (2.63) the equivalence between γ_α and γ_β follows, by director substitution,

$$\gamma_\beta = \dot{\gamma}_\beta t_\beta = \dot{\gamma}_\alpha \left(\frac{h_\alpha}{h_\beta}\right) t_\beta = \gamma_\alpha \tag{2.66}$$

where the original Ericksen's scaling relation $\dot{\gamma}_\alpha h_\alpha^2 = \dot{\gamma}_\beta h_\beta^2$ is satisfied in Eq. (2.61). It has been directly confirmed that the computational results obey the Ericksen scaling properties, but one should keep in mind that this scaling holds for a given preparation of the system (i.e., initial conditions, boundary conditions) but breaks down in the presence of stabilities leading to difference steady-state solutions (Ericksen, 1969; Han and Rey, 1994a–c; Rey, 1993a,b).

2.4.2 Wormlike Micellar Nematics

As mentioned above, wormlike micelles are observed when surfactant concentration is above 20–30 wt %. Their properties have received considerable attention in the last years due to their remarkable structural and complex viscoelastic rheological flow behavior (Cates, 1987; Larson and Doi, 1991; Spenley and Cates, 1994; Spenley et al., 1993, 1996; Yang, 2002). Nematic phases exhibit long-range orientational and translational order with no positional order, whereas hexagonal phases show both orientational and translational long-range order of the centers of mass of the micelles (Chen et al., 1977; Clawson et al., 2006; De Melo et al., 2007; Hendrix et al., 1983; Itri and Amaral, 1990, 1993; Larson, 1999; Lawson and Flautt, 1967; Leadbetter and Wrighton, 1979; Lukascheck et al., 1995; Muller et al., 1997; Onsager, 1949; Ostwald and Pieransky, 2000; Quist et al., 1992; Richtering et al., 1994; Schmidt et al., 1988; Semenov and Khokhlov, 1988).

2.4.2.1 Banded Patterns   Banded patterns are normal to the flow direction after cessation of flow for shear rates up to 50 s^{-1}, the pattern formation time is proportional to the square of the applied shear, and the pattern coarsening rate at sufficiently low preshear is well described by a diffusive process. The pattern is associated with relaxation of elastic energy. For shear rates greater than 50 s^{-1}, the material flow aligns. These features were found to be similar to those of lyotropic nematic polymers of the nonaligning type (Roux et al., 1995). Using given flow kinematics and a 1D spatial description, a full rheological phase diagram is based on the LdG for nonaligning nematics in terms of the length scale ratio R and the Ericksen number.

In Figure 2.23, the left schematic shows the rheological phase diagram in terms of the eight director modes across the cell thickness. At very small E, elasticity prevails and tumbling is arrested. The dotted parabola corresponding to small E contains the out-plane modes, where regions 3–5 display multistability.

The main features of these flow modes are summarized in the following explanation [Fig. 2.23 (left)]:

1. *In-Plane Elastic-Driven Steady State (EE)*   The steady state of this planar mode arises due to the long-rate order elasticity stored in the deformed tensor order parameter field. In this planar mode there is no orientation boundary layer behavior because there is no flow alignment in the bulk region.

2. *In-Plane Tumbling Wagging Composite State (IT)*   In this time-dependent planar mode, the director dynamic in the bulk region is rotational, and in the boundary layer it is oscillatory. The boundary between the bulk region and each boundary layer is characterized by the periodic appearance of the abnormal nematic state, which is characterized by two equal eigenvalues of the tensor order parameter (i.e., $\mu_k = \mu_r > \mu_l$) and follows a smoothly defect-free transition from the rotation bulk region to the fixed director anchoring at the surfaces by a director resetting mechanism. The insert in Figure 2.23 is discussed in full detail below and shows schematics of the existing stable solutions of Eq. (2.2).

3. *In-Plane Wagging State (IW)*   In this plane mode, the director dynamics over the entire flow geometry is periodic oscillatory with an amplitude decreasing from a maximum at the centerline to zero at the two bounding surfaces.

4. *In-Plane Viscous-Driven Steady State (IV)*   In this plane mode, the director profile shows a flow-aligning bulk region and two boundary layers. On traversing the boundary, the director rotates from the aligning angle to the flow direction at the walls.

5. *Out-of-Plane Elastic-Driven Steady State with Achiral Structure (OEA)*   In this nonplanar mode, the director shows steady twist structures, and the twist angle profiles are symmetric with respect to the centerline. The steady state arises due to the long-range order elasticity. Similar solutions are presented by the Leslie–Ericksen solutions. Following the bottom-to-top bounding surface, the net director twist rotation is nil.

6. *Out-of-Plane Elastic-Driven Steady State with Chiral Structure (OEC[n])*   In this nonplanar mode, the director shows steady twist structure, with $n\pi$ ($n = 1, 2$) radian difference between the anchoring angles at the lower and

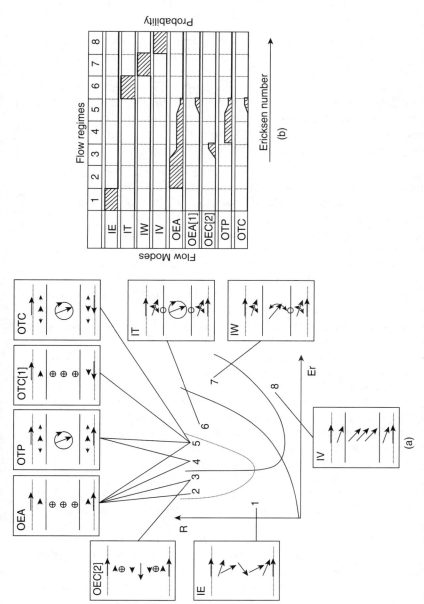

Figure 2.23 (a) Rheological phase diagram as a function of the ratio of short- to long-range elasticity (*R*) and the ratio of viscous flow to long-range elasticity effects (Er), and corresponding director configurations. There are eight flow regimes in the parametric space Er and *R* and nine flow modes. Lines represent flow regime transitions. The dotted line shows the transition between in-plane and out-of-plane modes. The arrows represent the director, and empty circles are the abnormal nematic state. (b) Shows flow modes observation probability as a function of the Ericksen number. The figure presents the concepts of flow regime transitions in terms of the change in the observation probability with increasing Er. [Adapted from Tsuji and Rey (1998).]

upper bounding surfaces, but without the presence of defects or disclinations. The different anchoring conditions are smoothly connected by the chiral director structure. A similar OEC[2] solution is predicted by the Leslie–Ericksen equations.

7. *Out-of-Plane Tumbling-Wagging Composite State with Periodic Chirality (OTP)* In this nonplanar mode, the bulk director dynamics is planar and rotational, and in the two boundary layers it is nonplanar and rotational, and in the two boundary layers it is nonplanar oscillatory. The spatial profiles of the periodic director motion are antisymmetric. The director field exhibits periodic chirality, that is, after a cycle of 2π rotation of the bulk director, the system periodically essentially recovers the spatially homogeneous director configurations [i.e., $n \cong (1, 0, 0)$] for $0 \leq y^* \leq 1$.

8. *Out-of Plane Tumbling-Wagging Composite State with π Chiral Structure (OTC)* The director dynamics is in-plane rotational in the bulk region and out-of-plane oscillatory in the boundary layers, the directors at the upper and lower bounding surfaces have opposite directions, and the system never recovers to the spatial homogeneous director configuration. Figure 2.23 is a schematic of the rheological phase diagram given in terms of R and Er, clearly indicating the parametric regions where the four planar modes and the five nonplanar modes are predicted. The dashed line containing regions 2–5 denotes the parametric envelope within which the out-of-plane modes exist and are stable; outside the dashed line envelope (regions 1, 6, 7, and 8) only stable planar modes exist. If R is sufficiently low, stable nonplanar modes do not exist at any Ericksen number. If R is sufficiently high, the stable nonplanar modes share a boundary with the planar elastic-driven steady-state mode (region 1) at lower Er and with a time-dependent tumbling-wagging composite mode (region 6) at high Er numbers. Thus out-of-plane modes arise when the R and Er numbers have significant magnitudes. In the nine inserts shown in Figure 2.23, the director orientation is denoted by thick arrows and the dashed lines denote the bounding surfaces ($y^* = 0, 1$); the fixed director surface orientation is denoted by the arrow lying on each dashed line. The symbol $\otimes$ denotes a director orientation along the vorticity axis. The main features of the spatial distributions are that two models (IE, OEC[2]) display one monotonic region behavior, while the other seven (IT, IW, IV, OEA, OEC[1], OTP, OTC) display a boundary layer and bulk region behavior. In the latter region the boundary layers are indicated by two thin lines parallel to the dashed lines. The dynamics of the director for the transient modes (IT, IW, OTP, OTC) are indicated as follows: the double arrowheads represents director wagging (oscillations), and the full circle with an arrow represents director tumbling (rotation). The director orientation of the lower surface of the inserts corresponding to OEC[1] and OTC shows that $n = -1, 0, 0$ and that the modes are chiral. The full twist shown in the insert for OEC[2] shows that this mode is also chiral. In addition, as shown below, the OTP mode exhibits periodic chirality.

Finally, observation probabilities of all flow modes for all the flow regimes are also shown in Figure 2.23 (right). In the figure, the probabilities are plotted for each flow mode, and thus for any Ericksen number the sum of the probabilities is 1. For example, for region I the probability of the IE mode is 1 and others are zero, and for region 4 the probability of OEA and OTP flow modes are almost 0.5, and the others are zero. The reason for the transition between regimes 2 and 3 arises from the loss of freedom in the director rotation direction. The boundary layer thickness increases with decreasing Er, and thus the two out-of-plane nucleation points become closer. When the points come closer than a certain correlation length, their rotation directions are also restricted by each other, and then the system loses the OEC[2] flow mode and changes from regime 3 to regime 2. Summarizing this section, the complete model given in Section (2.2) [Eq. (2.35)] predicts extensive multistability phenomena, involving planar, chiral achiral, steady, and time periodic nodes. The range and richness of the multistability is due to the presence of the two compatibilization mechanisms predicted by the complete theory (Tsuji and Rey, 1998).

2.4.2.2 Banded Textures   Banded texture predictions during flow and after cessation of flow have been compared with experimental data using LE and LdG models. The consistency between the two model predictions have been established (Tsuji and Rey, 1998). In the LE model of banded textures during flow, the pattern formation is driven by the OP mode, which nucleates a periodic array of elliptical splay–twist–bend inversion wall in the velocity/ velocity gradient plane with a wavelength close to the shear cell thickness. In the LdG model of banded textures, the nucleation and growth of OP modes give rise to a heterogeneous nonplanar orientation field that relaxes through the formation of a periodic texture. Figure 2.24 shows the out-of-plane component profile after cessation of flow, for $R = 0$ at the following dimensionless times: (a) $t^* = 2$, (b) 4, (c) 6, (d) 8, (e) 12, (f) 16, and (g) 20. The small source of the out-of-plane component near the surface is connected across the bulk region, and the banded texture is formed with almost the same director configuration as that during flow. Then, the texture relaxes through the shrinking of the director out-of-plane region.

Lastly it was observed (Santos and Amato, 1999) that for shear rates greater than 50 s^{-1}, a monodomain was observed. This is in full agreement with high shear rate flow-aligning region 8 of Figure 2.24 and with the predicted texture length-scale divergence for De > 1 (Tsuji and Rey, 1998).

2.4.2.3 Transient Shear Response   Transient shear rheology of cetyl-pyridinium chloride/hexanol (CPCL/Hex) indicates that wormlike micelles respond similarly as nonaligning lyotopric nematic polymers (Berret, 1997; Berret et al., 2001; Roux et al., 1995). In particular, the scaling properties of the transient shear stress, the effect of preshear, and the fact that the time to reach steady state is inversely proportional to shear rate were reported. The

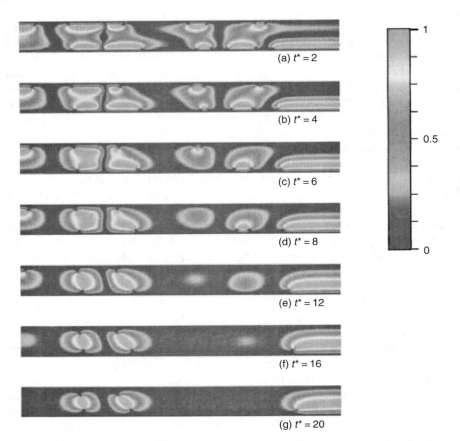

Figure 2.24 Director out-of-plane component n_z profile after cessation of flow, for $R = 0$ at the following dimensionless times: (a) $t^* = 2$, (b) 4, (c) 6, (d) 8, (e) 12, (f) 16, and (g) 20. Er = 1000 of the shear flow is applied until $t^* = 100$, and thus the result at $t^* = 100$ for Er = 1000 corresponds to that at $t^* = 0$ for Er = 0 (no flow). The color scale on the right indicates the correspondence of the magnitude of n_z and the shown colors. The relaxation of stored elastic energy drives the formation of a spatial pattern. [Adapted from Tsuji and Rey (1998).]

periodic oscillations in the shear stress after flow reversal were associated with the possibility that these wormlike micelles are nonaligning: $\lambda < 1$. From a theoretical point of view, the LE model of transient shear rheology predicts periodic damped stress oscillations for shear startup and flow reversal that are in perfect agreement with stress measurements for a low-molar-mass nonaligning thermotropic N_c material (Han and Rey, 1994a–c).

Figure 2.25 shows a characteristic flow reversal stress calculation. In flow reversal, the upper plate instantaneously reverses its moving direction at $t = t_{rv}$, and the corresponding apparent shear rate $\dot\gamma$ used here is

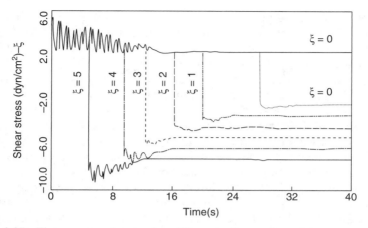

Figure 2.25 Flow reversal at six different times for shear startup flow $[\dot{\gamma} = 8[1 - 2H(t - t_v)](s^{-1})]$, $[\dot{\gamma} = 8 \rightarrow 8(s^{-1})$, $E = 5753$, temperature $= 35°C]$. H denotes the Heaviside unit step function. The corresponding flow reversal times are: $t_{rv} = 4.94$ s (solid line), $t_{rv} = 9.63$ s (dot-dash line), $t_{rv} = 12.50$ s (dash line), $t_{rv} = 16.35$ s (long dash line), $t_{rv} = 20.11$ s (triple dot-dash line), $t_{rv} = 27.79$ s (dotted line). As the time of the flow reversal is increased, the initial oscillations decrease, which is in agreement with the experimental observations. [Adapted from Han and Rey (1994a).]

$$\dot{\gamma} = 8[1 - 2H(t - t_v)](s^{-1}) \tag{2.67}$$

where H denotes the Heaviside unit step function. Figure 2.25 shows the shear stress τ as a function of time (s) during flow reversal, for six increasing reversal times $t_{rv} = 4.94$ s (solid line), 9.63 s (dot-dash line), 16.35 s (long dash line), 20.11 s (triple dot dash line), and 27.79 s (dotted line). The figure shows that (i) double peaks reappear for earlier reversal times, and (ii) when the flow is reversed at an earlier time, larger amplitude oscillatory behaviors are predicted.

2.5   CONCLUSIONS

A comprehensive review of the theory and computer simulation of lyotropic nematic phases, including micellar calamitic and discotic phases and wormlike micellar nematics, was presented based on the classical Leslie–Ericksen and Landau–de Gennes liquid crystal models. Emphasis was placed on the structure and dynamics of these systems. Important properties were discussed such as defects, texturing, and shear rheology using the tools, concepts, scaling laws, nonequilibrium phase diagrams, instability mechanisms, and pattern formation processes previously reported for calamitic and discotic thermotropic nematics and lyotropic nematic polymers. Strong similarities and analogous structural properties and dynamics behavior between micellar nematics and thermotropic nematics including flow birefringence, banded textures under magnetic fields, thermorehological responses across IN transitions, and shear-rheology scaling were discussed in detail.

Wormlike micelle rheology is a clear example of this, and experimental results for these systems present strong similarities with lyotropic nematic polymers, including banded textures after cessation of flow, nonaligning-to-aligning transition, damped oscillatory stress transients, and stress scaling properties of shear stress. Although not included in this review, some interesting biological manifestations, optical and biosensors, thermodynamics, converging and instabilities flows, bifurcations, fluctuations of nematic liquid crystals are available in the literature (Abukhdeir and Rey, 2008; de Luca and Rey, 2003, 2004, 2006a–c, 2007; Grecov and Rey, 2003a–c, 2004, 2005, 2006; Han and Rey, 1993a,b, 1994a–c, 1995a,b; Hwang and Rey, 2006a,b, 2007; Rey, 1990, 1993a,b, 1995, 2000a–c, 2001a,b, 2002, 2006, 2007, 2009, 2010; Rey and Denn, 1987, 1998a–c, 2002; Rey and Herrera-Valencia, 2010, 2012; Rey et al., 2004, 2005, 2011; Singh and Rey, 1995; Soule et al., 2009; Tsuji and Rey, 1997, 1998, 2000; Wang and Rey, 1997; Wincure and Rey, 2007; Yan and Rey, 2002, 2003; Zhou et al., 2006). Lastly, lyotropic chiral nematic shear rheology and texturing can also be described by the modeling methodology presented in this chapter.

ACKNOWLEDGMENTS

This work was supported, in part, by the U.S. Office of Basic Energy Sciences, Department of Energy, grant DE-SC0001412, and by the Petroleum Research Fund of the American Chemical Society. EHV gratefully acknowledges financial fellowship support from CONACY-MEXICO (Postdoctoral Grant/000000000124600 and Postdoctoral Grant/000000000147870).

REFERENCES

Abukhdeir, N. M., and Rey, A. D. (2008). Defect kinetics and dynamics of pattern coarsening in a two-dimensional smectic-A system. *Journal of New Physics*, *10*(6), 063035-1–17.

Acierno, D., La Mantia, F., Marrucci, G., and Titomanlio, G. (1976). A non linear viscoelastic model with structure dependent relaxation times. I. Basic formulation. *Journal of Non-Newtonian Fluid Mechanics*, *1*(2), 125–146.

Alves, V., Nakamatsu, S., Oliveira, E., Zappone, B., and Richetti, P. (2009). Anisotropic reversible aggregation of latex nanoparticles suspended in a lyotropic nematic liquid crystal: Effect of gradients of biaxial order. *Langmuir*, *25*(19), 11849–11856.

Amaral, L. Q. (1983). Magnetic orientation of nematic lyomesophases. *Molecular Crystal Liquid Crystal*, *100*, 85–91.

Auffret, Y., Roux, D., Kissi, N., Dunstan, D., and Pignot-Paintrand, I. (2009). Stress and strain controlled rheometry on a concentrated lyotropic lamellar phase of AOT/Water/Iso-octane. *Rheologica Acta*, *48*(4), 423–432.

Berni, M., Lawrence, C., and Machin, D. (2002). A review of the rheology of lamellar phase in surfactant systems. *Advances in Colloids and Interface Science*, *98*(2), 217–243.

Berret, J. F. (1997). Transient rheology of wormlike micelles. *Langmuir*, *13*(8), 2227–2234.

Berret, J. F., Séréro, Y., Winkelman, B., Calvet, D., Collet, A., and Viguier, M. (2001). Non-linear rheology of telechelic polymer networks. *Journal of Rheology*, *45*(2), 477–492.

Boden, N., Jackson, P. H., Mcmullen, K., and Holmes, M. C. (1979). Are nematic amphiphilic crystalline mesophases thermodynamically stable? *Chemical Physics Letters*, *65*(3), 476–479.

Boden, N., Radley, K., and Holmes, M. C. (1981). On the relationship between the micellar structure and the diamagnetic anisotropy of amphiphilic nematic mesophases. *Molecular Physics*, *42*, 493–496.

Boden, N., Bushby, R. J., and Hardy, C. (1985). New mesophases formed by water soluble discoidal amphiphiles. *Journal de Physique Letters*, *46*(7), 325–328.

Burghardt, W. R., and Fuller, G. G. (1991). Role of director tumbling in the rheology of polymer liquid crystals solutions. *Macromolecules*, *24*(9), 2546–2555.

Calderas, F., Sánchez-Solis, A., Maciel, A., and Manero, O. (2009). The transient flow of the PETPEN-Montmorillonite clay nanocomposites. *Macromolecular Symposia*, *283–284*(1), 354–360.

Cates, M. E. (1987). Reptation of living polymers: Dynamics of entangled polymers in the presence of reversible chain-scission reactions. *Macromolecules*, *20*(9), 2289–2296.

Cates, M. E., and Candau, S. J. (1990). Statics and dynamics of worm-like surfactants micelles. *Journal of Physics: Condensade Matter*, *2*(33), 6869–6892.

Chandrasekhar, S. (1992). *Liquid Crystals*, 2nd ed., Cambridge University Press, London.

Charvolin, J., Levelut, A. M., and Samulski, E. T. (1979). Lyotropic nematics: Molecular aggregation and susceptibilities. *Journal de Physique Lettres*, *40*, L-587–L-592.

Chen, D. M., Fujiwara, F. Y., and Reeves, L. W. (1977). Studies of behavior in magnetic-fields of some lyomesophase systems with respect to electrolyte additions. *Canadian Journal Chemistry*, *55*(12), 2396–2403.

Clawson, J. S., Holland, G. P., and Alam, T. P. (2006). Magnetic alignment of aqueous CTAB in nematic and hexagonal liquid crystalline phases investigated by spin-1-NMR. *Physical Chemistry Chemical Physics*, *8*(22), 2635–2641.

de Andrade Lima, L. R. P., and Rey, A. D. (2003a). Poiseuille flow of Leslie-Ericksen discotic liquid crystals: Solution multiplicity, multistability, and non-Newtonian rheology. *Journal of Non-Newtonian Fluid Mechanics*, *110*(2–3), 103–142.

de Andrade Lima, L. R. P., and Rey, A. D. (2003b). Computational modelling of ring textures in mesophase carbon fibbers. *Materials Research*, *6*(2), 285–293.

de Andrade Lima, L. R. P., and Rey, A. D. (2003c). Linear and non-linear viscoelasticity of discotic nematics under transient Poiseuille flows. *Journal of Rheology*, *47*(5), 1261–1282.

de Andrade Lima, L. R. P., and Rey, A. D. (2004a). Superposition and universality in the linear viscoelasticity of Leslie-Ericksen liquid crystals. *Journal of Rheolology*, *48*(5), 1067–1084.

de Andrade Lima, L. R. P., and Rey, A. D. (2004b). Assessing flow alignment of nematic liquid crystals through linear viscoelasticity. *Physical Review E*, *70*, 011701–0117013.

de Andrade Lima, L. R. P., and Rey, A. D. (2004c). Linear viscoelasticity of discotic mesophases. *Chemical Engineering Science*, *59*(18), 3891–3905.

de Andrade Lima, L. R. P., and Rey, A. D. (2004d). Computational modeling in processing flows of carbonaceous mesophases. *Carbon*, *42*(7), 1263–1268.

de Andrade Lima, L. R. P., and Rey, A. D. (2004e). Poiseuille flow of discotic nematic liquid crystals onion textures. *Journal of Non-Newtonian Fluid Mechanics*, *119*(1–3), 71–81.

de Andrade Lima, L. R. P., and Rey, A. D. (2005). Pulsatile flow of discotic mesophases. *Chemical Engineering Science*, *60*(23), 6622–6636.

de Andrade Lima, L. R. P., and Rey, A. D. (2006a). Back-flow in pulsatile flows of Leslie-Ericksen liquid crystals. *Liquid Crystal*, *33*(6), 711–712.

de Andrade Lima, L. R. P., and Rey, A. D. (2006b). Linear viscoelasticity of textured carbanaceous mesophases. *Journal of the Brazilian Chemical Society*, *17*(6), 1109–1116.

de Andrade Lima, L. R. P., and Rey, A. D. (2006c). Pulsatile flows of Leslie-Ericksen liquid crystals. *Journal of Non-Newtonian Fluid Mechanics*, *133*(1), 32–45.

de Andrade Lima, L. R. P., and Rey, A. D. (2006d). Superposition principles for small amplitude oscillatory shearing of nematic mesophases. *Rheologica Acta*, *45*(5), 1435–1528.

de Andrade Lima, L. R. P., Grecov, D., and Rey, A. D. (2006). Multiscale theory and simulation for carbon fiber precursors based on carbonaceous mesophases. *Plastics, Rubber and Composites*, *35*(617), 276–286.

de Gennes, P. G., and Prost, J. (1993). *The Physics of Liquid Crystals*, 2nd ed., Oxford University Press, London.

De Kee, D., and Fong, C. M. (1944). Rheological properties of structured fluids. *Polymer Engineering Science*, *34*(5), 438–445.

de Luca, G., and Rey, A. D. (2003). Monodomain and polydomain helicoids in chiral liquid-crystalline phases and their biological analogues. *European Physical Journal E*, *12*(2), 291–302.

de Luca, G., and Rey, A. D. (2004). Chiral front propagation in liquid crystalline materials: Formation of the planar domain twisted plywood architecture of biological fibrous composites. *Physical Review E*, *69*, 011706/1–13.

de Luca, G., and Rey, A. D. (2006c). Dynamics interactions between nematic point defects in the extrusion duct of spiders. *Journal of Chemical Physics*, *124*, 144904/1–8.

de Luca, G., and Rey, A. D. (2006b). Dynamical interactions between nematic point defects in the spinning extrusion duct of spiders. *Virtual Journal of Biological Physics Research*, *11*(8), 144904/1–8.

de Luca, G., and Rey, A. D. (2006a). Modeling of bloch inversion wall defects in nematic thin films. *Modeling and Simulation in Material Science and Engineering, 14*(8), 1397–1407.

de Luca, G., and Rey, A. D. (2007). Ring-like cores of cylindrically confined nematic point defects. *Journal of Chemical Physics, 126*(9), 094907/1–094907/11.

De Melo filho, A., Amadeu, N. S., and Fujiwara, F. Y. (2007). The phase diagram of the lyotropic nematic mesophase in the TTAB/NaBr/water system. *Liquid Crystal, 34*(6), 683–691.

Doi, M., and Ohta, T. (1991). Dynamics and rheology of complex interfaces, I. *Journal of Chemical Physics, 95*(2), 1242–1248.

Ericksen, J. L. (1969). A boundary-layer effect in viscometry of liquid crystals. *Transaction of the Society of Rheology, 13*, 9–15.

Erni, P., Cramer, C., Marti, I., Windhab, E. J., and Fisher, P. (2009). Continuous flow structuring of anisotropic biopolymer particle. *Advances in Colloids and Interface Science, 150*(1), 16–26.

Farhoudi, Y., and Rey, A. D. (1993a). Shear flow of nematics polymers. I. Orienting modes, bifurcations, and steady state rheological predictions. *Journal of Rheology, 37*(2), 289–314.

Farhoudi, Y., and Rey, A. D. (1993b). Shear flow of nematic polymers. II. Stationary regimes and start-up dynamics. *Journal of Non-Newtonian Fluid Mechanics, 49*(2–3), 175–204.

Farhoudi, Y., and Rey, A. D. (1993c). Ordering effects in shear flows of discotic polymers. *Rheologica Acta, 32*(3), 207–217.

Fernandes, P. R. G., and Figueiredo Neto, A. M. (1994). Flow birefrigence in lyotropic mixtures in the isotropic phase. *Physics Review E, 51*(1), 567–570.

Fernandes, P. R. G., Mukai, H., Honda, B. S. L., and Shibli, S. M. (2006). Induction of order in the isotropic phase of a lyotropic liquid crystals by pulsed light. *Liquid Crystal, 33*(3), 367–371.

Gelbart, W. M., Ben-Shaul, A., and Roux, D. (Eds.) (1994). *Micelles, Membranes, Microemulsions and Monolayers*. Springer, New York.

Giesekus, H. (1966). Die elastizitat von flussigkeiten. *Rheologica Acta, 5*(1), 29–35.

Giesekus, H. (1982). A simple constitutive equation for polymer fluids based on the concept of deformation-dependent tensorial mobility. *Journal of Non-Newtonian Fluid Mechanics, 11*(1–2), 69–109.

Giesekus, H. (1984). On configuration-dependent generalized Oldroyd derivatives. *Journal of Non-Newtonian Fluid Mechanics, 14*, 47–65.

Giesekus, H. (1985). Constitutive equation for polymer fluids based on the concept of configuration dependent molecular mobility: A generalized mean-configuration model. *Journal of Non-Newtonian Fluid Mechanics, 17*(3), 349–372.

Gobeaux, F., Belamie, E., Mosser, G., Davidson, P., Panine, P., and Giraud-Guille, M. M. (2007). Cooperative ordering of collagen triple helices in the dense state. *Langmuir, 23*(11), 6411–6417.

Golmohammadi, M., and Rey, A. D. (2009). Thermodynamic modelling of carbonaceous mesophase mixtures. *Liquid Crystal, 36*(1), 75–92.

Grecov, D., and Rey, A. D. (2003a). Shear-induced textural transitions in flow aligning liquid crystals polymers. *Physics Review E, 68*(6), 061704–061724.

Grecov, D., and Rey, A. D. (2003b). Theoretical computational rheology for discotic nematic liquid crystals. *Molecular Crystal Liquid Crystal*, *39*, 157–194.

Grecov, D., and Rey, A. D. (2003c). Transient shear rheology of discotic mesophases. *Rheologica Acta*, *42*(6), 590–604.

Grecov, D., and Rey, A. D. (2004). Impact of textures on stress growth of thermotropic liquid crystals subjected to step-shear. *Rheologica Acta*, *44*(2), 135–149.

Grecov, D., and Rey, A. D. (2005). Steady state and transient rheological behavior of mesophase Pitch: Part II. Theoretical. *Journal of Rheology*, *49*(1), 175–195.

Grecov, D., and Rey, A. D. (2006). Texture control strategies for flow-aligning, liquid crystal polymers. *Journal of Non-Newtonian Fluid Mechanics*, *139*(3), 197–208.

Gupta, G., and Rey, A. D. (2005). Texture rules for concentrated filled nematics. *Physics Review Letters*, *95*(12), 12785/1–4.

Han, W. H., and Rey, A. D. (1993a). Supercritical bifurcations in simple shear flow of a non-aligning nematic: Reactive parameter and director anchoring effects. *Journal of Non-Newtonian Fluid Mechanics*, *48*(1–2), 181–210.

Han, W. H., and Rey, A. D. (1993b). Stationary bifurcations and tricriticality in a creeping nematic polymer flow. *Journal of Non-Newtonian Fluid Mechanics*, *50*(1), 1–28.

Han, W. H., and Rey, A. D. (1994a). Orientation symmetry breakings in shear liquid-crystals. *Physics Review E*, *50*(2), 1688–1691.

Han, W. H., and Rey, A. D. (1994b). Dynamic simulations of shear-flow-induced chirality and twisted textures in a nematic polymer. *Physics Review E*, *49*(1), 597–613.

Han, W. H., and Rey, A. D. (1994c). Simulation and validation of nonplanar nemato-rheology. *Journal of Rheology*, *38*(5), 1317–1324.

Han, W. H., and Rey, A. D. (1995a). Theory and simulation of optical banded textures of nematic polymers during shear flow. *Macromolecules*, *28*(24), 8401–8405.

Han, W. H., and Rey, A. D. (1995b). Simulation and validation of temperature effects on the nematorheology of aligning and non-aligning liquid crystals. *Journal of Rheology*, *39*(2), 301–323.

Hendrix, V., Charvolin, J., Rawiso, M., Liebert, L., and Holmes, M. C. (1983). Anisotropic aggregates of amphiphilic molecules in lyotropic nematic phases. *Journal of Physical Chemistry*, *87*(20), 3991–3999.

Herrera, E. E., Calderas, F., Chávez, A. E., Manero, O., and Mena, B. (2009). Effect of random longitudinal vibrations pipe on the poiseuille flow of a complex liquid. *Rheologica Acta*, *48*(7), 779–800.

Herrera, E. E., Calderas, F., Chávez, A. E., and Manero, O. (2010). Study on the pulsating flow of worm-like micellar solutions. *Journal of Non-Newtonian Fluid Mechanics*, *165*(3–4), 174–183.

Hwang, D. K., and Rey, A. D. (2006a). Computational studies of optical texture of twist disclination loops in liquid crystals films using the finite-difference time-domain method. *Journal American Optics Society A*, *23*(2), 483–496.

Hwang, D. K., and Rey, A. D. (2006b). Optical modeling of liquid crystal biosensors. *Journal of Chemical Physics*, *125*, 174902/1–174902/9.

Hwang, D. K., and Rey, A. D. (2007). Computational modelling of texture formation and optical performance of liquid crystal films of patterned surfaces. *SIAM Journal on Applied Mathematics*, *67*(1), 214–234.

Itri, R., and Amaral, L. Q. (1990). Study of the isotropic-hexagonal transition in the system sodium dodecyl sulfate/water. *Journal of Physical Chemistry*, *94*(5), 2198–2202.

Itri, R., and Amaral, L. Q. (1993). Micellar-shape anisometry near isotropic-liquid crystal phase transition. *Physics Review E*, *47*(4), 2551–2557.

Khokhlov, A. R., and Semenov, A. N. (1981). Liquid-crystalline ordering in the solution of long persistent chains. *Physica A*, *108*(2–3), 546–556.

Kundu, S., Grecov, D., Ogale, A., and Rey, A. D. (2009). Shear flow induced microstructure of a synthetic mesophase pitch. *Journal of Rheology*, *53*(1), 85–113.

Kuzma, M. R., and Saupe, A. (1997). Structured and phase transitions of amphiphilic lyotropic liquid crystals. In P. Collings and J. S. Patel (Eds.), *Handbook of Liquid Crystals Research*, Oxford University Press, Oxford, England, Chapter 7.

Kuzma, M., Hui, Y. M., and Labes, M. M. (1989). Capillary viscometry of some lyotropic nematics. *Molecular Crystals Liquid Crystals*, *172*, 211–215.

Larson, R. (1999). *The Structure and Rheology of Complex Fluids*, Oxford University Press, New York.

Larson, R. G., and Doi, M. (1991). Mesoscopic domain theory for textured liquid crystalline polymers. *Journal of Rheology*, *35*(4), 539–563.

Lawson, K. D., and Flautt, T. J. (1967). Magnetically oriented lyotropic liquid crystalline phases. *Journal of the American Chemical Society*, *89*(21), 5489–5491.

Leadbetter, A., and Wrighton, P. (1979). Order parameters in Sa, Sc and N-phases by X-ray diffraction. *Journal de Physique Colloque*, *40*(C3), 234–242.

Lee, Y. S. (2008). *Self-Assembly and Nanotechnology: A Force Balance Approach*, Wiley, Hoboken, NJ.

Lhuillier, D., and Rey, A. D. (2004a). Nematic liquid crystals and ordered micropolar fluids. *Journal of Non-Newtonian Fluid Mechanics*, *120*(1–3), 169–174.

Lhuillier, D., and Rey, A. D. (2004b). Liquid cystalline nematic polymers revisited. *Journal of Non-Newtonian Fluid Mechanics*, *120*(1–3), 85–92.

Lu, C. Y. D., Chen, P., Ishii, Y., Komura, S., and Kato, T. (2008). Non-linear rheology of lamellar liquid crystals. *European Physics Journal E*, *25*(1), 91–101.

Lukascheck, M., Grabowski, D. A., and Schmidt, C. (1995). Shear-induced alignment of a hexagonal lyotropic liquid crystals as studied by rheo-NMR. *Langmuir*, *11*(9), 3590–3594.

Marrucci, G. (1985). Rheology of liquid crystalline polymers. *Pure and Applied Chemistry*, *57*(11), 1545–1552.

Moldenaers, P., Yanase, H., and Mewis, J. (1991). Flow-induced anisotropy and its decay in polymeric liquid crystals. *Journal of Rheology*, *35*(8), 1681–1699.

Muller, S., Fisher, P., and Schmidt, C. (1997). Solid-like director reorientation in sheared hexagonal lyotropic liquid crystals as studied by nuclear magnetic resonance. *Journal de Physique II France*, *7*(3), 421–432.

Murugesan, Y., and Rey, A. D. (2010). Structure and rheology of fiber-laden membranes via integration of nematodynamics and membranodynamic. *Journal of Non-Newtonian Fluid Mechanics*, *165*(1–2), 32–44.

Ostwald, P., and Pieransky, P. (2000). *Les Cristaux Liquides*, Gordon & Breach, Amsterdam.

Onsager, L. (1949). The effects of shape on the interaction of colloidal particles. *Annal of the New York Academy of Sciences*, *51*, 627–649.

Palangana, A. J., Depeyrot, J., and Figueiredo Neto, A. M. (1990). New thermal instability in a lyotropic uniaxial nematic liquid crystals. *Physics Review Letters*, *65*(22), 2800–2803.

Quist, P. O., Halle, B., and Fúro, I. (1992). Micelle size and order in lyotropic nematic phases from nuclear spin relaxation. *Journal of Chemical Physics*, *96*(5), 3875–3891.

Radley, K., and Reeves, L. W. (1975). Studies of ternary nematic phases by nuclear magnetic resonance. Alkali metal decyl sulfates/decanol/D_2O^1. *Canadian Journal of Chemistry*, *53*, 2998–3004.

Radley, K., Reeves, L. W., and Tracey, A. S. (1976). Effect of counterion substitution on the type and nature of nematic lyotropic phases from nuclear magnetic resonance studies. *Journal of Physical Chemistry*, *80*(2), 174–182.

Ramos, L. (2001). Scaling with temperature and concentration of the non-linear rheology of a soft hexagonal phase. *Physics Review E*, *64*(6), 061502–061509.

Rey, A. D. (1990). Defect mediated transition in a nematic flow. *Journal of Rheology*, *34*, 919–943.

Rey, A. D. (1993a). Analysis of shear-flow effects on liquid-crystalline textures. *Molecular Crystal Liquid Crystal*, *225*, 313–335.

Rey, A. D. (1993b). Rheological prediction of a transversely isotropic fluid model with extensible microstructure. *Rheologica Acta*, *32*(5), 447–456.

Rey, A. D. (1995). Bifurcational analysis of the isotropic-discotic nematic phase-transition in the presence of external flow. *Liquid Crystal*, *19*(3), 325–331.

Rey, A. D. (2000a). Theory of linear viscoelasticity in cholesteric liquid crystals. *Journal of Rheology*, *44*(4), 855–869.

Rey, A. D. (2000b). Theory of surface excess Miesowicz viscosities of planar nematic liquid crystal-isotropic fluid interfaces. *European Journal of Physics E*, *2*(2), 169–179.

Rey, A. D. (2000c). Mechanical theory of structural disjoining pressure in liquid crystal films. *Physics Review E*, *61*(4), 4632–4635.

Rey, A. D. (2001a). A rheological theory for liquid crystal thin films. *Rheologica Acta*, *40*(6), 507–515.

Rey, A. D. (2001b). Irreversible thermodynamics of liquid crystal interfaces. *Journal of Non-Newtonian Fluid Mechanics*, *96*(1–2), 45–62.

Rey, A. D. (2002). Simple shear small amplitude oscillatory rectilinear shear permeation flows of cholesteric liquid crystals. *Journal of Rheology*, *46*(1), 225–240.

Rey, A. D. (2006). Anisotropic fluctuation model for surfactant-laden liquid-liquid crystals interfaces. *Langmuir*, *22*(8), 3491–3493.

Rey, A. D. (2007). Capillary models for liquid crystals fibers, membranes, films and drops. *Soft Matter*, *2*, 1349–1368.

Rey, A. D. (2009). Flow and texture and modeling of liquid crystalline materials. *Rheology Review*, *2008*, 71–135.

Rey, A. D. (2010). Liquid crystals model of biological materials and processes. *Soft Matter*, *6*, 3402–3429.

Rey, A. D., and Denn, M. (1987). Analysis of converging and diverging flows of liquid crystal polymers. *Molecular Crystal Liquid Crystals*, *153*, 301–310.

Rey, A. D., and Denn, M. M. (1998a). Analysis of transient periodic textures in nematic polymers. *Liquid Crystal*, *4*(4), 409–419.

Rey, A. D., and Denn, M. M. (1998b). Converging flow of tumbling nematic liquid crystal. *Liquid Crystal*, *4*(3), 253–272.

Rey, A. D., and Denn, M. M. (1998c). Jeffrey-Hamel flow of Leslie-Ericksen nematic liquids. *Journal of Non-Newtonian Fluid Mechanics*, *27*, 375–401.

Rey, A. D., and Denn, M. M. (2002). Dynamical phenomena in liquid-crystalline materials. *Annual Reviews in Fluid Mechanics*, *34*, 233–266.

Rey, A. D., and Herrera-Valencia, E. E. (2010). Micromechanics model of liquid crystals anisotropic triple lines with applications to contact line self-assembly, *Lagmuir*, *26*(16), 13033–13037.

Rey, A. D., and Herrera-Valencia, E. E. (2012). Liquid crystal models of biological materials and silk spinning, *Biopolymers*, *97*(6), 374–396.

Rey, A. D., and Tsuji, T. (1998). Recent advances in theoretical liquid crystal rheology. *Macromolecular Theory and Simulation*, *7*(6), 623–639.

Rey, A. D., Grecov, D., and Das, S. K. (2004). Thermodynamics and flow modeling of meso and macrotextures in polymer liquid crystal material systems. *Industrial Engineering Chemistry Research*, *43*(23), 7343–7355.

Rey, A. D., Grecov, D., and de Andrade Lima, L. R. P. (2005). Multiscale simulation of flow-induced texture formation in polymer liquid crystals and carbonaceous mesophases. *Molecular Simulation*, *31*(2–3), 185–199.

Rey, A. D., Golmohammadi, M., and Herrera-Valencia, E. E. (2011). A model for mesophase wetting thresholds of sheets, fibers and fiber bundles. *Soft Matter*, *7*(10), 5002–5009.

Richtering, W. (2001). Rheology and shear induced structures in surfactants solutions. *Colloids and Interface Science*, *6*(5–6), 446–450.

Richtering, W., Laeuger, J., and Linemann, R. (1994). Shear orientation of a micellar hexagonal liquid crystalline phase: A rheo and small angle light scattering study. *Langmuir*, *10*(11), 4374–4379.

Roux, D. C., Berret, J. F., Porte, G., Peuvrel-Disdier, E., and Lindner, P. (1995). Shear-induced orientation and textures of nematic living polymers. *Macromolecules*, *28*(5), 1681–1687.

Santos, L., and Amato, M. A. (1999). Orientational diffusivities measured by Raleigh scattering in a lyotropic calamitic nematic (Nc) liquid crystal phase: The backflow problem revisited. *European Physical Journal B*, *7*(3), 393–400.

Sampaio, R., Palangana, A. J., Alves, F. S., and Simoes, M. (2005). Smectic A-cholesteric phase transition: A viscosity study. *Molecular Crystal Liquid Crystal*, *436*, 167/[1121]–176/[1130].

Savage, J. R., Caggioni, M., Spicer, P. T., and Cohen, I. (2010). Partial universality: Pinch-off dynamics in fluids with smetic liquid crystallie order. *Soft Matter*, *6*(5), 892–895.

Schmidt, C. (2006). Rheo-NMR spectroscopy. *Modern Magnetic Resonance*, 6, 1515–1521.

Schmidt, G., Muller, S., Lindner, P., Schmidt, C., and Richtering, W. (1988). Shear orientation of lyotropic hexagonal phases. *Journal of Physical Chemistry B*, 102(3), 507–513.

Semenov, A. N., and Khokhlov, A. R. (1988). Statistical physics of liquid-crystalline polymers. *Sovietic Physics—Uspekhi*, 31(11), 988–1014.

Siebert, H., Grabowski, D. A., and Schmidt, C. (1997). Rheo-NMR study of non-flow-aligning side chain liquid crystals polymer in nematic solution. *Rheologica Acta*, 36(6), 618–627.

Simoes, M., and Domiciano, S. M. (2003). Agreements and disagreements between theories and experiments in nematoviscosity. *Physical Review E*, 68, 011705–011712.

Simoes, M., Palangana, A. J., and Goncalves, A. E. (2000). Radial fluctuations and nonisotropic disclinations in nematic liquid crystals. *Physics Review E*, 61(53), 6007–6010.

Simoes, M., Palangana, A. J., Sampaio, A. R., de Campo, A., and Yamaguti, K. E. (2005). Viscosity peaks at the cholesteric-isotropic phase transitions. *Physics Review E*, 71(5), 051706-1–051706-5.

Singh, A. P., and Rey, A. D. (1995). Theory and simulation of extensional flow induced biaxiality in discotic mesophases. *Journal de Physique II*, 5(9), 1321–1348.

Sonin, S. (1987). Lyotropic nematics. *Sovietic Physics—Uspekhi*, 30(10), 875–896.

Soule, E. R., Abukhdeir, N. M., and Rey, A. D. (2009). Thermodynamics, transition dynamics, and texturing in polymer-dispersed liquid crystals with mesogens exhibiting a direct isotropic/smectic A transition. *Macromolecules*, 42(24), 9486–9497.

Spenley, N. A., and Cates, M. E. (1994). Pipe models for entangled fluids under strong shear. *Macromolecules*, 27(14), 3850–3858.

Spenley, N. A., Cates, M. E., and McLeish, T. C. B. (1993). Non-linear rheology of worm-like micelles. *Physics Review Letters*, 71(6), 939–942.

Spenley, N. A., Yuan, X. F., and Cates, M. E. (1996). Non-monotonic constitutive laws and the formation of shear banded flows. *Journal Physics II France*, 6(4), 551–571.

Srinivasarao, M. (1992). Rheology and rheo-optics of polymer liquid crystals: An overview of theory and experiment. *Chemtracts Macromolecular Chemistry*, 3, 149–178.

Stewart, W. I. (2004). *The Static and Dynamic and Cholesteric Liquid Crystals*, Taylor and Francis, London.

Tsuji, T., and Rey, A. D. (1997). Effect of long range order on sheared liquid crystalline materials. Part I. Compatibility between tumbling behavior and fixed anchoring. *Journal of Non-Newtonian Fluid Mechanics*, 73(1–2), 127–152.

Tsuji, T., and Rey, A. D. (1998). Orientation mode selection mechanisms for sheared nematic liquid crystalline materials. *Physical Review E*, 57(5), 5609–5625.

Tsuji, T., and Rey, A. D. (2000). Effect of long range order on sheared liquid crystalline materials, transition and rheological phase diagrams. *Physics Review E*, 62(6), 8141–8151.

Tschierske, C. (2007). Liquid crystals engineering—new complex mesophase structures and their relations to polymer morphologies, nanoscale patterning and crystals engineering. *Chemical Society Reviews, 36*(12), 1930–1970.

Vedenov, A. A., and Levchenko, E. B. (1983). Supermolecular liquid-crystalline structures in solutions of amphiphilic. *Soviet Physics Uspekhi, 26*, 747–774.

Wang, L., and Rey, A. D. (1997). Pattern formation and non-linear phenomena in stretched discotic nematic liquid crystal fibers. *Liquid Crystal, 23*(1), 93–111.

Wang, H. X., Zhang, G. Y., Feng, S. H., and Xie, X. L. (2005). Rheology as a tool for detecting mesophase transitions for a model nonyl phenol ethoxylate surfactant. *Colloids and Surface A: Physicochemical Engineering Aspects, 256*(1), 35–42.

Wincure, B., and Rey, A. D. (2007). Nanoscale analysis of defect shedding from liquid crystals interfaces. *Nano Letters, 7*(6), 1474–1479.

Yada, M., Yamamoto, J., and Yokoyama, H. (2003). Generation mechanism of shear yields stress for regular defect in water in cholesteric liquid crystals emulsions. *Langmuir, 19*(9), 3650–3655.

Yan, J., and Rey, A. D. (2002). Texture formation in carbonaceous mesophase fibers. *Physics Review E, 65*(1), 031713-1–031713-13.

Yan, J., and Rey, A. D. (2003). Modeling elastic and viscous effects on the texture of ribbon-shaped carbonaceous mesophase fibers. *Carbon, 41*(1), 105–121.

Yang, J. (2002). Viscoelastic wormlike micelles and their applications. *Colloids and Interface Science, 7*, 276–281.

Zimmer, J. E., and White, J. L. (1982). *Advances in Liquid Crystals*, Vol. 5. Academic Press, New York.

Zhou, J., Park, J. O., Luca, G., Rey, A. D., and Srinivasarao, M. (2006). Microscopic observations and simulations of bloch walls in nematic thin films. *Physics Review Letters, 97*(15), 157801/1–157801/4.

■■■■■■ CHAPTER 3

Dividing Planes of Hexagonal H$_{II}$ Mesophase

VESSELIN KOLEV

Casali Institute of Applied Chemistry, The Institute of Chemistry, The Hebrew University of Jerusalem, Jerusalem, Israel

Department of Chemical Engineering, Faculty of Chemistry, Sofia University, Sofia, Bulgaria

ABRAHAM ASERIN and NISSIM GARTI

Casali Institute of Applied Chemistry, The Institute of Chemistry, The Hebrew University of Jerusalem, Jerusalem, Israel

Abstract

Presented is a brief review of the cylindrical dividing surfaces (planes) of the hexagonal inverse H$_{II}$ phase—Luzzati, pivotal, and neutral. They are presented as interrelated surfaces through the general expression of elastic energy per molecule valid for the corresponding surface. For each of the planes, the routine for deriving the plane parameters—radius, area, and volume per molecule, as well as their role in terms of mechanical description the of H$_{II}$ phase—is described in detail. Special attention is paid to the numerical routines for parameter calculations from experimental data, using least-squares routines and the related methods of merit function optimization (minimization). In addition, cases of error estimation are presented along with discussions about the accuracy of used numerical routines.

Self-Assembled Supramolecular Architectures: Lyotropic Liquid Crystals, First Edition.
Edited by Nissim Garti, Ponisseril Somasundaran, and Raffaele Mezzenga.
© 2012 John Wiley & Sons, Inc. Published 2012 by John Wiley & Sons, Inc.

3.1 INTRODUCTION

Hexagonal inverse H$_{\text{II}}$ mesophase is the subject of large-scale theoretical research. It could be considered that research on H$_{\text{II}}$ topology started with the work of Luzzati and Husson in 1962. Their work is based purely on geometrical consideration inside a cylindrical coordinate system (i.e., axial symmetry is considered), which means the proposed structural design could be correct only if H$_{\text{II}}$ phase rods are treated as axial-infinite structures. Such an approximation of a finite structure with an infinite one is possible since the longitudinal dimension of the rods is much larger than the transverse one. Following that simplification, Luzzati and Husson published simple and easy-to-use formulas for deriving the radius of the water channel and the cross-section area of a lipid molecule at that position. Their formulas include, as a major external parameter, the cylinder-center–cylinder-center distance, which in turn is related to the experimentally observed values of first-order Bragg spacing—the relation between structural parameters theory and the attendant experiment. In addition, the lipid volume fraction has to be taken into account. Using the result of their work, every scientist can calculate the radius of the water channel that defines the position of the Luzzati plane by using only spectroscopic data and the calculated volume lipid fraction. The described radius of the water cylinder is involved in the description of the volume of lipid molecule, V_l, which is considered to be the volume of a right truncated cone equal to that of a lipid molecule calculated on the base of its density (Fig. 3.1). The height of the cone is defined using the position of the Luzzati

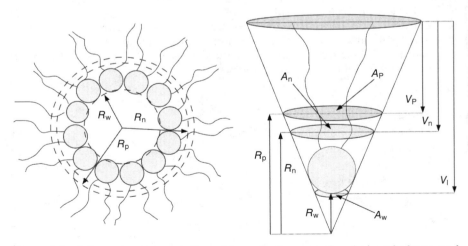

Figure 3.1 Schematic representation of Luzzati, the neutral and pivotal planes, and the related variables: R_w, A_w, radius and area at Luzzati plane; R_n, A_n, V_n, radius, area, and volume related to the neutral plane; R_p, A_p, V_p, radius, area, and volume, related to the pivotal plane. Note that $R_p > R_n > R_w$, $A_p > A_n > A_w$, and $V_l > V_n > V_p$.

plane—the radius of the water cylinder. See the respective formulas in Kulkarni et al. (2010) about the minimal and maximal lengths of the lipid molecule that are also involved in the determination of the cone height.

But when it comes to a mechanical description of the hexagonal inverse H_{II} phase, the abstraction of the Luzzati plane is insufficient because it cannot be directly connected to the mechanical characteristics of the rod structure—the monolayer bending modulus and the spontaneous curvature radius. When the rod structure deformation depends on the experimental technique used, the monolayer bending modulus and the spontaneous curvature radius definitions have to be taken into account. Their definitions are related to a specific abstracted plane—the pivotal plane (Fig. 3.1). If the deformation does not depend on experimental technique, both the monolayer bending modulus and the spontaneous curvature radius must be defined relative to another abstract plane—the natural plane (Fig. 3.1). Both the pivotal and natural planes are connected to each other—the position of the natural plane could be calculated only if that of the pivotal plane is known. In turn, the position of the pivotal plane (the radius of the pivotal plane) depends on the lattice constant of the H_{II} mesophase. The invariant values of the monolayer bending modulus and the spontaneous curvature radius are those at the natural plane.

The aim of the present chapter is to help the reader make the required sequence of assumptions and calculations that leads to deriving the radii of the three planes, their related areas of the lipid molecule, and the corresponding mechanical parameters. It also draws attention to the correct statistical treatment of the data and plane models.

3.2 DETERMINATION AND ROLE OF THE PLANES

Inside the lipid structure of the H_{II} phase, it is possible to define three cylindrical Gibbs dividing planes/surfaces (ordered by their distance from the water channel): *Luzzati* plane, *neutral* plane, and *pivotal* plane (Fig. 3.1). According to Gibbs terminology, the defined planes should have *zero surface excess of water and lipid*. Kozlov and Winterhalter (1991a,b) proposed the following expression of the elastic energy per lipid molecule, at any of the above listed dividing surfaces of H_{II} phase:

$$F \approx \frac{1}{2} A_0 E_{RR} \left(\frac{1}{R} - \frac{1}{R_0} \right)^2 + E_{AR} (A - A_0) \left(\frac{1}{R} - \frac{1}{R_0} \right) + \frac{1}{2} E_{AA} \frac{(A - A_0)^2}{A_0} \qquad (3.1)$$

where E_{RR}, E_{AR}, and E_{AA} are the corresponding elastic moduli. The values and definitions of A and R depend on the selection of the dividing surface. It is very important to note that the neutral and pivotal planes are well defined only for *small* deformations within the limits of validity of the quadratic free energy expansion (Leikin et al., 1996). In other words, Eq. (3.1) defines the curvature-independent energy of the interstices only for *small* deviations from the spontaneous curvature $1/R_0$. Equation (3.1) has the specific form for each of the given planes that is due to the specific physical description of the plane.

3.2.1 Luzzati Plane

The position of the Luzzati plane is defined by the radius of the water cylinder, R_w (Luzzati and Husson, 1962):

$$R_w = \alpha\sqrt{\frac{\sqrt{3}}{2\pi}(1 - \phi_l)} \qquad \text{(see Fig. 3.1)} \tag{3.2}$$

Here α is the lattice parameter of the H_{II} phase and ϕ_l is the lipid volume fraction (Kulkarni et al., 2010). For the hexagonal phase, α is the cylinder-center–cylinder-center distance and is related to the measured first-order Bragg spacing of the hexagonal phase, d_{hex}:

$$\alpha = \frac{2}{\sqrt{3}} d_{hex} \tag{3.3}$$

The corresponding area per molecule at the Luzzati surface (Fig. 3.1) is given by the formula (Luzzati and Husson, 1962)

$$A_w = \frac{4\pi R_w V_l}{\alpha^2 \phi_l \sqrt{3}} \tag{3.4}$$

where V_l is the lipid molecular volume. The position of the Luzzati plane is somewhere inside the hydrated structure of the polar lipid heads. It is easy to calculate R_w and A_w using only Eqs. (3.2) and (3.4) and an appropriate experimental data set (d_{hex}, ϕ_l). Table 3.1 contains a set of calculated values of R_w and A_w for different lipids at different temperatures and water mole ratios.

Although Eq. (3.3) does not show any dependence between d_{hex} and ϕ_l, it is definitely there. Figure 3.2 shows the experimentally determined dependence $d_{hex}(\phi_w)$, where $\phi_w = 1 - \phi_l$ is the water volume fraction. The dependence between d_{hex} and ϕ_l exists up to the full hydration of the polar lipid heads (Fig. 3.2a). Figure 3.2b shows only the part with clear dependence between parameters (below full hydration). The easiest way to express the connection between the lattice parameter and the lipid volume fraction is through the pivotal plane parameters (Section 3.2.2). In addition, there is a model that explains in some detail the contribution of the second lipid in the lattice parameter α (Lafleur et al., 1996):

$$R_w = \frac{\alpha}{2} - \alpha^*_{H_{II}} - \alpha_{PH} \tag{3.5}$$

Here, $\alpha^*_{H_{II}}$ is the sum of the contributions of the lipid chain and includes the structural influence of the second component, while α_{ph} is the contribution

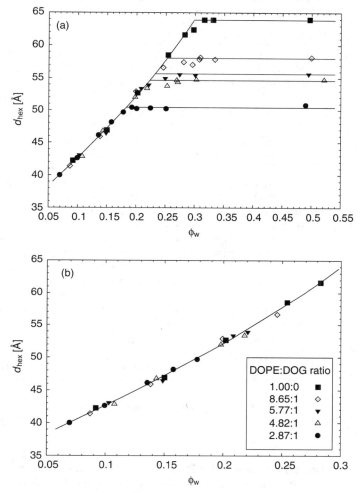

Figure 3.2   Experimental dependence of d_{hex} on the water volume fraction, ϕ_w at different dioleoyl phosphatidylethanolamine : dioleoyl glycerol (DOPE : DOG) ratios: (a) the horizontal lines describe the dependence at fully hydrated lipid heads and (b) the same dependence at less than full hydration. [Data source: Leikin et al. (1996).]

of the polar lipid heads (it can be estimated on the basis of the structure of the lipid head). Table 3.2 contains a measured data set for α and $\alpha_{H_{II}}^*$ in 1-palmitoyl-2-oleoyl-sn-phopshatidylethandolamine/1-palmitoyl-2-oleoyl-sn-phopshatidylcholine (POPE/POPC) mixtures. Unfortunately, $\alpha_{H_{II}}^*$ is not connected to an appropriate theoretical model—it is a product of ^2H nuclear magnetic resonance (NMR) measurements.

According to the classical approach (Kulkarni et al., 2010), the lipid molecular volume could be calculated by using the density of the used lipid. There is

TABLE 3.1 Calculated Data for R_w and A_w (at the Luzzati Plane) and R_p and A_p (at the Pivotal Plane), for Different Lipids at Different Temperatures and Water Mole Ratios, n_w[a]

Lipid	T (°C)	n_w	R_p (Å)	R_w (Å)	A_p (Å^2)	A_w (Å^2)	V_p/V_l
$(18:1c\Delta^9)_2$PE	10	18.0	37.8	22.8	78.5 ± 2.9	47.3	0.215 ± 0.086
	15	17.5	34.2	22.1	73.2 ± 1.7	47.4	0.399 ± 0.044
	20	16.9	29.1	21.5	63.9 ± 1.6	47.1	0.650 ± 0.034
	20	19.5	30.6	22.6	69.8 ± 0.3	50.6	0.616 ± 0.050
	22	19.1	30.0	22.1	70.3 ± 1.4	51.8	0.604 ± 0.024
	22	20.3	31.0	22.6	72.9 ± 1.3	53.9	0.562 ± 0.027
	22	18.1	28.9	21.6	66.3 ± 0.8	50.2	0.649 ± 0.015
	25	16.6	28.4	21.0	64.3 ± 1.6	47.5	0.659 ± 0.033
	30	16.2	27.8	20.5	64.6 ± 2.0	47.6	0.665 ± 0.040
	35	16.0	27.3	20.0	65.5 ± 2.1	48.0	0.663 ± 0.041
	40	15.6	26.2	19.5	64.2 ± 1.9	47.9	0.697 ± 0.037
	45	15.2	25.3	19.1	63.4 ± 1.9	47.8	0.719 ± 0.035
	50	14.9	24.5	18.6	63.0 ± 2.1	48.0	0.738 ± 0.035
	55	14.7	24.2	18.3	63.7 ± 2.2	48.3	0.735 ± 0.037
	60	14.5	23.0	17.9	64.5 ± 2.4	48.5	0.774 ± 0.039
	65	14.2	22.8	17.5	63.3 ± 2.5	48.6	0.763 ± 0.039
	70	14.0	22.2	17.2	63.0 ± 2.0	48.8	0.776 ± 0.031
	75	13.8	21.7	16.8	63.3 ± 1.8	49.1	0.782 ± 0.028
	80	13.7	21.6	16.6	64.5 ± 1.8	49.5	0.773 ± 0.032
	85	13.7	21.3	16.4	65.0 ± 2.0	50.0	0.777 ± 0.029
	90	13.6	21.3	16.2	66.3 ± 2.9	50.5	0.767 ± 0.043
$(20:0)_2$PE	99	15.7	25.3	21.1	55.7 ± 3.3	46.4	0.845 ± 0.038
$(O-12:0)_2$PE	135	15.8	26.4	16.0	104.0 ± 4.0	63.2	0.114 ± 0.090

Source: Marsh (2011).

[a]The last column contains the calculated ratios between the pivotal volume, V_p, to the respective volume of molecule, V_l.

TABLE 3.2 Comparison between the Intercylinder Distance, Derived from NMR and X-ray Diffraction, for Different Mixtures of POPE/POPC and Different Temperatures[a]

Mixture and Temperature	α (X ray) (Å)	α (NMR) (Å)	α_{II}^* (Å)
POPE:POPC (91:09); 30°C	91.2	89.3	12.0
POPE:POPC (91:09); 40°C	86.7	85.7	11.8
POPE:POPC (91:09); 60°C	79.8	80.7	11.4
POPE:POPC (82:18); 20°C	103.0	102.1	12.3
POPE:POPC (82:18); 30°C	96.5	95.7	12.0
POPE:POPC (82:18); 40°C	91.7	90.7	11.7
POPE:POPC (82:18); 50°C	87.9	88.3	11.5
POPE:POPC (82:18); 60°C	84.5	85.0	11.3
POPE:POPC (82:18); 70°C	81.7	82.4	11.2
POPE:POPC (68:32); 40°C	104.8	107.0	11.9

Source: Lafleur et al. (1996).

[a]The values in the third column are used in Eq. (3.5).

another approach, introduced by Leikin et al. (1996), that insists that an effective lipid molecule volume should be used:

$$V_1 = V_A + x V_B \qquad (3.6)$$

Here V_A is the molar volume of lipid that builds the rods and V_B is the molecular volume of the second lipid. Equation (3.6) is used to standardize the analysis of lipid mixture and to evaluate the pivotal parameters (see Section 3.2.2).

3.2.2 Pivotal Plane

The pivotal plane is the plane that has a molecular cross-sectional area invariant upon isothermal bending. The position of the pivotal plane (Fig. 3.1) depends on the relationship between area compressibility and bending of the monolayer (Leikin et al., 1996). In other words, the pivotal plane position and elastic constants are specific to a particular deformation. Especially in the H_{II} phase, the pivotal plane is the surface at which the area remains constant as the curvature in the phase is changed by varying the water content. The last definition connects the area at the pivotal plane to the process of hydration of lipid head groups, but it is not very usable because it does not incorporate the natural plane position (Marsh, 2011).

For the pivotal plane, Eq. (3.1) could be transformed as follows (Helfrich, 1973; Kirk et al., 1984):

$$F \approx \frac{1}{2} A_p k_{cp} \left(\frac{1}{R_p} - \frac{1}{R_{0p}} \right)^2 \qquad (3.7)$$

where A_p is the area per molecule at the pivotal surface, k_{cp} is the monolayer bending modulus, R_p is the radius of the plane, and $1/R_{0p}$ is the spontaneous curvature.

3.2.2.1 Bicontinuous Inverse Cubic Phases

Note: This section is included only for comparison purposes. It is not a part of the study of H_{II} phase and its goal is to help the reader to better understand the process of data management.

The models of bicontinuous inverse cubic phases are based on the infinite periodic minimal surface model (IPMS). The minimal surface lies at the bilayer midplane, and it has the property that the mean curvature, defined as half the sum of the principal curvatures, is zero everywhere on the surface. The pivotal surfaces of each monolayer are displaced from the midplane at a distance l_p (midplane is defined by IPMS), and the area per molecule at the pivotal plane is given by the expression (Templer et al., 1998b):

$$A_p = \frac{2 V_1}{a \phi_1} \left[\sigma_0 + 2\pi \chi \left(\frac{l_p}{a} \right)^2 \right] \qquad (3.8)$$

Here, a is the cubic lattice constant, σ_0 is the dimensionless area per cubic unit cell of the minimal surface, V_l is the volume of lipid molecule, and χ is the Euler characteristic of the surface, per cubic unit cell. It should be noted that χ has a role of a *topology index* of the surface. The volume fraction between pivotal and minimal surfaces can be expressed as follows (Marsh, 2011; Turner et al., 1992):

$$\frac{V_p}{V_1}\phi_1 = 2\sigma_0 \frac{l_p}{a} + \frac{4\pi}{3}\chi\left(\frac{l_p}{a}\right)^3 \tag{3.9}$$

Here V_p is the pivotal volume of the molecule and V_1 is the volume of the molecule. Equation (3.8) is used primarily to calculate l_p and its only physically meaningful solution is (Marsh, 2011)

$$l_p = a\sqrt{\frac{2\sigma_0}{\pi|\chi|}}\cos\left(\frac{\pi+\vartheta}{3}\right) \qquad \vartheta = \arccos\left(\phi_1\frac{V_p}{V_1}\sqrt{\frac{9\pi|\chi|}{8\sigma_0^3}}\right) \tag{3.10}$$

Twice the value of l_p determines the separation of the pivotal surfaces in bicontinuous cubic phases—$2l_p$. By combining Eqs. (3.8) and (3.10) it is easy to derive the expression for the lattice parameter of bicontinuous inverse cubic phases:

$$a = \frac{2\sigma_0 V_1}{A_p\phi_1}\left[1 - 4\cos\left(\frac{\pi+\vartheta}{3}\right)\right] \tag{3.11}$$

The most common use of Eq. (3.11) in data management is to fit a set of experimental data (a, ϕ_1) and to find the corresponding optimal values of σ_0, A_p, V_p, and χ (using the nonlinear least-squares method). Due to the large number of fit parameters (four), a large set of experimental data must be used. The nonlinear optimization, based on the nonlinear least-squares method, could be transferred into an economizing one by reducing the number of the variable parameters by one. Such a reduction is possible because a has a linear dependence on $1/A_p$. Therefore, $1/A_p$ can be calculated (as an internal/local solution) at each step of the nonlinear routine for finding σ_0, V_p, and χ. This type of hybrid optimization reduces the computational error and speeds up the entire process of optimization.

Due to the importance of the optimization design for the processing of experimental data, a brief technical description is presented. If the given set of data (a, ϕ_1) consists of N data pairs, and $\sigma^2(a_i)$ is the standard error of the ith measurement of a $(i = 1, \ldots, N)$, then the merit function can be defined as

$$\Phi\left(\frac{1}{A_p}, V_p, \sigma_0, \chi\right) = \sum_{i=1}^{N}\frac{1}{\sigma^2(a_i)}\left\{a_i - \frac{2\sigma_0 V_1}{A_p\phi_{1,i}}\left[1 - 4\cos^2\left(\frac{\pi+\vartheta_i(V_p,\sigma_0,\chi)}{3}\right)\right]\right\}^2 \tag{3.12}$$

where

$$\vartheta_i(V_p, \sigma_0, \chi) = \arccos\left(\phi_{1,i} \frac{V_p}{V_1} \sqrt{\frac{9\pi|\chi|}{8\sigma_0^3}}\right) \tag{3.13}$$

Differentiating Eq. (3.12) by $1/A_p$, an internal solution for A_p can be derived for a given set of values (V_p, σ_0, χ):

$$A_p(V_p, \sigma_0, \chi) = 2\sigma_0 V_1 \frac{\sum_{i=1}^{N} \frac{1}{\sigma^2(a_i)\phi_{1,i}^2} \left[1 - 4\cos^2\left(\frac{\pi + \vartheta_i(V_p, \sigma_0, \chi)}{3}\right)\right]^2}{\sum_{i=1}^{N} \frac{a_i}{\sigma^2(a_i)\phi_{1,i}} \left[1 - 4\cos^2\left(\frac{\pi + \vartheta_i(V_p, \sigma_0, \chi)}{3}\right)\right]} \tag{3.14}$$

The correctness of the result could be proved by setting $N = 1$ and comparing the derived expression to Eq. (3.11). By means of Eq. (3.14), the merit function set of parameters can be reduced by one:

$$\Phi(V_p, \sigma_0, \chi) = \sum_{i=1}^{N} \frac{1}{\sigma^2(a_i)} \left\{a_i - \frac{2\sigma_0 V_1}{A_p(V_p, \sigma_0, \chi)\phi_{1,i}} \left[1 - 4\cos^2\left(\frac{\pi + \vartheta_i(V_p, \sigma_0, \chi)}{3}\right)\right]\right\}^2 \tag{3.15}$$

The process of deriving V_p, σ_0, and χ is a subject of the chosen routine for nonlinear optimization, most often one of the following: downhill simplex algorithm, Powell's method, conjugate gradient (CG) algorithm, Broyden–Fletcher–Goldfarb–Shanno (BFGS), and Newton-conjugate-gradient (Newton-CG) (Press et al., 2007).

3.2.2.2 *Hexagonal Inverse Phase*   In hexagonal inverse phase (H_{II}), the position of the pivotal plane is given by (Marsh, 2011)

$$R_p = \alpha \sqrt{\frac{\sqrt{3}}{2\pi}\left(1 - \frac{V_p}{V_1}\phi_1\right)} \tag{3.16}$$

Here V_p is the pivotal volume (Fig. 3.1). The position of the pivotal plane, R_p, is offset from the position of the lipid–water interface, R_w:

$$R_p = R_w \sqrt{\frac{1}{1 - \phi_1}\left(1 - \frac{V_p}{V_1}\phi_1\right)} \tag{3.17}$$

It is easy to predict that if $V_p = V_1$ then $R_p = R_w$. The corresponding area per lipid molecule at the pivotal surface is

$$A_p = \frac{4\pi R_p V_1}{\sqrt{3}\alpha^2\phi_1} \tag{3.18}$$

By combining Eqs. (3.16) and (3.18), an expression for α can be derived (Templer et al., 1998b):

$$\alpha = \frac{2V_1}{A_p \phi_1} \sqrt{\frac{2\pi}{\sqrt{3}} \left(1 - \frac{V_p}{V_1} \phi_1\right)} \tag{3.19}$$

Equation (3.19) can be used to construct a target (merit) function and to fit a set of experimental data (α, ϕ_1) in order to obtain the values of A_p and V_p. The idea behind the optimization routine here is similar to those applied for bicontinuous cubic phases before [Eqs. (3.11)–(3.15)]. The corresponding merit function takes the form

$$\Phi\left(\frac{1}{A_p}, V_p\right) = \sum_{i=1}^{N} \frac{1}{\sigma^2(\alpha_i)} \left(\alpha_i - \frac{2V_1}{A_p \phi_1} \sqrt{\frac{2\pi}{\sqrt{3}} \left(1 - \frac{V_p}{V_1} \phi_{1,i}\right)}\right)^2 \tag{3.20}$$

The respective internal solution for A_p, for a given value of V_p, could be expressed as

$$A_p(V_p) = V_1 \sqrt{\frac{8\pi}{\sqrt{3}} \frac{\sum_{i=1}^{N} \frac{1}{\sigma^2(\alpha_i)\phi_{1,i}^2} \left(1 - \frac{V_p}{V_1}\phi_{1,i}\right)}{\sum_{i=1}^{N} \frac{\alpha_i}{\sigma^2(\alpha_i)\phi_{1,i}} \sqrt{1 - \frac{V_p}{V_1}\phi_{1,i}}}} \tag{3.21}$$

and it can help to reduce by one the number of parameters of the merit function:

$$\Phi(V_p) = \sum_{i=1}^{N} \frac{1}{\sigma^2(\alpha_i)} \left(\alpha_i - \frac{2V_1}{A_p(V_p)\phi_1} \sqrt{\frac{2\pi}{\sqrt{3}} \left(1 - \frac{V_p}{V_1}\phi_{1,i}\right)}\right)^2 \tag{3.22}$$

The reduced expression, Eq. (3.22), represents a merit function with only one parameter and its minimum can be easily calculated by using the golden section method (Press et al., 2007). Figure 3.3 gives an idea of the quality of a similar optimization using the nonreduced model from Eq. (3.20). Table 3.2 contains calculated values of R_p and A_p for different lipids at different temperatures and water mole ratios, as well as the V_p/V_1 ratio. The routine explained here could be extended by adding V_1 to the list of the fit parameters when necessary.

There is another way to reduce the number of fit parameters by setting V_1/A_p and V_p/V_1 as independent parameters in Eq. (3.20) (March, 2011).

There is an alternative method for obtaining R_p, A_p, and V_p that is applicable to the H_{II} phase. It is simpler, but works well only if the dependence between A_w^2 and A_w/R_w is linear (Fuller et al., 2003; Leikin et al., 1996):

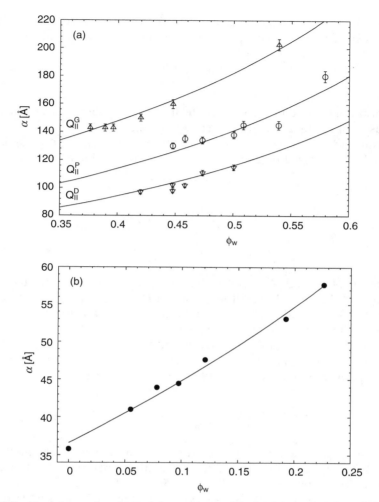

Figure 3.3 Experimental measured dependence of the lattice parameter as a function of water volume part, ϕ_w: (a) Q_{II} phases of 2:1 LA/DLPC measured between 35 and 45°C and (b) H_{II} phase of 2:1 MA/DMPC at 60°C. The continuous lines depict the performed least squares fits for the pivotal planes with (a) Eq. (3.11) and (b) Eq. (3.18). [Data source: Templer et al. (1998a).]

$$A_w^2 = A_p^2 - 2V_p \frac{A_w}{R_w} \tag{3.23}$$

So the first task, before using the model, is to confirm statistically that the used data of the set $(A_w^2, A_w/R_w)$ have linear dependence at some reasonable significance level. Then the pivotal volume value can be determined from the slope, and the pivotal surface can be estimated from the calculated intercept of the linear fit. Equation (3.23) could be rewritten in a more general form:

$$\frac{A_w^2}{V_1^2} = \frac{A_p^2}{V_1^2} - 2\frac{V_p}{V_1}\left(\frac{A_w}{V_1 R_w}\right) \tag{3.24}$$

This form can be used to standardize the linear fit routine in the case of simultaneous fit of multiple sets of data (Leikin et al., 1996).

Another model for determining both the pivotal radius and pivotal cross-sectional area is based on geometrical considerations (Fuller et al., 2003):

$$R_p = R_w\sqrt{1 + \frac{\phi_1}{1 - \phi_1}\frac{V_p}{V_1}} \tag{3.25}$$

$$A_p = A_w\sqrt{1 + \frac{\phi_1}{1 - \phi_1}\frac{V_p}{V_1}} \tag{3.26}$$

In terms of A_p and V_p, it is possible to formulate another (recursive-like) definition of the pivotal plane for H_{II} phase: The pivotal plane is defined as a surface inside the lipid phase such that both A_p and V_p are constants when the distance between the rod axes of H_{II} varies.

The radius of the pivotal plane is incorporated in the process of determining the elastic parameters at the pivotal plane of the lipid mixture from osmotic stress experiments (Gruner et al., 1986; Rand et al., 1990):

$$\Pi R_p^2 = 2k_{cp}\left(\frac{1}{R_p} - \frac{1}{R_{0p}}\right) = -2k_{cp}\frac{1}{R_{0p}} + 2k_{cp}\frac{1}{R_p} \tag{3.27}$$

where Π is the measured osmotic stress. The described linear model ΠR_p^2 versus $1/R_p$ can be easily adapted to process the experimental data from H_{II} phase analysis. First, the initial data set (Π, d_{hex}, ϕ_1) is taken from the experiment and used to derive the set (R_p, ϕ_1). Next, the final data set $(\Pi R_p^2, 1/R_p)$ has to be determined by using a fit routine and Eq. (3.27). Figure 3.4 illustrates the quality of the performed linear fit. The estimated value of the slope is twice the value of k_{cp}. To obtain the value of the spontaneous curvature at the pivotal plane, $1/R_{0p}$, the calculated intercept is divided by the slope value.

The measured values of the spontaneous curvature are properties of the lipid mixture. As the mixture consists of two components (the first one builds the hexagonal rods), $1/R_{0p}$ should depend on the quantity of the two components separately. For example, if the mixture consists of DOPE and DOG (an intensively studied system), the spontaneous curvature could be expressed as (Leikin et al., 1996):

$$\frac{1}{R_{0p}} = (1 - m_{DOG})\frac{1}{R_{0p}^{DOPE}} + m_{DOG}\frac{1}{R_{0p}^{DOG}} \tag{3.28}$$

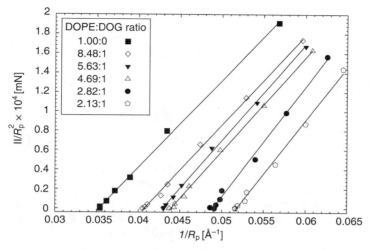

Figure 3.4 Determination of the monolayer bending modulus, k_{cp}, and the spontaneous curvature at the pivotal plane, $1/R_{0p}$, using an appropriate linear fit [Eq. (3.24)]. The experimental data set is for the H_{II} phase of DOPE : DOG at different mole ratios. [Data source: Leikin et al. (1996).]

where m_{DOG} is the DOG mole fraction and $1/R_{0p}^{DOPE}$ and $1/R_{0p}^{DOG}$ are the specific radii of curvature for each of the components. The respective dependencies of the spontaneous curvature at the pivotal plane, $1/R_{0p}$, and the monolayer bending modulus, k_{cp}, on the mole ratio of the second compound (DOG) are shown in Figure 3.5.

3.2.3 Neutral Plane

The neutral plane (Fig. 3.1) is defined as that surface where the bending and stretching (compression) deformations are energetically uncoupled, so that its position does not depend on the deformation or on the experimental techniques (Kozlov and Winterhalter, 1991a,b; Leikin et al., 1996). The position of the neutral surface is determined only by the mechanical properties of the lipid and is *independent of the deformation* as long as the elastic response remains linear (Leikin et al., 1996). The neutral plane should be near the interface between the hydrophilic and hydrophobic regions of the monolayer (Siegel and Kozlov, 2004), and the process of hydration leads to deformation of the neutral surface (Leikin et al., 1996). The spontaneous curvature and bending modulus defined for the neutral plane are *unique*, spontaneous properties of the lipid. For bending deformation of symmetrical bilayers in the absence of bilayer stretching and when the monolayers can freely slide along each other, the pivotal plane coincides with the neutral surface.

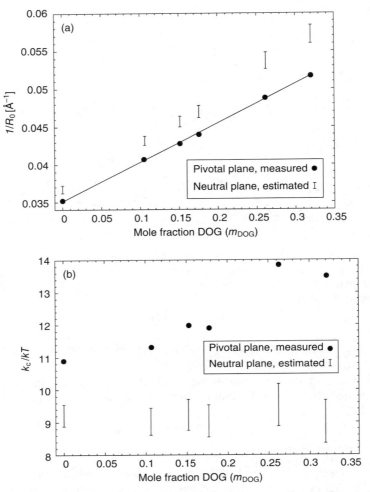

Figure 3.5 Dependence of the radius of spontaneous curvature, $1/R_0$, on mole fraction of DOG, for (a) pivotal and neutral planes and (b) the respective dependence of the monolayer bending modulus, k_c. All data refer to DOPE/DOG mixture. The used value of K is about 100–150 dyn/cm (see text for explanation). [Adapted from Leikin et al. (1996).]

The expression of the elastic energy could be derived from Eq. (3.1) by taking into account that at the position of the neutral plane, the elastic modulus E_{AR} is exactly zero (Kozlov and Winterhalter, 1991b):

$$F \approx \frac{1}{2} A_{0n} k_{cn} \left(\frac{1}{R_n} - \frac{1}{R_{0n}} \right)^2 + \frac{1}{2} K \frac{(A_n - A_{0n})^2}{A_{0n}} \tag{3.29}$$

Here, R_n is the radius of the neutral plane, $1/R_{0n}$ is the spontaneous curvature, A_n is the lipid area at the neutral surface, A_{0n} is the relevant spontaneous area, and K is the lateral compressibility modulus of the monolayer. Unlike the pivotal plane, the position of the neutral surface does not depend on the deformation or on the used experimental technique, so it is impossible to simply use some variant of the model applied previously for the pivotal plane. When applying Eq. (3.29) it should be kept in mind that the neutral and pivotal planes are well defined only for *small deformations* within the limits of validity of Eq. (3.1). That limitation has to be compatible with the experimental methods used for obtaining data for the analysis.

Due to the specificity described above, the neutral plane could be studied in terms of the deformation of H_{II} phase upon hydration. The first step is to derive the spontaneous parameters at the pivotal plane—$1/R_{0p}$ and k_{cp}. The spontaneous parameters at the neutral plane—$1/R_{0n}$ and k_{cn}—can be calculated by using the following set of equations (Leikin et al., 1996):

$$\frac{1}{R_{0n}} = \frac{1}{R_{0p}} \sqrt{\frac{1+\gamma}{1-\gamma}} \tag{3.30}$$

$$k_{cn} = k_{cp} \frac{(1-\gamma)^{5/2}}{(1+\gamma)^{3/2}} \tag{3.31}$$

$$\gamma \equiv \frac{k_{cn}}{K} \frac{1}{R_{0n}^2} \tag{3.32}$$

Here, γ is the compression/bending ratio, and it can be estimated only if the value of K—the lateral compressibility modulus of the monolayer—is known (from an independent source). In addition,

$$A_{0n} \approx A_p \sqrt{\frac{1-\gamma}{1+\gamma}} \tag{3.33}$$

$$A_n \approx A_{0n} \left[1 - \gamma R_{0n} \left(\frac{1}{R_n} - \frac{1}{R_{0n}} \right) \right] \tag{3.34}$$

The neutral volume, V_n, could be estimated by means of the relation (Leikin et al., 1996)

$$V_n = V_p - A_{0n} R_{0n} \frac{\gamma}{1-\gamma} \tag{3.35}$$

or

$$R_n A_n = 2V_n + 2V_1 \frac{1-\phi_1}{\phi_1} \tag{3.36}$$

Using the calculated set of data $[R_n A_n/2, (1 - \phi_l)/\phi_l]$ and an appropriate linear fit routine, it is possible not only to prove (or disapprove) the validity of Eq. (3.36) but eventually to prove the validity of the assumptions of the entire model [Eqs. (3.30–3.36)] and their validity regarding the studied system. First, the values of V_n, obtained by Eq. (3.35), and the linear fit [Eq. (3.36)], should be compared to each other. Second, the obtained value of V_l from the slope should be compared to the calculated value used to derive R_p, A_p, and their spontaneous analogs.

Figure 3.5 shows the calculated dependence of the monolayer bending modulus, k_{cn}, and the radius of spontaneous curvature, $1/R_{0n}$, at the neutral plane position, for the mole ratio of the second component of the lipid mixture—DOG (for DOPE/DOG mixture). It is shown (Fig. 3.5b) that the values of the monolayer bending moment *do not depend* on the mole ratio of DOG, while the ones of the radius of spontaneous curvature *do depend*.

It is very important to explain the origin of the error of k_{cn} and $1/R_{0n}$. In Figure 3.5 both values are drawn with bars instead of points. It is obvious that their estimation errors are high and the reason for that is the magnitude of the sampling error of K, which seems to be a general problem of the method. Some improvement could be obtained if the value of K is evaluated using appropriate statistics.

3.3 CONCLUDING REMARKS

The described models of the dividing planes of the hexagonal H$_{II}$ phase have a strong physical background and are widely developed. The corresponding methods of data processing are also well developed and deployed in analytical practice. Despite their relative simplicity, they provide a sufficiently accurate tool set for studying the position and structure of the dividing planes of the H$_{II}$ phase and the related physical quantities. Unfortunately, such simplicity does not guarantee the correctness of the produced results and that peculiarity must be taken into account. To produce the correct results, appropriate statistics must be applied: first to the input data and second to the regression model.

It is important to use well-estimated values of the lipid volume ratios and accurately measured densities of the used lipid and water. The most critical point during the calculation process is the goodness of the proposed fit routines because they determine the values of the mechanical characteristics of hexagonal inverse H$_{II}$ phase. The total error of the routines is hard to determine exactly because of its additive sense. On the other hand, it is clear that it must include the error of the model approximation and the error of the (method of) measurement. The second is more difficult to control because it is a random variable and depends on a lot of factors. So the most important task is (i) to prove statistically that the used theoretical model adequately describes the current experimental data and (ii) to publish the corresponding statistical description along with the data (Bevington, 1969). It should be mentioned that the merit functions are related to χ^2 distribution, and the corresponding probability should be taken in account when assessing the reliability of the results. For example, fitting data without taking into account the statistical description of the performed fit (the variance in the estimates of the fit parameters, the relation between the linear correlation coefficient, and a goodness-of-fit measure, 95% confidence/prediction interval, etc.) is a common problem in scientific practice. It does not allow the researcher or reader to conclude whether the description of data throughout the used model is adequate. Another common problem is the use of unweighted data, that is, to set $\sigma = 1$ in merit function. In that case, data may lie perfectly on the fit line, but the calculated parameters might be unreliable or, at worst, completely wrong. When reliable results have to be obtained, the propagation of the error must be always taken into account.

REFERENCES

Bevington, P. R. (1969). *Data Reduction and Error Analysis for the Physical Sciences*, McGraw-Hill, New York, Chapters 1–4.

Fuller, N., Benatti, C. R., and Rand, R. P. (2003). Curvature and bending constants for phosphatidylserine-containing membranes. *Biophysics Journal*, *85*, 1667–1674.

Gruner, S. M., Parsegian, V. A., and Rand, R. P. (1986). Directly measured deformation energy of phospholipid H_{II} hexagonal phases. *Faraday Discussions Chemical Society*, *81*, 29–37.

Helfrich, W. (1973). Elastic properties of lipid bilayers: Theory and possible experiments. *Zeitschrift fur Naturforschung*, *28C*, 693–703.

Kirk, G. L., Gruner, S. M., and Stein, D. L. (1984). A thermodynamic model of the lamellar to inverse hexagonal phase transition of lipid membrane-water system. *Biochemistry*, *23*, 1093–1102.

Kozlov, M. M., and Winterhalter, M. (1991a). Elastic moduli for strongly curved monolayers. Analysis of experimental results. *Journal de Physique, France II*, *1*, 1085–1100.

Kozlov, M. M., and Winterhalter, M. (1991b). Elastic moduli for strongly curved monolayers. Position of the neutral surface. *Journal de Physique, France II*, *1*, 1077–1084.

Kulkarni, V. C., Wachter, W., Iglesias-Salto, G., Engelskirchen, S., and Ahualli, S., (2010). Monoolein: A magic lipid? *Physical Chemistry Chemical Physics*, *13*, 3004–3021.

Lafleur, M., Bloom, M., Eikenberry, E. F., Gruner, S. M., Han, Y., and Cullis, P. (1996). Correlation between lipid plane curvature and lipid chain order. *Biophysics Journal*, *70*, 2747–2757.

Leikin, S., Kozlov, M. M., Fuller, N. L., and Rand, R. P. (1996). Measured effects of diacylglycerol of phospholipid membranes. *Biophysics Journal*, *71*, 2623–2632.

Luzzati, V., and Husson, F. (1962). X-ray diffraction studies of lipid-water systems. *Journal of Cell Biology*, *12*, 207–219.

Marsh, D. (2011). Pivotal surfaces in inverse hexagonal and cubic phases of phospholipids and glycolipids. *Chemistry and Physics of Lipids*, *164*, 177–183.

Press, W. H., Teukolsky, S. A., Vetterling, W. T., and Flannery, B. P. (2007). *Numerical Recipes: The Art of Scientific Computing*, 3rd ed., Cambridge University Press, New York.

Rand, R. P., Fuller, N. L., Gruner, N. L., and Parsegian, V. A. (1990). Membrane curvature, lipid segregation, and structural transitions for phospholipids under dual-solvent stress. *Biochemistry*, *29*, 76–87.

Siegel, D. P., and Kozlov, M. (2004). The Gaussian curvature elastic modulus of N-monomethylated dioleoylphosphatidylethanolamine: Relevance to membrane fusion and lipid phase behavior. *Biophysics Journal*, *87*, 366–374.

Templer, R. H., Khoo, B. J., and Seddon, M. J. (1998a). Gaussian curvature modulus of an amphiphilic monolayer. *Langmuir*, *14*, 7427–7434.

Templer, R. H., Seddon, J. M., Warrender, N. A., Syrykh, A., Huang, Z., Winter, R., and Erbes, J. (1998b). Inverse bicontinuous cubic phases in 2:1 fatty acid/phosphatidylcholine mixtures. The effects of chain length, hydration and temperature. *Journal of Physical Chemistry B*, *102*, 7251–7261.

Turner, D. C., Wang, Z. G., Gruner, S. M., Mannock, D. A., and McElhaney, R. N. (1992). Structural study of the inverted cubic phases of di-dodecyl alkyl-β-D-glucopyranosyl-rac-glycerol. *Journal de Physique, France II*, *2*, 2039–2063.

CHAPTER 4

Nanocharacterization of Lyotropic Liquid Crystalline Systems

PATRICK G. HARTLEY and HSIN-HUI SHEN

CSIRO Materials Science and Engineering, Clayton, Victoria, Australia

Abstract

The development of techniques for the characterization of the nanostructure of lyotropic liquid crystalline systems have been crucial in enabling increased understanding of fundamental properties and efficacy in technological applications. This chapter provides a review of a number of commonly used and emerging nanocharacterization techniques, with a view to providing a resource for academic and industrial researchers joining the field.

Self-Assembled Supramolecular Architectures: Lyotropic Liquid Crystals, First Edition.
Edited by Nissim Garti, Raffaele Mezzenga, and Ponisseril Somasundaran.
© 2012 John Wiley & Sons, Inc. Published 2012 by John Wiley & Sons, Inc.

4.1 INTRODUCTION

Lyotropic liquid crystalline (LLC) phases result from the self-assembly of amphiphilic molecules such as lipids and surfactants in water. Classical examples of LLCs are the lamellar, hexagonal, and cubic phases, as shown in Figure 4.1. LLCs often possess long-range orientational order (in the range of nanometers to millimeters), resulting in unique hydrophobic–hydrophilic "nanoenvironments," which has led to their development in a number of technological areas such as personal care products and pharmaceutics (Spicer, 2005a). Concomitant with this has been the need for detailed analysis of LLC properties, including their thermodynamic equilibrium and metastable phase behavior, and nanostructure. This latter need has in turn driven the development of new techniques for LLC nanocharacterization.

In this chapter, we review the current techniques that are employed for LLC nanocharacterization. Section 4.2 discusses characterization of "bulk" LLC nanostructure, while Section 4.3 deals with dispersed (i.e., colloidal) LLCs. Section 4.4, meanwhile, explores recent developments in the characterization of LLC interfacial properties. The aim is to give an overview of key techniques for researchers joining this exciting subfield of "soft-matter nanotechnology".

4.2 BULK LYOTROPIC LIQUID CRYSTALS: NANOCHARACTERIZATION AND PHASE BEHAVIOR

The key determinant of LLC equilibrium nanostructure is the thermodynamic phase behavior of the amphiphile–water system. Since phase behavior is also

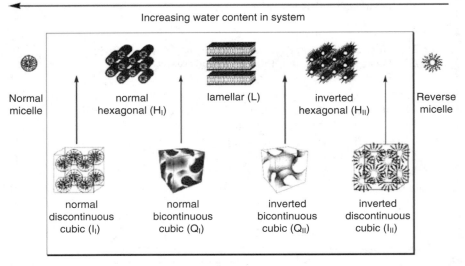

Figure 4.1 Representations of different lyotropic liquid crystal phases (Gin et al., 2008).

often critical to technological applications, there exists a considerable body of LLC research focused on understanding phase behavior.

Phase behavior data is most commonly reported in the form of the temperature–binary composition (i.e., water–amphiphile) phase diagram (Laughlin, 1994). Phase diagrams are generated by producing specific compositions and carefully analyzing the phases that result without causing compositional alterations. Classically, this is achieved by sealing precise compositions within glass ampoules.

LLC phases manifest themselves in different regions of the phase diagram depending on the amphiphile being studied, and in certain cases they share a phase boundary with excess water, which results in the LLCs being "dilutable in excess water," an important property when it comes to producing LLC dispersions, which are discussed later.

There are several characterization techniques that can be employed in determining temperature–composition phase diagrams of LLC systems. These range from extremely "low tech" (e.g., direct visual observation) to extremely "high tech" (employing, e.g., synchrotron X-ray radiation). We first introduce optical microscopy, which allows the simple and direct observation of various LLC properties as temperature is varied and is often the first technique applied in qualitative studies of LLC phase behavior. When applied in combination with spectroscopic measurements, as employed in diffusive interfacial transport (DIT)–near-infrared microspectroscopy, quantitative information regarding the position of compositional phase boundaries can also be obtained. Rheological measurements of LLC phases are discussed in Section 4.2.2. Scattering techniques (particularly X-ray scattering) have become a powerful tool in determining LLC nanostructure and are discussed in Section 4.2.3. Pulsed field gradient nuclear magnetic resonance (NMR) is a complementary technique to small-angle X-ray scattering (SAXS) which will be introduced in Section 4.2.4. Recently, analysis of free volume obtained from position annihilation lifetime spectroscopy (PALS) has also been found to show consistency with phase and rheological behavior, and this technique is briefly discussed in Section 4.2.5. Electron microscopy techniques, which allow the direct observation of phase nanostructure, are also discussed in Section 4.2.6.

4.2.1 Polarized Optical Microscopy

Friedrich Reinitzer discovered liquid crystals in 1888 and observed their ability to rotate the polarization of light (Reinitzer, 1888). Since then, polarized light microscopy has been a standard technique for observing liquid crystalline phases, including lyotropic liquid crystals.

The observation of LLCs using light and polarized light microscopy is a simple technique for basic phase behavior characterization. Birefringent anisotropic phases (such as the lamellar and hexagonal phases) are readily detected under cross-polarized light (Laughlin, 1994; Rosevear, 1968), with specific optical textures allowing such phases to be distinguished from each other, as shown in Figure 4.2.

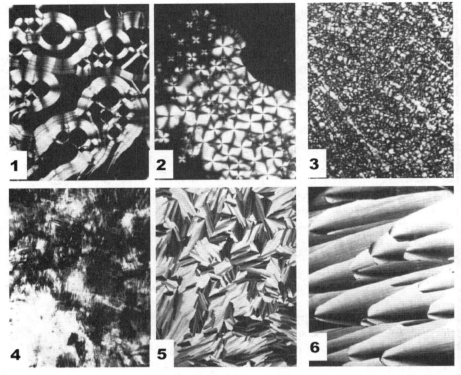

Figure 4.2 Optical textures observed under crossed polarizers for lyotropic liquid crystals from the classical work of Rosevear (Rosevear, 1968): (1–3) lamellar phases (denoted "neat" phases in the original work) and (4–6) hexagonal phases (denoted "middle" phases in the original work).

Isotropic phases (e.g., cubic and micellar phases) do not alter light polarization and hence appear dark under cross polarizers. In such systems, it is often still possible to distinguish phases from one another based on qualitative observation of phase viscosity by, for example, observing the appearance of entrained bubbles. Specifically, high-viscosity phases (e.g., cubic phases) typically inhibit bubbles from adopting a fully spherical geometry.

A particularly useful implementation of polarized light microscopy is the *water penetration* or *water flooding* experiment. Here, a small sample of the amphiphile of interest is placed on a microscope slide, and a cover slip is used to form a sandwiched disk of amphiphile. A small quantity of water is then introduced at the edge of the cover slip, and capillary force draws water toward the sample. The microscope is focused at a position where the water contacts the sample. A surfactant–water concentration gradient is established at this interface, which results in the appearance of different phases depending on

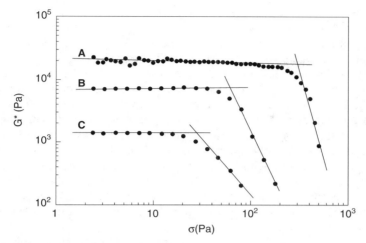

Figure 4.3 Complex modulus G^* versus shear stress of liquid crystal samples: A = cubic phase, B = lamellar phase, and C = hexagonal phase. The critical shear stress is determined from the intersect between the linear and nonlinear viscoelastic regions. [Adapted from An et al. (2006)].

the water content. While not quantitative in terms of precise water–surfactant composition, this simple measurement allows a "snapshot" of the major elements of the phase diagram to be recorded, without the need for laborious phase diagram studies. This is particularly useful when searching for the existence of specific phases of interest.

Efforts to make the flooding technique more quantitative with respect to phase composition have led to the development of a family of *diffusive interfacial transport* techniques (Laughlin and Munyon, 1987). Here, the flooding experiment is combined with a spectroscopic technique that allows water content to be determined within each of the observed phases. However, experimental complexity has prevented widespread uptake of these techniques.

4.2.2 Rheology

Different LLC phases exhibit different rheological properties. Thus quantitative rheological studies of LLC systems are supportive of other phase behavior observations.

LLC phases typically show shear thinning behavior, while measurements of the critical shear stress, at which LLC structure breaks down, can be related to phase behavior (Cordobes et al., 2005). As shown in Figure 4.3, An et al. (2006) have shown that the critical shear stress decreases in the order cubic liquid crystal (330 Pa) > hexagonal liquid crystal (64 Pa) > lamellar (27 Pa), consistent with the greater interconnectivity (and hence resistence to shear) of the cubic LLCs in particular.

Oscillatory shear measurements typically employ a concentric cylinder geometry, where an oscillatory shear stress (denoted variously as σ or τ, with units of pascals) is induced across an annular volume of sample, whose deformation (shear strain, γ) is measured. The loss (viscous) modulus (G'') and storage (elastic) modulus (G') of the material can be derived from the amplitudes and phase offset (δ) of the applied stress and measured strain oscillations. The loss factor, tan δ, is given by the ratio of viscous to elastic response (G''/G') and is particularly useful in characterizing liquid-to-gel transitions. Finally, varying the shear rate in a *frequency sweep* allows the identification of characteristic deviations between viscous and elastic behaviors, such as yielding and shear thinning behavior. Mezzenga et al. (2005) have suggested that signature rheological properties exist for different LLC phases. Their work involved identifying characteristic relaxation times based on a crossover between G' and G''-dominated behaviors derived from frequency sweep measurements, as shown in Figure 4.4.

A recent development in rheological studies of LLC systems has been the advent of particle tracking microrheology (PTM). Here, hard colloidal spheres in the submicrometer size range are introduced into the LLC formulation, and the dynamics of their motion (in the form of a mean–square displacement, MSD) are calculated from intensity autocorrelation functions measured using diffusing wave spectroscopy (DWS) (Alam and Mezzenga, 2011; Mason et al., 1997). These dynamics are related to the viscoelastic properties of the LLC mesophase within which the particles are embedded.

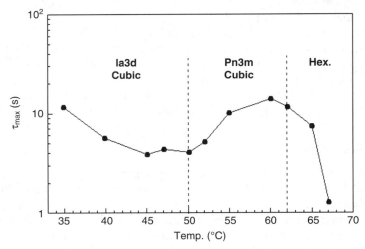

Figure 4.4   Relaxation time (τ_{max}) versus temperature for LLCs composed of 80 wt % Dimodan U/J and 20 wt % water. Time τ_{max} is obtained by the crossover of G' and G''. A negative slope of τ_{max} with temperature is associated with the Ia3d phase region, while positive slope characterizes the Pn3m region. The drop of τ_{max} beyond 62°C is consistent with the appearance of the hexagonal phase, as measured at 65°C by SAXS (Mezzenga et al., 2005).

4.2.3 Scattering Techniques

Scattering techniques provide a powerful method for obtaining quantitative information on size, shape, and nanostructure within LLC phases. Three kinds of radiation are typically used in LLC scattering studies, light, X-ray, and neutron studies. The principles of small-angle X-ray and neutron scattering are broadly the same and have been discussed in a large number of articles. (Glatter and Kratky, 1982; King et al., 1999). In this section, we focus on X-ray and neutron scattering approaches to the characterization of LLC phase behavior.

4.2.3.1 Small-Angle X-ray Scattering Small-angle X-ray scattering (SAXS) is the most recognized method used to study the nanostructural features of lyotropic liquid crystals, both in their bulk and dispersed forms. (Pabst et al., 2000; Rappolt et al., 2000). Cubic lyotropic liquid crystals were first recognized using X-ray scattering techniques by Luzzati and Husson in 1962 (Luzzati and Husson, 1962), and several structural models were proposed. (Luzzati et al., 1968a,b). Around the same time, the temperature–composition phase diagram of monoglycerides in water was published by Lutton (1965). Since then, a great diversity of amphiphile systems have been studied using SAXS, including simple and complex natural/synthetic surfactants, phospholipids, glycolipids, synthetic and natural carbohydrate surfactants, dispersed LLCs (see later), and many more (Bäverback et al., 2009; Briggs et al., 1996; Mencke and Caffrey, 1991).

Figure 4.5 is a schematic diagram of a SAXS experiment. A collimated beam of X-ray radiation of wavelength λ (typically ~1 Å) is incident upon a thin solution sample, typically millimetres or less in thickness.

The radiation passes through the sample (i.e., is transmitted), is absorbed, or is scattered (elastically/inelastically). The intensity of scattering as a function of angle is measured relative to the incident beam using an X-ray detector, such as a charge-coupled device (CCD) camera.

Non-sample-related scattering (background) is subtracted from data files, and the scattered intensity is normalized relative to sample transmission.

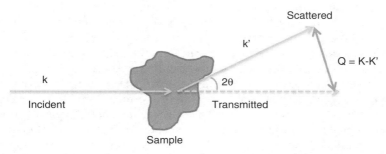

Figure 4.5 Wave vector geometry in a small-angle scattering measurement.

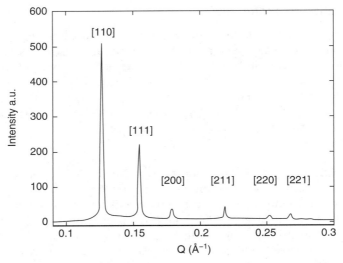

Figure 4.6 Radially integrated SAXS profile for the surfactant phytantriol in excess water at 20°C recorded at the Australian Synchrotron SAXS/WAXS beam line by the authors. The corresponding Miller indices for the Pn3m cubic phase are identified from the relative peak spacings as shown in Table 4.1.

Radial integration of the resulting two-dimensional (2D) intensity versus angle data yields the one-dimensional scattering function $I(Q)$, where Q is the length of the scattering vector, defined as the difference between the wave propagation vectors for incident (k) and scattered (k') radiation:

$$|Q| = |k - k'| = \frac{4\pi \sin(\theta / 2)}{\lambda}$$

with λ being the wavelength of the radiation and 2θ the scattering angle.

The long-range orientational order possessed by many LLC systems results in classical intense peaks in the $I(Q)$ vs. Q plot, as shown in Figure 4.6. These peaks are precisely analogous to the Bragg peaks measured for crystalline systems using X-ray diffraction; however, the peak positions found for LLCs generally occur at much lower Q values (corresponding to lower scattering angles), since liquid crystalline structure is supramolecular, with typical unit cell dimensions in the nanometer rather than Angstrom range.

In general, for LLC systems, the position of the peaks in the $I(Q)$ vs. Q plot can be used to identify the phase nanostructure (crystallographic space group) and unit cell dimensions. The crystallographic space group is derived from matching the relative positions (in Q) between measured Bragg peaks with the allowable Miller indices (h,k,l) for different phases. Typically, three or more such peaks are required for unequivocal identification. This relationship between relative Q spacing and Miller index is summarized for a number of

TABLE 4.1 Bragg Peak Spacing Ratios and Corresponding Miller Indices (h,k,l) for Double Diamond Cubic (Pn3m), Primitive Cubic (Im3m), Gyroid Cubic (Ia3d), Lamellar (L_α), and Hexagonal ($H_{I/II}$) LLC Phases

Peak Spacing Ratio:	1	$\sqrt{2}$	$\sqrt{3}$	$\sqrt{4}$	$\sqrt{6}$	$\sqrt{7}$	$\sqrt{8}$	$\sqrt{9}$
Pn3m (hkl)	—	110	111	200	211	—	220	221
Im3m (hkl)	—	110	—	200	211	—	220	—
Ia3d (hkl)	—	—	—	—	211	—	220	—
L_α (hkl)	110	—	—	220	—	—	—	300
$H_{I/II}$ (hk)	10	—	11	20	—	21	—	30

typical LLC structures in Table 4.1. The mean lattice parameter, a, may also be calculated from the measured spacings, d, between scattering planes (given by $d = 2\pi/q$), using established relationships (Winter, 2002). Such analysis allows extremely subtle changes in unit cell dimensions to be followed as a function of composition and/or sample environment for instance (Seddon et al., 2006).

The advent of synchrotron X-ray sources has reduced SAXS data collection times down to the order of milliseconds, allowing studies of the kinetics of LLC nanostructural rearrangements in real time. Many synchrotron-based SAXS/wide-angle X-ray scattering (WAXS) beam lines have been employed in studying LLC systems. These include beam lines such as ID02 at the European Synchrotron Radiation Facility (Grenoble, France), ChemMatCARS at the Advanced Photon Source (Chicago, IL), and the SAXS/WAXS beam line at the Australian Synchrotron, to name but a few.

Time-resolved X-ray scattering has been employed to study phase transitions between LLC phases in stopped-flow experiments (Gradzielski and Narayanan, 2004) and to deduce the dynamics of order–disorder transitions in response to pressure and temperature jumps (Caffrey and Hing, 1987; Gruner, 1987; Kriechbaum et al., 1989; Seddon et al., 2006). Such studies have, for example, revealed that fast phase transitions in phospholipid systems proceed without complete disruption of liquid crystalline order.

4.2.3.2 Small-Angle Neutron Scattering (SANS)

Neutron scattering relies on the interactions between a neutron beam and atomic nuclei within a sample. This is in contrast to X-ray scattering, which results from interactions with electron density. As such, neutron scattering varies nonsystematically as a function of isotopic composition, rather than systematically with atomic number, as is the case for X-ray scattering. This nonsystematic isotope dependence allows neutron scattering to be employed to study spatial distrubutions of specific chemical species within a sample. Through the use of isotopic substitution, variable contrasts in neutron scattering length density can be created within samples. In particular, the large difference in scattering length density between hydrogen and deuterium can be employed in *contrast-matching* experiments, where the scattering due to different phases in a sample can be enhanced or masked to allow a specific phase of interest to be studied. An advantage of neutron scattering in the study of structure and dynamics at the microscopic level is that neutrons are nondestructive, relative to X-rays, and hence radiation damage rarely occurs in a neutron-based experiment. The relatively slow data acquisition times for neutron-scattering experiments, and the requirement for a suitable source (i.e., nuclear reactor), however, limits their applicability in the study of dynamic phenomena.

While neutron reflectivity has been used extensively in the study of amphiphile adsorption at air–liquid or solid–liquid interfaces (Crowley et al., 1990), SANS has been employed to elucidate amphiphile aggregation pheomena in bulk aqueous solutions (Knoll et al., 1981). SANS employing contrast matching has proven particularly effective in characterizing mixing/demixing phenomena in lipid LLC mixtures (Winter, 2002). A key finding has been the potential for nonideal mixing associated with coexistence regions in mixed LLC phase diagrams. Combining SANS with rheological measurements has also allowed nanostructural alignment of LLCs in shear fields to be demonstrated. (Singh et al., 2004).

4.2.4 Nuclear Magnetic Resonance

The diffusion coefficients of both lipid–surfactant and water in LLCs is accessible using pulsed field gradient NMR measurements. Such measurements were the first to elucidate the structure of the bicontinuous cubic phase (Lindblom et al., 1979). This is distinct from discontinuous cubic phases, such as the Fd3m micellar cubic phase, where the closed nature of the aggregate units results in lower diffusion coefficients (Eriksson and Lindblom, 1993; Lindblom and Wennerstrom, 1977; Oradd et al., 1995).

The basis of the technique is to use NMR peak attenuation to measure the displacement (translational diffusion) of spins due to specific nucleii in the lipid–surfactant and water over a defined period in a magnetic field gradient. Attenuation is measured as a function of field gradient pulse duration, and the results can be fitted to yield the diffusion coefficient (Oradd et al., 1995).

In general, the NMR spectra of cubic LLC phases exhibit narrow, well-resolved peaks. This is due to the fact that the molecular motions are isotropic. NMR employing different isotopes (e.g., ^{2}H, ^{4}N, and ^{31}P) also allows diffusion coefficients of specific species within mixtures to be measured directly without using probe molecules and without model-dependent assumptions (Cullis and De Kruijff, 1979).

For example, Eriksson and Lindblom (1993) used NMR to detect lipid diffusion in cubic phases formed from mixed systems. In the three-component system: monoolein (MO)/dioleoylphosphatidylcholine (DOPC)/D_2O, the ^{1}H-NMR spin echo spectrum consisted of major peaks that were specifically attributable to the different lipid constituents. The attenuation of these peaks was able to show that the diffusion of both lipids was reduced when DOPC

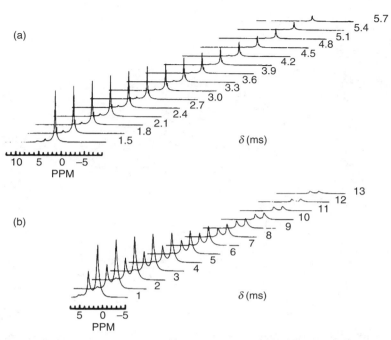

Figure 4.7   NMR diffusion experiment in cubic phases showing peak attenuation as duration of field gradient pulse (δ) is increased. Sample compositions: (a) MonooleinO/D$_2$O 70:30 wt/wt and (b) monoolein/DOPC/D$_2$O 42.5:42.5:15 wt/wt (Eriksson and Lindblom, 1993).

was added to the system (Fig. 4.7), which has important implications when considering multicomponent biological membranes.

4.2.5   Positron Annihilation Lifetime Spectroscopy

PALS is a well-established technique used for investigating chain packing and nanostructure in polymer (Ata et al., 2009; Misheva et al., 2000) high-viscosity liquid (Jain, 1995) and recently lyotropic liquid crystals (Dong et al., 2009). It is used to provide an atomic-scale probe of the dimensions of free volume within materials, based on the lifetime of orthopositronimum (oPs) localized in inter- and intramolecular spaces.

PALS has promising potential as a novel technique for characterizing the phase behavior of lyotropic liquid crystals. However, there is a lack of systematic correlation of PALS parameters with molecular structure.

Dong et al. (2009) have compared the relationship between the oPs lifetime and rheology of mesophases summarized in Figure 4.8. They found that the oPs lifetime is correlated with rheological properties, lipid chain packing (*d* spacing), and mobility in various mesophases.

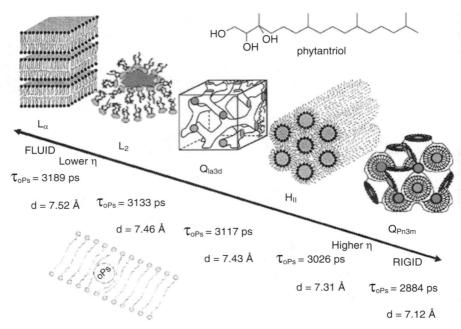

Figure 4.8 Orthopositronimum lifetimes and rheological properties of different mesophases of phytantriol (Dong et al., 2009).

4.2.6 Electron Microscopy

Electron microscopy (EM) techniques offer the potential for direct visualization of LLCs at <5-nm resolution in both bulk and dispersed form (see later). Transmission (TEM) and scanning (SEM) techniques have been used to explore different aspects of LLC behavior. TEM, with its high resolution and ability to image the internal structure of materials, has been extensively employed to study LLC phase nanostructure.

The general approach to TEM analysis is to prepare a sample as a thin film such that an electron beam can pass through it. The image is collected with contrast resulting from differential absorption and scattering from different regions of the sample. These differences may be natural, or enhanced via the use of heavy-metal stains (e.g., osmium tetroxide), to generate either negative or positive contrast. Many EM-based techniques originally developed for the study of biological specimens have been applied to LLC systems. As with biological specimens, however, the risk of image artifacts due to sample preparation and the requirement for vacuum during imaging requires careful consideration (Bangham and Horne, 1964; Delacroix et al., 1996; Moren et al., 2000).

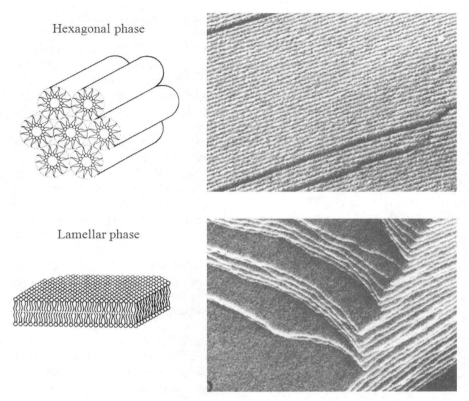

Hexagonal phase

Lamellar phase

Figure 4.9 Freeze–Fracture transmission electron micrographs of LLC structures formed by phosphatidylethanolamine lipids (Verkleij, 1984).

Freeze–fracture electron microscopy (FFEM) has been extensively applied to the study hydrated lipid phases (Andersson et al., 1995; Larsson, 1989; Verkleij, 1984). In this technique, frozen samples are cleaved under vacuum. The fracture plane typically follows the interior (hydrophobic) domains of membrane-like structures. A thin layer of electron conductive material (e.g., carbon or platinum) is then deposited obliquely on the exposed surface to form a replica that is imaged in the TEM. Here, image contrast results from variations in thickness of the deposited conductive layer due to shadowing effects during deposition as shown in Figure 4.9 (Severs, 2007; Verkleij, 1984).

A key limitation of conventional EM techniques is the necessity to expose samples to a vacuum, which is required to minimize scattering of the incident electron beam due to molecules of any gases present. This presents a particular problem for the study of LLCs since their phase behavior is intrinsically susceptible to both the in vacuo drying and cooling conditions normally required.

For this reason, FFEM studies of lipid and other LLC systems were initially dogged by concerns regarding the preservation of mesophase structure after freezing, and this led to the development of "fast-quenching" cryofixation procedures (e.g., using liquid ethane), which offer a reduced risk of ice crystal formation or other sample perturbation during the freezing process (Gulik-Krzywicki, 1994). Simultaneous studies using X-ray scattering and electron microscopy were critical in validating such approaches (Costello and Gulik-Krzywicki, 1976; Gulik-Krzywicki and Costello, 1978).

Combining cryogenic preparation techniques with conventional TEM has led to the development of cryo-TEM. Here, a thin film of sample is produced on a 200-mesh copper TEM grid coated with perforated carbon film, which then is plunged into a cooling medium, such as liquid ethane just above its freezing point. The film very rapidly vitrifies, without ice or sample crystallization. Frozen grids are stored in liquid nitrogen until required. The grid with the vitrified film is then transferred to the microscope and examined at liquid nitrogen temperature in the transmission mode. The structures captured in the vitrified film are thus observed without dehydration, and hence amphiphile assemblies can be imaged directly in their aqueous state. Cryo-TEM has contributed significantly during the last 30 years to understanding the variety of self-assembled nanostructures formed by amphiphilic molecules in dilute aqueous solutions. The majority of these studies has focused on dispersed LLC systems and are described in Section 4.3.2, since bulk viscous/gel phases are not amenable to the sample preparation approaches described above (Moren et al., 2000).

4.3 DISPERSED LYOTROPIC LIQUID CRYSTALLINE CHARACTERIZATION

The earliest description of dispersed lamellar phase systems (vesicles and liposomes) involved electron microscopy techniques in the 1960s (Bangham and Horne, 1964). Since then, the preparation of LLCs in dispersed form has been explored extensively from both fundamental scientific and technological viewpoints (Spicer, 2005a). Dispersed lamellar phases in the form of unilamellar and multilamellar vesicles and liposomes have structures reminiscent of biological membranes, and hence have received considerable attention (Genty et al., 2003).

In 1994 Landh proposed the use of amphiphlic triblock copolymers to colloidally stabilize submicron dispersions of nonlamellar LLCs of monoglycerides, now known as cubosomes and hexasomes (Amar-Yuli et al., 2007). While most of the techniques mentioned in the previous section are broadly applicable to the characterization of dispersed LLCs, their advent has also spawned new characterization approaches (Sagalowicz et al., 2006b). In this section, we will focus on the use of electron microscopic and scattering techniques in characterizing LLC dispersions.

4.3.1 Small-Angle X-ray Scattering in Dispersed LLC Systems

Synchrotron SAXS is particularly useful in the study of LLC dispersions since the high-intensity X rays that are produced overcome many of the difficulties in collecting scattering data from these intrinsically dilute (i.e., low scattering) samples.

SAXS can be employed to study bilayer thickness and interlamellar spacings in dispersed lamellar LLCs (Balgavy et al., 2001), as well as phase transitions resulting from lipid exchange/solubilization processes (Shen et al., 2010) and interactions with drugs (Schutze and Muller-Goymann, 1998) or biomolecules such as DNA (deoxyribonucleic acid) (Koynova et al., 2006).

SAXS has also been extensively applied to studying phase behavior and transitions in nonlamellar LLC dispersions (Siekmann et al., 2002). In particular, the method has evolved as an important tool for the investigation of cubosomes (Andersson et al., 1995; Boyd et al., 2007) and hexasomes (Almgren, 2003; Salonen et al., 2008).

For instance, Yaghmur and co-workers (2005) have shown that the addition of tetradecane (TC) to monolinolinein-dispersed LLCs results in a phase progression from bicontinuous cubic Pn3m (cubosomes) to inverted hexagonal H_2 (hexosomes) to discontinuous micellar cubic Fd3m (micellar cubosomes) particles (Yaghmur et al., 2005). A new form of particle denoted emulsified microemulsion (EME) was also identified in this system. (See Fig. 4.10.) Mixtures of different members of this family of dispersions (which the authors denote as isasomes) were found to equilibrate with each other by material transfer (Moitzi et al., 2007).

4.3.2 Electron Microscopy Techniques for Dispersed LLC Systems

4.3.2.1 *Cryogenic Transmission Electron Microscopy (Cryo-TEM)*
Cryo-TEM has been used to probe the shape, defects, and morphology of dispersed LLC particles. Studies of vesicle solubilization and amphiphile exchange phenomena in lamellar LLC particles have also been undertaken (Heerklotz, 2008; Silvander et al., 1996). Cryo-TEM images and SAXS data are highly complementary. Indeed, much as in SAXS, fast Fourier transform (FFT) analysis of cryo-TEM images can be used to determine the internal crystallographic properties of dispersed LLC particles (Barauskas et al., 2005; Gustafsson et al., 1996, 1997). Gustafsson et al. (1996, 1997) were the first to characterize cubosome morphology by cryo-TEM. They found faceted particles and internal peridocities consistent with SAXS *d*-spacing measurements. Many other dispersed LLC particle morphologies have since been studied as shown in Figure 4.11.

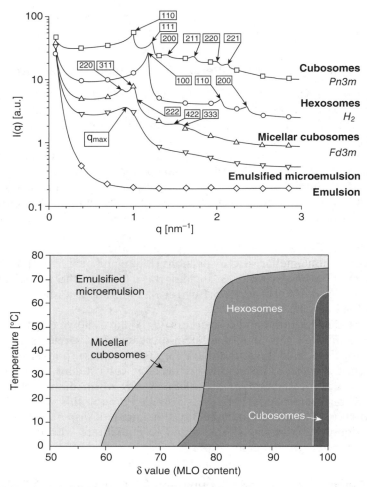

Figure 4.10 SAXS data and a putative "phase diagram" showing LLC phases found within aqueous monolinolinein (MLO) dispersions as a function of % MLO relative to decane (Moitzi et al., 2007).

Cryo-TEM image analysis often finds that cubosomes are observed in coexistence with vesicles or vesiclelike structures (Fig. 4.11a). Hexosomes show two morphologies: one comprising faceted rigid particles as shown in Figure 4.11 and others exhibiting curved striations (Sagalowicz et al., 2006a). Spicer and Hayden (2001) also employed cryo-TEM extensively in developing a low-energy dilution approach to cubosome manufacture.

4.3.2.2 Tilt-Angle Cryo-TEM In conventional cryo-TEM imaging as described above, crystallographic analysis (e.g., confirmation of three-dimensional

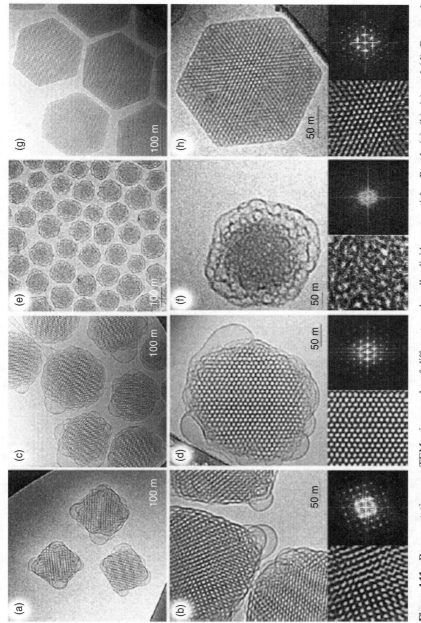

Figure 4.11 Representative cryo-TEM micrographs of different nonlamellar lipid nanoparticles. Panels (a), (b), (c), and (d): Reversed bicontinuous cubic phase particles viewed along [001] [(a) and (b)] and [111] [(c) and (d)] directions. Panels (e) and (f): Monodisperse "sponge" phase nanoparticles. Panels (g) and (h): Reversed hexagonal monocrystalline particles. Fourier transforms of magnified areas in panels (b), (d), (f), and (h) show the structural periodicity of the different nanoparticles consistent with the mesophase structures indicated above (Barauskas et al., 2005).

113

(3D) space group) is limited by the 2D data obtained. Tilt-angle cryo-TEM overcomes this limitation by imaging multiple sample orientations for single particles, allowing the alignment and study of different crystallographic planes relative to the incident electron beam, as shown in Figure 4.11. (Barauskas et al., 2005; Sagalowicz et al., 2006, 2007). Control of the tilt-angle allows greater distinction between different dispersed LLC particles (liposomes, cubosomes, micellar cubosomes, and hexosomes), and recording images at two to five different tilting angles allows sufficient data to perform crystallographic analysis to be obtained in order to differentiate between Pn3m or Im3m cubic symmetries, for example (Sagalowicz et al., 2006, 2007).

4.3.2.3 Scanning Electron Microscopy and Cryo-Field Emission Scanning Electron Microscopy (Cryo-FESEM)

Unlike TEM, in SEM, an electron beam is raster scanned across the surface of a sample, and the secondary electrons, or X-rays, are "back-scattered" from a conductive coating applied to the sample surface. This method generates a lower resolution image (typically 10–100 nm) but allows the direct mapping of surface features and can even be used for elemental analysis. As such, SEM allows surface morphological characterization of dispersed LLC systems but suffers from many of the sample preparation problems discussed earlier with respect to noncryogenic TEM techniques and hence has been less widely applied.

Spicer et al. (2002) employed a spray-drying method to produce dry powder cubosome precursors encapsulated with hydrophobically modified starch. SEM indicated uniform encapsulation of the monoolein by the starch and interrelationship between powder particle size and cubosome particle size distributions. Using the same method, cubic phase particles were used as templates for infiltration of silica precursor solutions. Figure 4.12 shows the SEM

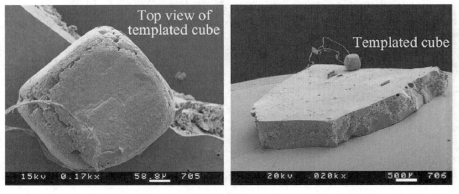

Figure 4.12 SEM image of a silica cube templated by coating a cubosome with a silica sol, drying, and then dissolving the cubosomes (Spicer, 2005b).

image of a 100-μm cube formed by coating a large cubosome with silica sol, evaporating the water to consolidate the silica, and dissolving the cubosome using an organic solvent (Spicer, 2005b).

Analagous to cryo-TEM, cryo-FESEM is a high-resolution SEM technique capable of visualizing the surface morphology of vitreous frozen surfactant mesophases at nanometer resolution.

Recently, Boyd and co-workers (Boyd et al., 2007; Rizwan et al., 2007) have used the cryo-FESEM technique to directly investigate the 3D morphology and surface structures of dispersed cubosome and hexosome particles. They found that the 3D cubosome structure enclosing aqueous water channels agreed well with earlier mathematical models. Both cryo-TEM and cryo-FESEM images displayed cubic morphology with coexisting vesicular structures (Fig. 4.13).

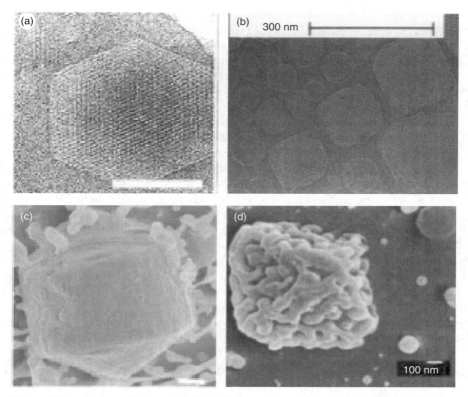

Figure 4.13 Images of phytantriol-based cubosomes and hexosomes obtained using cryo-TEM [(a) and (b)] and cryo-FESEM images [(c) and (d)]. The smaller white particles coexisting with cubosomes in the cryo-FESEM image were thought to be stabilizer (Pluronic F127) particles (Boyd et al., 2007).

4.4   SURFACE AND INTERFACE CHARACTERIZATION

Since the first description by Luzzati in the early 1960s (Luzzati and Husson, 1962) studies of lyotropic liquid crystalline systems have predominantly focused on the investigation of their interior nanostructure in solution, as described in the earlier sections. However, in both biology and technological applications, it is interfacial interactions that define their properties (e.g., membrane interactions and solute transport). As such, it is perhaps surprising that relatively little attention has been given to interfacial characterization of LLC systems. The application of advanced surface characterization techniques would therefore seem to be a major opportunity in understanding and manipulating LLC behavior. Herein, we will briefly discuss two such techniques that have recently been applied in this field, atomic force microscopy and neutron reflectivity (Rittman et al., 2010; Vandoolaeghe et al., 2008, 2009a,b).

4.4.1   Atomic Force Microscopy

Characterization of cubic phases uses scattering techniques (SAXS and SANS) allows the discrimination of different cubic phase symmetries and provides information regarding dispersion particle size. However, SAXS is analogous to powder diffraction in that it averages scattering over randomly oriented micron-sized domains (i.e., dispersion particles or particle subdomains. The data therefore provide little information regarding the interface of the cubic phase with water or of the boundary between adjacent domains. Similarly, many groups have imaged discrete single domains of cubic phase (Q_{II}) as cubosomes, using cryo-TEM (Gustafsson et al., 1996). However, it is not clear to what extent they are representative of domains within a bulk polydomain sample since they are stabilized by a co-surfactant that forms an additional layer at the cubosome surface. The properties of domain boundaries and defects are likely to strongly affect technological applications, viscoelastic properties, and phase transition kinetics, yet understanding their nature has been an enduring challenge both experimentally and theoretically (Andersson et al., 1995; Boyd et al., 2007).

Atomic force microscopy (AFM) has the potential to address these challenges. It gives direct visualization and has sufficient spatial resolution to image individual water channels within a single domain, thus providing information on domain orientation and epitaxy and domain boundaries. It can be carried out under ambient conditions, in a temperature range consistent with equilibrium lyotropic phase formation, under water or in air, and thus has potential for structural analysis of LLC phases and dispersed particles, both at equilibrium and during phase transitions.

AFM is a member of the family of scanned probe microscopies and was first developed by Binning, Quate, and Gerber in the 1980s (Strausser and Heaton, 1994). The technique is based upon raster scanning a microscopically sharp probe (a typical tip radius of curvature is ~10–50 nm) across a sample

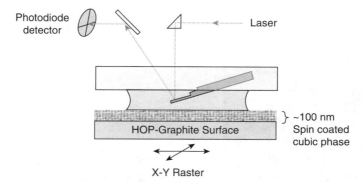

Figure 4.14 Schematic of AFM setup used in study by Rittman et al. (2010). The dotted line shows the path of the laser used in the optical lever detection system (Rittman et al., 2010).

surface, and monitoring deflection, or force exerted on the tip due to changes in surface interactions and/or sample topography. Tip displacements can be measured at subnanometer resolution using an optical lever technique (see Figure 4.14), thereby affording scanned probe microscopies the potential for atomic scale imaging. Advanced imaging modes such as intermittent/ dynamic/tapping mode (Dufrene, 2008) and soft contact imaging (Fleming and Wanless, 2000) have been developed to minimize tip sample forces, allowing soft materials such as surfactant aggregates to be imaged with minimal perturbation.

Neto et al. (1999) were the first to use AFM to image dispersed LLC particles under liquid at ambient temperatures. More recently, Rittman et al. (2010) have demonstrated for the first time high-resolution imaging of LLC interfaces with water using tapping-mode AFM. Thin films of monoolein and phytantriol were spin coated onto highly oriented pyrolytic graphite and allowed to hydrate within the AFM cell. Individual water channels within a single domain were imaged, thus providing information on domain orientation and epitaxy, as well as domain boundaries. As shown in Figure 4.15, the images show periodic arrays of water channels with spacing and symmetry consistent with published SAXS data on the bulk materials.

4.4.2 Neutron Reflectometry

As discussed above, the interactions between LLC phases are important technologically with respect to the development of nanocapsule delivery systems and from a fundamental biological perspective since they define interactions with and between biological lipid membranes in membrane fusion. (Dong et al., 2006; Vandoolaeghe et al., 2009a). Neutron reflectometry (NR) is ideally suited to characterizing interactions at lipid bilayer surfaces (Penfold, 2002),

Figure 4.15 AFM non-contact-mode images of Q$_{II}$ phases in excess water. All are height images, except (b) and (e), which show phase. (a) and (d) are from monoolein and (b), (c), and (e) from phytantriol. Unit cell lattices have been drawn. Using the right-hand axis convention, with lattice vectors listed anticlockwise about the obtuse angle, the dimensions of these lattices are as follows: (a) 14.4 nm, 12.8 nm, 120°; (b) 8.3 nm, 6.9 nm, 120°; (c) 14.2 nm, 14.2 nm, 92°; (d) 25.0 nm, 15.3 nm, 115°; and (e) 19.7 nm, 13.6 nm, 100° (Rittman et al., 2010).

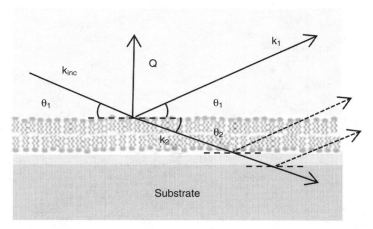

Figure 4.16 Simplified geometry of neutron reflection from a supported lipid bilayer system.

and recent studies have shown how LLC dispersion-supported lipid bilayer interactions can be characterized using this technique (H.-H. Shen et al., 2011; Vandoolaeghe et al., 2008, 2009a).

The basic theory of NR with respect to a lipid bilayer deposited on smooth planar interface is illustrated in Figure 4.16. When a beam of neutrons is incident on a planar interface, it is both reflected and refracted. Specular reflection is defined as reflection with the angle of reflection equal to the angle of incidence (θ_1). At all other angles the signal is referred to as off-specular reflection.

The relative proportion of the reflected and refracted waves depends on the nature of the surface, in particular on the relative neutron interaction potentials of the interfaces. The signal in neutron reflection is determined by the perpendicular variations in this potential through the surface, based on the scattering length densities of the constituent materials, effectively their neutron refractive indices.

Experimentally, the intensity of reflected neutrons as a function of the total momentum transfer perpendicular to the interface (Q) is determined in the experiment, given by

$$Q = \frac{4\pi \sin \theta}{\lambda}$$

Detailed modeling of the measured reflectivity profile is used to extract layer arrangement thicknesses, roughness, and scattering length densities for different components within the layered system.

Vandoolaeghe et al. (2008, 2009b,c) have published several papers on cubosome interactions with a silica surface with and without supported bilayers

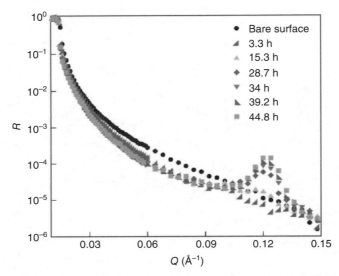

Figure 4.17 Neutron reflectivity profiles at different time points for GMO cubosome adsorption on an SiO$_2$ surface at a concentration of 0.05 mg/ml (Vandoolaeghe et al., 2009b).

using neutron reflectometry (NR). Figure 4.17 shows evolving reflectivity profiles for silica surfaces interacting with glyceryl monooleate (GMO) cubosomes recorded in D$_2$O at different incubation times. The appearance of a distinctive Bragg diffraction peak at Q-0.12 Å^{-1} demonstrates the presence of an organized repeating structure perpendicular to the plane of the surface, which is thought to correspond to the ordered arrangement of lipid molecules in the cubosomes. Different models were used in an attempt to optimize the fits to the experimental reflectivity profiles, all of them consisting of a single layer adjacent to the surface with an adjoining layer with a repeating structure with a total layer thickness of around 1500, which is close to the size of the cubosomes.

They further used neutron reflectometry to study exchange mechanisms between cubosomes and lipid bilayers. A lipid exchange mechanism was proposed that involved an initial absorption of material from the GMO cubosomes to form a mixed bilayer, and a subsequent release of material/particles from the surface, as the uptake of GMO reached a maximum (Vandoolaeghe et al., 2009).

Recently, Shen et al. (2010, 2011) studied phytantriol-based cubosomes interactiong with 1-palmitoyl-2-oleoyl-sn-glycero-3-phosphocholine (POPC) bilayers. This system also showed growth of a Bragg peak even at high coverage of the POPC bilayer (98%), which is in contrast to the GMO study, where the Bragg peak only appeared at low (55%) bilayer coverage. It was concluded that the difference was due to incorporation of bilayer phospholipids within the internal cubic structure of the nanoparticles, and hence the phytantriol-based cubosomes appear to have a stronger interaction with model membrane compared to GMO. This behavior was correlated with biological data, which demonstrates that phytantriol cubosomes have a higher cytotoxicity at 40 μg/ml than GMO cubosomes. These observations point to a potential cytotoxicity mechanism based on relative membrane fusogenicities that might have important implications for the design of in vivo drug delivery systems in particular (Shen et al., 2011).

4.5 SUMMARY AND OUTLOOK

It is some 50 years since the first systematic steps were taken to identify and explain nanostructural diversity in lyotropic liquid crystals at equilibrium (Luzzati and Husson, 1962). Since then, both the biological and technological importance of these systems has been recognized, studied, and exploited. This progress has been facilitated in large part by the development of characterization techniques that allow precise nanostructural information to be recorded in "close to native" environmental conditions.

As these techniques have matured, so have the questions that scientists seek to answer. Two key trends emerge as we survey the field today. The first is that new generations of instrumentation are focused on probing the *dynamic* behavior of lyotropic liquid crystalline systems in response to stimuli that are physical (e.g., light, pressure, temperature), chemical (e.g., catalytic), or biological (e.g., interactions with biomacromolecules, biological fluids, and cell membranes). These studies will doubtless extend knowledge regarding the fundamental biological role of LLC systems, as well as their technological utility.

The second scientific challenge is to better characterize the *interfacial interactions* of LLC systems. Previously, such studies have been limited by the lack of availability of techniques that can probe surface nanostructure in aqueous environments. The two interfacial characterization techniques highlighted in this chapter, atomic force microscopy and neutron reflectivity, may represent the vanguard of techniques that will be focused on such studies in coming years. Again, understanding these interactions will likely lead to advances in our understanding of the role of lyotropic liquid crystals in lipid membrane biophysics in particular.

REFERENCES

Alam, M. M., and Mezzenga, R. (2011). Particle tracking microrheology of lyotropic liquid crystals. *Langmuir, 27*, 6171–6178.

Almgren, M. (2003). Alexander lecture 2003: Cubosomes, vesicles, and perforated bilayers in aqueous systems of lipids, polymers, and surfactants. *Australian Journal of Chemistry, 56*(10), 959–970.

Amar-Yuli, I., Wachtel, E., Ben Shoshan, E., Danino, D., Aserin, A., and Garti, N. (2007). Hexosome and hexagonal phases mediated by hydration and polymeric stabilizer. *Langmuir*, *23*, 3637–3645.

An, Y., Xu, J., Zhang, J., Hu, C. G., Li, G. Z., Wang, Z. N., Wang, Z. N., Zhang, X. Y., and Zheng, L. Q. (2006). Studies on the phase properties of lyotropic liquid crystals of BriJ35/sodium oleate/oleic acid/water system: By means of polarizing microscope, SAXS, 2 H-NMR and rheological methods. *Science in China Series B—Chemistry*, *49*(5), 411–422.

Andersson, S., Jacob, M., Lidin, S., and Larsson, K. (1995). Structure of the cubosome—a closed lipid bilayer aggregate. *Zeitschrift für Kristallographie*, *210*(5), 315–318.

Ata, S., Muramatsu, M., Takeda, J., Ohdaira, T., Suzuki, R., Ito, K., Kobayashi, Y., and Ougizawa, T. (2009). Free volume behavior in spincast thin film of polystyrene by energy variable positron annihilation lifetime spectroscopy. *Polymer*, *50*(14), 3343–3346.

Balgavy, P., Dubnickova, M., Kucerka, N., Kiselev, M. A., Yaradaikin, S. P., and Uhrikova, D. (2001). Bilayer thickness and lipid interface area in unilamellar extruded 1,2-diacylphosphatidylcholine liposomes: A small-angle neutron scattering study. *Biochimica et Biophysica Acta*, *1512*, 40–52.

Bangham, A. D., and Horne, R. W. (1964). Negative staining of phospholipids and their structural modification by surface-active agents as observed in the electron microscope. *Journal of Molecular Biology*, *8*, 660–668.

Barauskas, J., Johnsson, M., and Tiberg, F. (2005). Self-assembled lipid superstructures: Beyond vesicles and liposomes. *Nano Letters*, *5*(8), 1615–1619.

Bäverback, P., Oliveira, C. L. P., Garamus, V. M., Varga, I., Claesson, P. M., and Pedersen, J. S. (2009). Structural properties of β-dodecylmaltoside and C12E6 mixed micelles. *Langmuir*, *25*, 7296.

Boyd, B. J., Rizwan, S. B., Dong, Y. D., Hook, S., and Rades, T. (2007). Self-assembled geometric liquid-crystalline nanoparticles imaged in three dimensions: Hexosomes are not necessarily flat hexagonal prisms. *Langmuir*, *23*(25), 12461–12464.

Briggs, J., Chung, H., and Caffrey, M. (1996). The temperature-composition phase diagram and mesophase structure characterization of the monoolein/water system. *Journal de Physique II*, *6*(5), 723–751.

Caffrey, M., and Hing, F. S. (1987). A temperature gradient method for lipid phase diagram construction using time-resolved X-ray diffraction. *Biophysical Journal*, *51*, 37–46.

Cordobes, F., Franco, J. M., and Gallegos, C. (2005). Rheology of the lamellar liquid-crystalline phase in polyethoxylated alcohol/water/heptane systems. *Grasas y Aceites*, *56*(2), 96–105.

Costello, M. J., and Gulik-Krzywicki, T. (1976). Correlated X-ray diffraction and freeze-fracture studies on membrane model systems perturbations induced by freeze-fracture preparative procedures. *Biochimica et Biophysica Acta*, *455*, 412–432.

Crowley, T. L., Lee, E. M., Simister, E. A., Thomas, R. K., Penfold, J., and Rennie, A. R. (1990). The application of neutron reflection to the study of layers adsorbed at liquid interfaces. *Colloids and Surfaces*, *52*, 85–106.

Cullis, P. R., and De Kruijff, D. (1979). Lipid polymorphism and the functional roles of lipids in biological membranes. *Biochimica et Biophysica Acta*, *559*, 399–420.

Delacroix, H., Gulik-Krzywicki, T., and Seddon, J. M. (1996). Freeze fracture electron microscopy of lyotropic lipid systems: Quantitative analysis of the inverse micellar cubic phase of space group Fd3m (Q227). *Journal of Molecular Biology, 258*, 88–103.

Dong, Y. D., Larson, I., Hanley, T., and Boyd, B. J. (2006). Bulk and dispersed aqueous phase behavior of phytantriol: Effect of vitamin E acetate and F127 polymer on liquid crystal nanostructure. *Langmuir, 22*(23), 9512–9518.

Dong, A. W., Pascual-Izarra, C., Pas, S. J., Hill, A. J., Boyd, B. J., and Drummond, C. J. (2009). Positron annihilation lifetime spectroscopy (PALS) as a characterization technique for nanostructured self-assembled amphiphile systems. *Journal of Physical Chemistry B, 113*(1), 84–91.

Dufrene, Y. F. (2008). Towards nanomicrobiology using atomic force microscopy. *Nature Microbiology, 6*, 674–680.

Eriksson, P. O., and Lindblom, G. (1993). Lipid and water diffusion in bicontinuous cubic phases measured by NMR. *Biophysical Journal, 64*(1), 129–136.

Fleming, B. D., and Wanless, E. J. (2000). Soft-contact atomic force microscopy imaging of adsorbed surfactant and polymer layers. *Microscopy and Microanalysis, 6*, 104–112.

Genty, M., Couarraze, G., Laversanne, R., Degert, C., Navailles, C., Gulik-Krzywicki, T., and Grossiord, J.-L. (2003). Characterization of a complex dispersion of multilamellar vesicles. *Colloid and Polymer Science, 282*, 32–40.

Gin, D. L., Pecinovsky, C. S., Bara, J. E., and Kerr, R. L. (2008). Functional lyotropic liquid crystal materials. In K. Takashi (Ed.), *Liquid Crystalline Functional Assemblies and Their Supramolecular Structures*, Springer, Berlin, pp. 181–222.

Glatter, O., and Kratky, O. (1982). *Small Angle X-ray Scattering*, Academic, London.

Gradzielski, M. G., I., and Narayanan, T. (2004). Dynamics of structural transitions in amphiphilic systems monitored by scattering techniques. *Progress in Colloid and Polymer Science, 129*, 32–29.

Gruner, S. M. (1987). Time-resolved X-ray diffraction of biological materials. *Science, 238*, 305–312.

Gulik-Krzywicki, T. (1994). Electron microscopy of cryofixed biological specimens. *Biology of the Cell, 80*, 161–163.

Gulik-Krzywicki, T., and Costello, M. J. (1978). The use of low temperature X-ray diffraction to evaluate freezing methods used in freeze-fracture electron microscopy. *Journal of Microscopy, 112*(1), 103–113.

Gustafsson, J., Ljusberg-Wahren, H., Almgren, M., and Larsson, K. (1996). Cubic lipid-water phase dispersed into submicron particles. *Langmuir, 12*(20), 4611–4613.

Gustafsson, J., Ljusberg-Wahren, H., Almgren, M., and Larsson, K. (1997). Submicron particles of reversed lipid phases in water stabilized by a nonionic amphiphilic polymer. *Langmuir, 13*(26), 6964–6971.

Heerklotz, H. (2008). Interactions of surfactants with lipid membranes. *Quarterly Reviews of Biophysics, 41*(3/4), 205–264.

Jain, P. C. (1995). Positron-annihilation spectroscopy and micellar systems. *Advances in Colloid and Interface Science, 54*, 17–54.

King, S. M., Pethrick, R. A., and Dawkins, J. V. (1999). Small-angle neutron scattering. In *Modern Techniques for Polymer Characterisation*. Wiley, Chichester.

Knoll, W., Ibel, K., and Sackmann, E. (1981). Small-angle neutron scattering study of lipid phase diagrams by the contrast variation method. *Biochemistry*, *20*, 6379–6383.

Koynova, R., Wang, L., and MacDonald, R. (2006). An intracellular lamellar–nonlamellar phase transition rationalizes the superior performance of some cationic lipid transfection agents. *Proceedings of the National Academy of Sciences*, *103*(39), 14373–14378.

Kriechbaum, M., Rapp, G., Hendrix, J., and Laggner, P. (1989). Millisecond time resolved x ray diffraction on liquid-crystalline phase transitions using infrared laser T jump technique and synchrotron radiation. *Review of Scientific Instruments*, *60*(7), 2541–2544.

Landh, T. (1994). Phase behavior in the system pine oil monoglycerides-poloxamer 407–water at 20°C. *Journal of Physical Chemistry*, *98*, 8453–8467.

Larsson, K. (1989). Cubic lipid-water phases—structures and biomembrane aspects. *Journal of Physical Chemistry*, *93*(21), 7304–7314.

Laughlin, R. G. (1994). *The Aqueous Phase Behavior of Surfactants*, Academic, London.

Laughlin, R. G., and Munyon, R. L. (1987). Diffusive interfacial transport: A new approach to phase studies. *Journal of Physical Chemistry*, *91*, 3299–3305.

Lindblom, G., and Wennerstrom, H. (1977). Amphiphile diffusion in model membrane systems studied by pulsed NMR. *Biophysical Chemistry*, *6*(2), 167–171.

Lindblom, G., Larsson, K., Johansson, L., Fontell, K., and Forsen, S. (1979). The cubic phase of monoglyceride-water systems arguments for a structure based upon lamellar bilayer units. *Journal of the American Chemical Society*, *101*(19), 5465–5470.

Lutton, E. S. (1965). Phase behavior of aqueous systems of monoglycerides. *Journal of the American Oil Chemists Society*, *42*(12), 1068–1070.

Luzzati, V., and Husson, F. (1962). Structure of liquid-crystalline phases of lipid-water systems. *Journal of Cell Biology*, *12*(2), 207–219.

Luzzati, V., Gulikkrz, T., Rivas, E., Reisshus, F., and Rand, R. P. (1968a). X-ray study of model systems—structure of lipid-water phases in correlation with chemical composition of lipids. *Journal of General Physiology*, *51*(5p2), S37–S43.

Luzzati, V., Tardieu, A., Gulikkrz, T., Rivas, E., and Reisshus, F. (1968b). Structure of cubic phases of lipid-water systems. *Nature*, *220*(5166), 485–488.

Mason, T., Gang, H., and Weitz, D. A. (1997). Diffusing-wave-spectroscopy measurements of viscoelasticity of complex fluids. *Journal of the Optical Society of America A*, *14*(1), 139–149.

Mencke, A. P., and Caffrey, M. (1991). Kinetics and mechanism of the pressure-induced lamellar order-disorder transition in phosphatidylethanolamine—a time-resolved X-ray-diffraction study. *Biochemistry*, *30*(9), 2453–2463.

Mezzenga, R., Meyer, C., Servais, C., Romoscanu, A. I., Sagalowicz, L., and Hayward, R. C. (2005). Shear rheology of lyotropic liquid crystals: A case study. *Langmuir*, *21*, 3322–3333.

Misheva, M., Mihaylova, M., Djourelov, N., Kresteva, M., Krestev, V., and Nedkov, E. (2000). Positron annihilation lifetime spectroscopy studies of irradiated poly(propylene-co-ethylene)/poly(ethylene-co-vinyl acetate) blends. *Radiation Physics and Chemistry*, *58*(1), 39–47.

Moitzi, C., Guillot, S., Fritz, G., Salentinig, S., and Glatter, O. (2007). Phase reorganization in self-assembled systems through interparticle material transfer. *Advanced Materials*, *19*, 1352–1358.

Moren, A. K., Regev, O., and Khan, A. (2000). A Cryo-TEM study of protein-surfactant gels and solutions. *Journal of Colloid and Interface Science*, *222*(2), 170–178.

Neto, C., Aloisi, G., Baglioni, P., and Larsson, K. (1999). Imaging soft matter with the atomic force microscope: Cubosomes and hexosomes. *Journal of Physical Chemistry B*, *103*(19), 3896–3899.

Oradd, G., Lindblom, G., Fontell, K., and Ljusberg-Wahren, H. (1995). Phase diagram of soybean phosphatidylcholine diacylglycerol water studied by X-ray diffraction and 31P and pulsed field gradient ^{1}H-NMR: Evidence for reversed micelles in the cubic phase. *Biophysical Journal*, *68*, 1856–1863.

Pabst, G., Rappolt, M., Amenitsch, H., Bernstorff, S., and Laggner, P. (2000). X-ray kinematography of temperature-jump relaxation probes the elastic properties of fluid bilayers. *Langmuir*, *16*(23), 8994–9001.

Penfold, J. (2002). Neutron reflectivity and soft condensed matter. *Current Opinion in Colloid & Interface Science*, *7*(1–2), 139–147.

Rappolt, M., Pabst, G., Rapp, G., Kriechbaum, M., Amenitsch, H., Krenn, C., Bernstorff, S., and Laggner, P. (2000). New evidence for gel-liquid crystalline phase coexistence in the ripple phase of phosphatidylcholines. *European Biophysics Journal with Biophysics Letters*, *29*(2), 125–133.

Reinitzer, F. (1888). Beiträge zur Kenntniss des Cholesterins. *Monatshefte für Chemie (Wien)*, *9*(1), 421–441.

Rittman, M., Frischherz, M., Burgmann, F., Hartley, P. G., and Squires, A. (2010). Direct visualisation of lipid bilayer cubic phases using atomic force microscopy. *Soft Matter*, *6*(17), 4058–4061.

Rizwan, S. B., Dong, Y. D., Boyd, B. J., Rades, T., and Hook, S. (2007). Characterisation of bicontinuous cubic liquid crystalline systems of phytantriol and water using cryo field emission scanning electron microscopy (cryo FESEM). *Micron*, *38*(5), 478–485.

Rosevear, F. B. (1968). Liquid crystals: The mesomorphic phases of surfactant compositions. *Journal of the Society of Cosmetic Chemists*, *19*, 581–594.

Sagalowicz, L., Mezzenga, R., and Leser, M. E. (2006a). Investigating reversed liquid crystalline mesophases. *Current Opinion in Colloid & Interface Science*, *11*(4), 224–229.

Sagalowicz, L., Michel, M., Adrian, M., Frossard, P., Rouvet, M., Watzke, H. J., Yaghmur, A., De Campo, L., Glatter, O., and Leser, M. E. (2006b). Crystallography of dispersed liquid crystalline phases studied by cryo-transmission on electron microscopy. *Journal of Microscopy—Oxford*, *221*, 110–121.

Sagalowicz, L., Acquistapace, S., Watzke, H. J., and Michel, M. (2007). Study of liquid crystal space groups using controlled tilting with cryogenic transmission electron microscopy. *Langmuir*, *23*(24), 12003–12009.

Salonen, A., Muller, F., and Glatter, O. (2008). Dispersions of internally liquid crystalline systems stabilized by charged disklike particles as Pickering emulsions: Basic properties and time-resolved behavior. *Langmuir*, *24*(10), 5306–5314.

Schutze, W., and Muller-Goymann, C. (1998). Phase transformation of a liposomal dispersion into a micellar solution induced by drug loading. *Pharmaceutical Reearch, 15*(4), 538–543.

Seddon, J. M., Squires, A. M., Conn, C. E., Ces, O., Heron, A. J., Mulet, X., Shearman, G. C., and Templer, R. H. (2006). Pressure-jump X-ray studies of liquid crystal transitions in lipids. *Philosophical Transactions of the Royal Society A—Mathematical Physical and Engineering Sciences, 364*(1847), 2635–2655.

Severs, N. J. (2007). Freeze-fracture electron microscopy. *Nature Protocols, 2*(3), 547–576.

Shen, H. H., Crowston, J. G., Huber, F., Saubern, S., McLean, K. M., and Hartley, P. G. (2010). Influence of dipalmitoyl phosphatidylserine on phase behavior of and cellular response to lyotropic liquid crystalline dispersions. *Biomaterials, 31*(36), 9473–9481.

Shen, H.-H., Hartley, P. G., James, M., Nelson, A., Defendi, H., and McLean, K. M. (2011). The interaction of cubosomes with supported phospholipid bilayers using neutron reflectometry and QCM-D. *Soft Matter, 7*, 8041–8049.

Siekmann, B., Bunjes, H., Koch, M. M. H., and Westesen, K. (2002). Preparation and structural investigations of colloidal dispersions prepared from cubic monoglyceride-water phases. *International Journal of Pharmaceutics, 244*(1–2), 33–43.

Silvander, M., Karlsson, G., and Edwards, K. (1996). Vesicle solubilization by alkyl sulfate surfactants: A cryo-TEM study of the vesicle to micelle transition. *Journal of Colloid and Interface Science, 179*, 104–113.

Singh, M., Agarwal, V., De Kee, D., McPherson, G., John, V., and Bose, A. (2004). Shear-induced orientation of a rigid surfactantMesophase. *Langmuir, 20*, 5693–5702.

Spicer, P. T. (2005a). Cubosome processing: Industrial nanoparticle technology development. *Transactions of the Institution of Chemical Engineers, Part A, Chemical Engineering Research and Design, 83*(A11), 1283–1286.

Spicer, P. T. (2005b). Progress in liquid crystalline dispersions: Cubosomes. *Current Opinion in Colloid & Interface Science, 10*(5–6), 274–279.

Spicer, P. T., and Hayden, K. L. (2001). Novel process for producing cubic liquid crystalline nanoparticles (cubosomes). *Langmuir, 17*, 5748–5756.

Spicer, P. T., Small, W. B., Lynch, M. L., and Burns, J. L. (2002). Dry powder precursors of cubic liquid crystalline nanoparticles (cubosomes). *Journal of Nanoparticle Research, 4*(4), 297–311.

Strausser, Y. E., and Heaton, M. G. (1994). Scanning probe microscopy—technology and recent innovations. *American Laboratory, 26*(6), 20.

Vandoolaeghe, P., Rennie, A. R., Campbell, R. A., Thomas, R. K., Hook, F., Fragneto, G., Tiberg, F., and Nylander, T. (2008). Adsorption of cubic liquid crystalline nanoparticles on model membranes. *Soft Matter, 4*(11), 2267–2277.

Vandoolaeghe, P., Barauskas, J., Johnsson, M., Tiberg, F., and Nylander, T. (2009a). Interaction between lamellar (vesicles) and nonlamellar lipid liquid-crystalline nanoparticles as studied by time-resolved small-angle X-ray diffraction. *Langmuir, 25*(7), 3999–4008.

Vandoolaeghe, P., Campbell, R. A., Rennie, A. R., and Nylander, T. (2009b). Adsorption of intact cubic liquid crystalline nanoparticles on hydrophilic surfaces: Lateral organization, interfacial stability, layer structure, and interaction mechanism. *Journal of Physical Chemistry C, 113*(11), 4483–4494.

Vandoolaeghe, P., Rennie, A. R., Campbell, R. A., and Nylander, T. (2009c). Neutron reflectivity studies of the interaction of cubic-phase nanoparticles with phospholipid bilayers of different coverage. *Langmuir, 25*(7), 4009–4020.

Verkleij, A. J. (1984). Lipidic intramembranous particles. *Biochimica et Biophysica Acta, 779*, 43–63.

Winter, R. (2002). Synchrotron X-ray and neutron small-angle scattering of lyotropic lipid mesophases, model biomembranes and proteins in solution at high pressure. *Biochimica et Biophysica Acta, 1595*, 160–184.

Yaghmur, A., de Campo, L., Sagalowicz, L., Leser, M. E., and Glatter, O. (2005). Emulsified microemulsions and oil-containing liquid crystalline phases. *Langmuir, 21*(2), 569–577.

CHAPTER 5

Self-Assembly in Lipidic Particles

ANAN YAGHMUR

Department of Chemistry, University of Graz, Graz, Austria

Department of Pharmacy, School of Pharmaceutical Sciences, Faculty of Health and Medical Sciences, University of Copenhagen, Copenhagen, Denmark

OTTO GLATTER

Department of Chemistry, University of Graz, Graz, Austria

Abstract

The present contribution summarizes our previous investigations on the formation of emulsions, whose particles consist of a self-assembled inverted-type liquid crystalline phase or an inverted-type microemulsion. In this context, the main focus was on replacing either the dispersed oil droplets in normal O/W emulsions, or the kinetically stabilized internal W/O emulsion in double W/O/W emulsions, by an inverted-type liquid crystalline phase or an inverted-type microemulsion system. Owing to the physico-chemical properties of their internal nanostructures, these unique aqueous dispersions are superior to conventional emulsions and double emulsions. They are attractive as nanonreactors and as host systems for solubilizing active molecules (drugs, flavors, and vitamins) in the cosmetics, pharmaceutical, and food industries. This chapter describes the effect of varying temperature and solubilizing oil on the reversible structural transitions of the internal nanostructures of these lipidic dispersions.

Self-Assembled Supramolecular Architectures: Lyotropic Liquid Crystals, First Edition.
Edited by Nissim Garti, Ponisseril Somasundaran, and Raffaele Mezzenga.
© 2012 John Wiley & Sons, Inc. Published 2012 by John Wiley & Sons, Inc.

5.1   INTRODUCTION

In the literature, the biological relevance of lipidic nonlamellar structures in various cells has been pointed out (Almsherqi et al., 2006; de Kruijff, 1997; Deng et al., 2002; Ellens et al., 1989; Luzzati, 1997; Patton and Carey, 1979). As an example, Patton and Carey (1979) observed the formation of bicontinuous cubic phases during fat digestion. These self-assembled nanostructures were also explored in a variety of cells as a result of cell stress, starvation, or lipid and protein alterations (Almsherqi et al., 2006; Deng et al., 2002). Moreover, there is a growing body of evidence concerning the crucial role of peptides (such as viral peptides) and proteins for inducing the lamellar to nonlamellar structural transitions in biological cells (Colotto and Epand, 1997; Ellens et al., 1989; Siegel, 1999). For instance, these transitions are considered as a keystone element in the first steps of the fusion process (Colotto and Epand, 1997; Ellens et al., 1989).

The characterization of real biological systems is considered a difficult task due to their complexity. Therefore, the physicochemical properties of model, biologically relevant, binary or ternary lamellar and nonlamellar phases (de Kruijff et al., 1997; Hyde et al., 1997; Lewis et al., 1998; Lindblom and Rilfors, 1989; Ortiz et al., 1999; Yaghmur et al., 2007) as well as their corresponding nanostructured dispersions have received considerable attention (Angelova et al., 2005; Barauskas et al., 2005; Gustafsson et al., 1996; Larsson, 2000; Rizwan et al., 2007; Spicer, 2004). In this context, numerous investigations have been carried out on the physicochemical properties and the phase behavior of biologically relevant surfactant-like lipid–water systems (Barauskas and Landh, 2003; de Campo et al., 2004; Larsson, 1983; Lindblom et al., 1979; Lutton, 1965; Qiu and Caffrey, 1998, 1999), particularly the temperature–water content phase diagrams of various binary monoglyceride–water systems (Larsson, 1983; Lindblom et al., 1979; Lutton, 1965). Among these systems, we have recently investigated the monolinolein (MLO)–water binary mixture (de Campo et al., 2004), which forms reverse isotropic micellar solution (L_2), lamellar (L_α), inverted-type hexagonal (H_2), and cubic (V_2) liquid crystalline phases within a feasible experimental temperature range (i.e., 20–94°C) (Fig. 5.1). V_2 is a three-dimensional (3D) bicontinuous phase composed of bilayers that separate aqueous channel networks, and this phase has an infinite periodic minimal surface (IPMS). An MLO–water system has two types of bicontinuous cubic phases depending on the water content, that is, the cubic assemblies with the Ia3d (the gyroid type, C_G) and the Pn3m (the diamond type, C_D) symmetries. The H_2 consists of hydrophilic cylinders in a continuous nonpolar matrix composed of the MLO hydrocarbon tails.

Recent studies (de Campo et al., 2004; Guillot et al., 2006; Moitzi et al., 2007; Sagalowicz et al., 2006; Salonen et al., 2007; Yaghmur and Glatter, 2009; Yaghmur et al., 2005, 2006a,b) have been focused on the formation and characterization of various emulsions in which the droplets consisted of well-defined nanostructures. This required an exchange of either the kinetically stabilized oil droplets in normal oil-in-water (O/W) emulsions or the internal

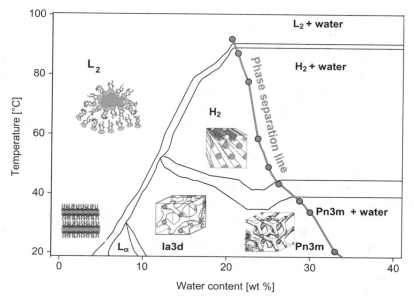

Figure 5.1 Identity and location in the temperature–composition phase diagram of MLO–water as determined by small-angle X-ray scattering (SAXS) in the heating direction between 20 and 94°C (de Campo et al., 2004). The black guidelines bring into evidence the phase transitions and regions of coexistence between the mesophases. The gray line corresponds to the phase separation line between the single-phase regions (left) and the biphasic systems: self-assembled nanostructures coexisting with excess water (right).

water-in-oil (W/O) emulsion in double W/O/W emulsions, by kinetically dispersed droplets confining thermodynamically stable nanostructures. In other words, we aimed to detect the influence of various parameters on *ISAsomes* (*internally self-assembled particles or somes*), which are aqueous dispersions of a family of colloidal droplets (cubosomes, hexosomes, micellar cubosomes, and emulsified microemulsions) with internal self-assembled nanostructures. Owing to their unique physicochemical properties, these nonviscous emulsion nanoobjects are superior to conventional dispersions (emulsions, double emulsions, and liposomes).

There is a huge surge of interest in utilizing nanostructured emulsions in a variety of applications due to their possible biological relevance, their high interfacial area, and their capacity to solubilize hydrophilic, amphiphilic, and hydrophobic active guest molecules with biological and pharmaceutical relevance (Boyd, 2005; Drummond and Fong, 2000; Larsson, 2009; Leser et al., 2006; Malmsten, 2006; Mezzenga et al., 2005; Siekmann et al., 2002; Yaghmur and Glatter, 2009). For instance, potential applications include nanoreactors or host nanoparticulate delivery systems in the cosmetic, pharmaceutical, and food industries.

This chapter describes the effect of variations in temperature and solubilizing oil on the reversible structural transitions of the nanoscale interior self-assembled domains of the kinetically stabilized MLO-based droplets. Intriguingly, our findings demonstrated that the formation of the confined nanoscaled tunable hierarchical structures was driven only by the principles of self-assembly. Our results revealed the possibility to emulsify W/O microemulsion systems—even at room temperature—by adding an oil such as tetradecane (TC) to the binary MLO–water system. The formed interior nanostructures were thermodynamic equilibrium structures, and the emulsified particles showed a swelling–deswelling behavior during the heating/cooling cycles (denoted as "breathing mode"). We also demonstrated that the interior self-assembled nanostructures of these dispersed droplets could be modulated by varying the lipid composition (Yaghmur et al., 2006b). This could be achieved by replacing a certain amount of MLO by a surfactant favoring the formation of the L_α structure. For instance, we found that diglycerol monooleate (DGMO, a nonionic polar lipid with a large headgroup) had a significant impact on the internal nanostructures of MLO-based aqueous dispersions (Yaghmur et al., 2006b).

Our main concern was to prove that both the oil-loaded and the oil-free self-assembled nanostructures were preserved in the presence of the stabilizer after application of a high-energy input during the dispersing procedure. To achieve this objective, it was important to carry out detailed systematic investigations on both the dispersed and the nondispersed bulk phases. In this chapter, we describe also the temperature-induced direct liposomes–cubosomes transitions in the binary monoelaidin (ME)–water system.

The interested reader is also directed to the contribution of Glatter and Kulkarni (Chapter 6) in this book. They summarized further recent studies on these nanostructured aqueous dispersions.

5.2 EMULSIFICATION OF MLO–WATER NANOSTRUCTURES

Various studies have been performed to prepare stable lipidic colloidal dispersions of V_2 and H_2 phases, and subsequently explore their internal structure (Angelova et al., 2005; Barauskas et al., 2005; Gustafsson et al., 1996; Larsson, 2000; Rizwan et al., 2007; Spicer, 2004). This section describes investigations devoted mainly to characterizing the temperature-induced structural transitions of the various confined nanostructures in the dispersions in order to explore the temperature-dependent reversible formation of self-assembled internal structures, as well as to report on the formation of what we termed an emulsified L_2 phase (ELP) at high temperatures.

In order to shed light on the confined nanostructure of the MLO–water system in a colloidal emulsified state, MLO was dispersed in large amounts of excess water by means of ultrasonication, where the block copolymer F127 was used as stabilizer. This led to the formation of submicron-sized cubosome

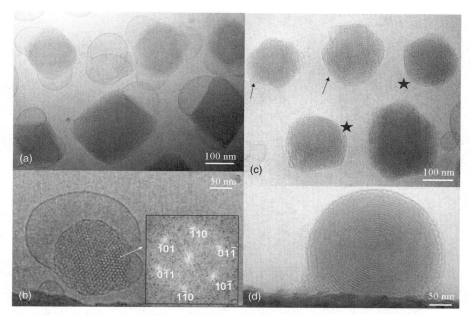

Figure 5.2 Cryo-TEM images of a dispersion consisting of 4.625 wt % MLO, 0.375 wt % F127, and 95.0 wt % water (de Campo et al., 2004). At 25°C [images (a) and (b)], the internal structure of the cubosome particle [inset in image (b)] showed a hexagonal arrangement with interplanar distances d of approximately 6 nm. This was compatible with a cubic structure of Pn3m symmetry with a lattice parameter, a, of 8.5 nm (a good agreement with the SAXS analysis). At 55°C [images (c) and (d)], certain of the nanostructured hexosome particles exhibited an internal hexagonal symmetry [arrows in image (c)], whereas others exhibited curved striations [stars in images (c)].

particles with nanostructured interiors as shown in the cryo-TEM (transmission electron microscopy) images in Figure 5.2. The coexistence of cubosomes and vesicles (Figs. 5.2a and 5.2b) has also been observed in previous studies on dispersed monoglycerides (Barauskas et al., 2005; Gustafsson et al., 1996; Larsson, 2000).

Figure 5.3 shows the temperature-dependent SAXS curves of the aqueous dispersion. It is clear that the internal structures of this MLO-based dispersion underwent a transition from Pn3m, denoted Q^{224} (cubosomes, emulsified cubic phase, ECP), via H_2 (hexosomes, emulsified hexagonal phase, EHP), to the fluid L_2 phase (emulsified L_2 phase, ELP) as the temperature was increased. The ELP was of particular interest since it had to our knowledge never before been described in the literature. This emulsion consisted of submicron-sized droplets with a nanostructured, nonviscous fluid isotropic micellar water–MLO-rich interior and had no long-range order.

Figure 5.4 shows SAXS patterns for the MLO dispersion (thick lines) as well as the corresponding nondispersed fully hydrated sample (thin lines) at

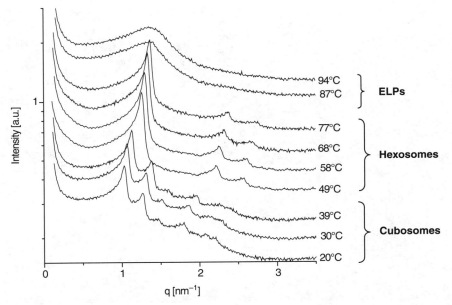

Figure 5.3   Temperature dependence of the SAXS curves for the same MLO-based dispersion described in Figure 5.2. [Reprinted from de Campo et al. (2004).]

three temperatures: 25, 60, and 98°C. Increasing the temperature [as discussed in de Campo et al. (2004) and Yaghmur et al. (2005)] was expected to induce a dehydration of the MLO headgroup. This would lead to an increase in the kink states in the lipid acyl chains and thereby their *effective* volume would also increase. As a consequence, the negative spontaneous curvature would become enhanced.

Our results clearly indicated a remarkable resemblance in the SAXS patterns; the peak positions for the bulk and the dispersion were identical. The confined nanostructure was preserved in all investigated temperatures. Owing to this identical nanostructure in both the dispersed and nondispersed phases, the water content in the dispersed droplets could be readily calculated from the maximum water solubilization capacity of the fully hydrated nondispersed systems (their structures were independent of the water content) (de Campo et al., 2004; Yaghmur et al., 2005). An additional interesting point is the significant decrease in water solubilization capacity with increasing temperature. For instance, the internal nanostructure of the cubosome particles contained 33 wt % water at 20°C, whereas the ELP contained only 20 wt % water at 94°C, in analogy with the nanostructure of the corresponding nondispersed bulk phase (as shown in Fig. 5.1).

As mentioned above, the symmetries of the internal phase were preserved for all investigated temperatures. Only our results revealed that the internal

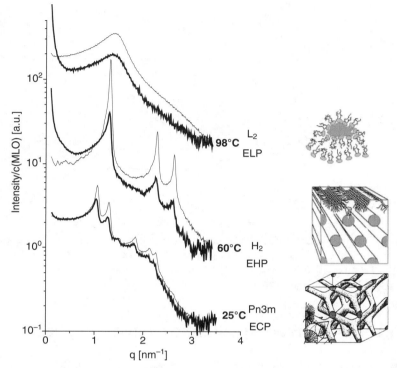

Figure 5.4   Comparison of SAXS curves of the dispersed and the nondispersed fully hydrated sample with excess water (40 wt % water) at three temperatures (de Campo et al., 2004): The thick lines correspond to the dispersion and the thin lines to the bulk phase. The intensities were normalized by the respective MLO concentration.

phase transition from H_2 to L_2 seemed to take place already at slightly lower temperatures in the dispersion (de Campo et al., 2004). This good agreement between the results obtained for both the dispersed and the nondispersed systems shows that the internal structure of the dispersions at each temperature actually corresponded to the equilibrium structure of MLO with excess water. Moreover, only a small part of F127 seemed to be incorporated into the internal structure of the dispersed particles. In other words, F127 was very effective as a stabilizer and covered mainly the outer surface of the emulsified particles.

The temperature-induced transition from cubosomes to hexosomes was observed by cryo-TEM (Figs. 5.2a and 5.2b). At 25°C, the fast Fourier transforms (FFTs) of the internal structure of the particles were compatible with the cubic Pn3m symmetry. At a higher temperature (55°C), the hexosome particles displayed internal arrangements of a hexagonal symmetry (arrows in Fig. 5.2c) and/or curved striations (Figs. 5.2c and 5.2d).

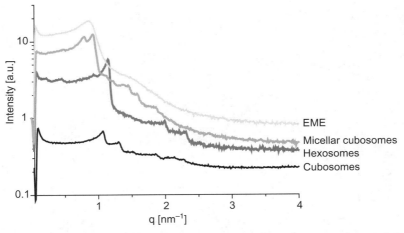

Figure 5.5 Effect of solubilized TC content on the scattering curves of MLO-based emulsified systems at 25°C. In these dispersions, the α ratio {[(mass of oil)/(mass of MLO)] × 100} varied from 0 (cubosomes) to 28 (hexosomes), to 40 (micellar cubosomes), and to 60 (EME). [Reprinted from Yaghmur et al. (2006a).]

5.3 EMULSIFICATION OF OIL-LOADED NANOSTRUCTURES

In the previous section, we found an indication for the formation of an ELP at high temperature. As a next step, our attention was mainly directed to studying the oil-loaded aqueous dispersions in an attempt to significantly decrease the structure transition temperatures and to prove that it was possible to form droplets with a confined internal oil-loaded self-assembled microemulsion system at room temperature. This experimentally ensures that the self-assembled nanostructure is preserved after applying the dispersing procedure.

In our studies, we investigated the impact of TC solubilization on the internal nanostructure of MLO-based aqueous dispersions. Figure 5.5 shows the SAXS scattering curves at 25°C obtained from dispersions in which the α ratio value, defined as [(mass of oil)/(mass of MLO) × 100], ranged between 0 and 60. In the absence of oil (i.e., $\alpha = 0$), cubosomes with an internal cubic Pn3m phase were formed. For the TC-loaded dispersion with $\alpha = 28$, the scattering curve displayed three peaks in the characteristic ratio for a H_2 phase (hexosomes). Intriguingly, a further increase of the TC concentration in the dispersion ($\alpha = 40$) gave rise to a pattern with more than seven characteristic peaks of a discontinuous micellar cubic structure of the Fd3m type, referred to the number Q^{227} (micellar cubosomes, emulsified micellar cubic phase, EMCP). The mean lattice parameter, a, of the internal nanostructures of cubosomes, hexosomes, and micellar cubosomes were 8.6, 6.3, and 21.7 nm, respectively. At higher TC concentrations ($\alpha \geq 60$), the scattering curve showed only one broad peak, which was typical for a thermodynamically stable microemulsion phase (emulsified W/O microemulsion system, EME).

The formation of EMEs is very attractive in both basic and applied research since these unique nanostructured droplets have superior physicochemical properties to conventional emulsions and double emulsions. These effective nanoparticulate systems are of great interest for various pharmaceutical, food, and cosmetic applications.

Our study revealed that the addition of TC increased the flexibility of the surfactant film, thereby destabilizing the internal H_2 phase in favor of an inverted microemulsion phase. It was thus evident that it was possible for the formation of droplets confining W/O microemulsion system at room temperature to take place (Yaghmur et al., 2005).

Our results also indicate that the structural transformation from hexosomes to EME was not direct but rather took place in a certain TC concentration range in which micellar cubosomes with the internal space group Fd3m (also referred to by the symbol I_2) were formed between the hexosomes and EMEs. However in the absence of TC (Fig. 5.1), the results with respect to both the dispersed and the nondispersed systems were similar to those from other previous investigations carried out on the impact of temperature on the phase behavior of binary monoglycerides–water systems (Barauskas & Landh, 2003; de Campo et al., 2004; Larsson, 1983; Lindblom et al., 1979; Lutton, 1965; Qiu & Caffrey, 1998, 1999): None of these studies displayed any indication of the existence of the Fd3m phase when the temperature was varied from 20 to 94°C.

It was important to prove that confined oil-loaded self-assembled nanostructures could be well preserved during the formation of the emulsions. We also wished to show that there was no *leaking* process in the internal nanostructure, which could occur as a result of either TC or water being expelled from these structures to the surrounding aqueous phase. Our approach was to simply compare the structural transitions of the TC-loaded dispersed MLO particles to those that occurred in the corresponding fully hydrated nondispersed ternary MLO–TC–water systems. It is also noteworthy that the addition of TC decreased the water solubilization capacity. For instance, in the absence of oil ($\alpha = 0$), the water content was determined to be approximately 32 wt % (de Campo et al., 2004). Upon increasing the TC content to $\alpha = 19$, 75, and 110, the water content decreased to less than 25, 20, and 15 wt %, respectively (Yaghmur et al., 2005).

We found in all investigated dispersions (of which two examples are presented in Fig. 5.6) that the internal structures were well preserved: The peak positions were identical to those of the corresponding nondispersed fully hydrated samples. Figure 5.6a shows an oil-loaded H_2 phase, and Figure 5.6b displays an example of an EME with an internal nanostructure identical to that of the corresponding nondispersed W/O microemulsion coexisting with excess water (Winsor-II-type microemulsion system). We were thus, in principle, able to form EMEs without concomitant loss of the W/O microemulsion's integrity. The average hydrodynamic radius of these droplets, as determined by dynamic light scattering (DLS), was in the range of 120–200 nm.

The submicron-sized dispersed EME particles were also observed by cryo-TEM. Figure 5.7 shows that the confined W/O nanostructures were clearly

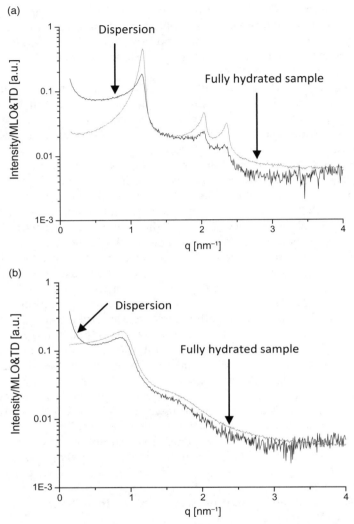

Figure 5.6   Scattering curves of two TC-loaded MLO emulsified systems at 25°C and those of the corresponding fully hydrated nondispersed bulk samples with excess water. The corresponding α ratios were (a) 19 and (b) 75. [Reproduced from Yaghmur et al. (2005).]

different from those previously obtained for the hexosomes and the cubosomes (Barauskas et al., 2005; Gustafsson et al., 1996; Larsson, 2000). No long-range order was observed. The outer shape of the dispersed droplets was similar to that found for the conventional O/W or W/O emulsion system. However, the FFTs of the EME droplets indicated the presence of an internal arrangement within the particles.

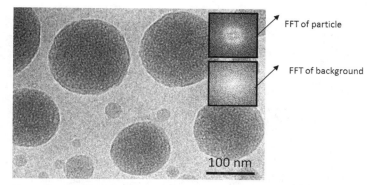

Figure 5.7 Cryo-TEM observation of EME droplets ($\alpha = 75$). The top insert represents an FFT of a particle revealing its internal nanostructure. [Reproduced from Yaghmur et al. (2005).]

5.4 STABILITY OF THE INTERNALLY OIL-LOADED NANOSTRUCTURES

In order to prove that the composition of the oil-loaded internal nanostructures did not change with time, the dispersions' stability against ageing was studied. Two examples of hexosomes and EMEs are presented in Figure 5.8. Intriguingly, there was no change in the internal nanostructures of these dispersions after 4 months of storage at room temperature. The presented data show that the oil molecules were maintained in the internal nanostructure of the particles. There was no experimental evidence found for a leakage with time, that is, the solubilized oil did not move from the internal nanostructures to the continuous aqueous phase. In other words, there was no indication of a formation of TC-rich normal O/W emulsion droplets in addition to the MLO-rich nanostructured dispersion.

These findings support our hypothesis that the internal structure of the dispersed particles was formed only through self-assembly principles. On this basis, there was no evidence of a reorganization of the lipid (and lipid–oil) molecules during the preparation of the dispersions. Thus, these constituents behaved as pseudocomponents at all investigated temperatures: The solubilized oil (and water) was maintained in the MLO nanostructure during the dispersing procedure. In other words, our results showed no indication of any significant change in the composition at the interfacial film when a high-energy input was applied in the presence of the polymeric stabilizer F127. It should be pointed out that the reorganization of the lipid (and oil) molecules could be induced by replacing the primary surfactant MLO by any surfactant that would not remain bound with oil and water during the dispersing procedure.

(a)

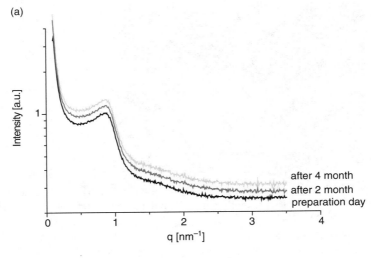

(b)

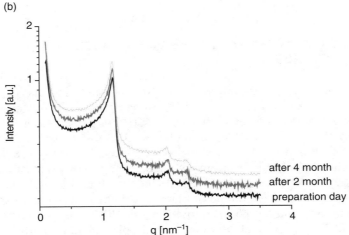

Figure 5.8 Stability against aging: Scattering curves of MLO-based ISAsomes with α ratios of (a) 75 and (b) 19 are shown. The measurements were carried out at 25°C: after preparation (black line), after 2 months, and after 4 months. [Reprinted from Yaghmur et al. (2005).]

For instance, we found that the Winsor-II-type system based on a TC-loaded W/O microemulsion system of the well-known sodium bis(2-ethylhexyl) sulfosuccinate (AOT) lost its stability when it was emulsified in the presence of F127, thus forming a structureless TC-in-water emulsion instead of an EME (data not shown).

5.5 THE BREATHING MODE

The tuning of the inner particle nanostructures could be readily controlled by altering the system's composition and/or varying the temperature. Figure 5.9 summarizes the impact of temperature and TC loading on the MLO-based dispersed and nondispersed systems.

SAXS investigations on both oil-free and oil-loaded MLO dispersions were carried out in order to elucidate the structural changes of these internal nano-structures that occur during heating–cooling processes. In the absence of oil ($\alpha = 0$), the MLO dispersion was measured at three temperatures in a heating–cooling cycle (Fig. 5.10a). As discussed above, heating of the dispersion led to structural transitions of the internal nanostructure in the order V_2 (Pn3m) $\rightarrow$ H_2 $\rightarrow$ ELP. Figure 5.10a shows also that cooling the dispersion induced the inverse order of the structural transitions without a hysteresis or other changes after application of the heating–cooling cycles. Therefore, the respective scattering curves of the dispersion at 25°C (cubosomes) and 58°C (hexosomes) were found to coincide. This meant that the internal nanostructure, similar to its counterpart in the fully hydrated nondispersed phase, depended only on the current temperature, irrespective of whether it was reached by heating or cooling.

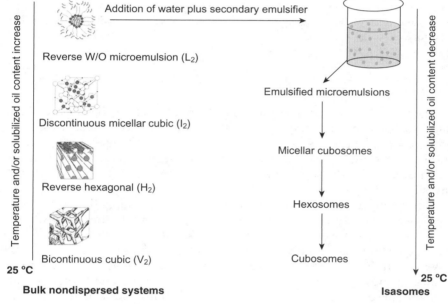

Figure 5.9 Formation of ISAsomes from MLO-based nanostructures.

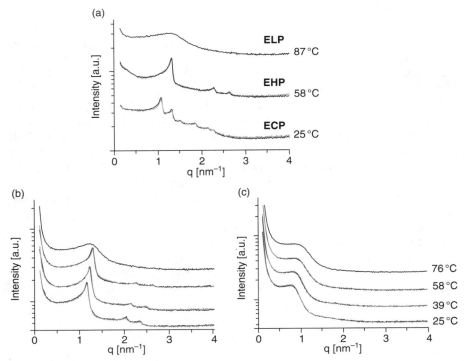

Figure 5.10 Reversibility: Scattering curves of three dispersions with different α values: (a) 0, (b) 19, and (c) 110. The dispersions were measured during heating (black lines), and during cooling (gray lines). The curves have been shifted by a constant arbitrary factor for the sake of visibility. [Reprinted from de Campo et al. (2004) and Yaghmur et al. (2005).]

The same study was also performed on oil-loaded dispersions. Figures 5.10b and 5.10c show the scattering curves of two examples of TC-loaded dispersions (with $\alpha = 19$ and $\alpha = 110$, respectively) as functions of temperature (from 25 to 76°C). Our results revealed the structural reversibility of the confined nanostructures. It is worth noting that the scattering curves were identical at each investigated temperature.

Our results indicate that the internal nanostructure in all dispersions (oil-free and oil-loaded phases) was identical to its counterpart in the fully hydrated bulk system. Moreover, it was, at a certain temperature, independent of the thermal history, that is, either heating or cooling to the required temperature gave rise to the same structure. This fact was clear evidence that the formed nanostructures inside the kinetically stabilized particles were thermodynamic equilibrium structures in analogy with that in the fully hydrated nondispersed bulk phases. It is also worth noting that the emulsified particles showed a water swelling–deswelling behavior during the heating–cooling cycles [denoted as *breathing mode* (de Campo et al., 2004)]. This temperature-dependent behavior indicated a reversible exchange of water into and out of the confined internal particle structures during the cooling and heating cycles. For instance, in oil-free samples, the internal structures follow the phase separation line in Figure 5.1; the particles expelled water upon heating and reabsorbed water upon cooling in a reversible way.

It should be pointed out that the slope of the separation line in the MLO–water phase was negative. It was, therefore, similar to that observed in most investigated monoglyceride–water systems. However, in certain cases, such as in a monolaurin–water mixture, this line has a positive slope (Lutton, 1965), thus suggesting that also a reverse breathing mode of dispersed monolaurin particles could be feasible.

5.6   MODULATION OF THE INTERNAL NANOSTRUCTURES

This section describes, in two parts, the modulation of the internal self-assembled nanostructure of the kinetically stabilized ISAsomes by varying the lipid composition, as well as how to tune back the oil-loaded internal nanostructure from the H_2 (hexosomes) or the W/O microemulsion system (EME) to the V_2 phases (cubosomes). The underlying idea is based on a partial replacement of the primary lipid MLO, which is favoring the formation of nonlamellar phases, by a surfactant favoring the formation of the L_α phases (such as diglycerol monooleate, DGMO, or soybean phosphatidylcholine, PC) (Yaghmur et al., 2006b).

5.6.1   Oil-Free MLO–DGMO- and MLO–PC-Based Systems

The impact of DGMO on the internal structure of the MLO-based aqueous dispersions was investigated in the absence of oil ($\alpha = 0$). Figure 5.11 shows the SAXS curves at 25°C obtained from dispersions in which the β-ratio value, defined as [[mass of DGMO (or mass of PC)]/(mass of MLO) × 100], was in the range of 0–25. At low DGMO concentrations (at β = 5 and 10), the scattering curves still showed the characteristic peaks for a cubic Pn3m structure, as well as a peak at low q values. The appearance of this additional peak suggested the formation of two phases within the confined internal nanostructures of our dispersed particles: the Pn3m cubic phase (diamond type C_D) coexisting with the Im3m cubic phase (primitive type C_P). The Bonnet ratios for these dispersed particles with the confined Pn3m–Im3m coexisting phases were in the range of 1.3–1.34. These values corresponded well with the theoretical value of 1.279 (Hyde, 1996). At a higher DGMO concentration, the three observed peaks had ratios characteristic of a cubic phase with the symmetry Im3m; a scattering curve for β = 25 is shown as an example in Figure 5.11. The cryo-TEM micrographs obtained for this dispersion are shown in Figure 5.12.

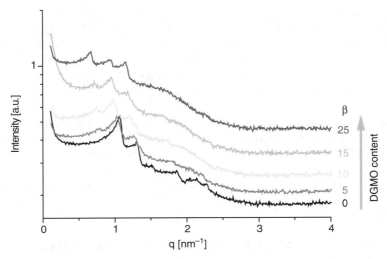

Figure 5.11   Effect of DGMO content on scattering curves of MLO-based aqueous dispersions at 25°C for an α ratio of 0 and with β values in the range of 0–25 (Yaghmur et al., 2006b). The intensities have been shifted by a constant arbitrary factor for the sake of visibility.

Figure 5.12a displays vesicles attached to the formed cubosomes, and in Figures 5.12b and 5.12c the electron beam is aligned with the [100] direction of the dispersed particle, and the {110} and {200} planes contribute to the appearance of a square motif. In Figures 5.12d and 5.12e the electron beam is oriented along [111] and a hexagonal motif, formed by the {110} planes, is observed. It should be pointed out that the particle was tilted in order to discriminate between the Im3m and Pn3m space group (Sagalowicz et al., 2006, 2007). Tilting by approximately 30° enabled access to the [110] axis of the observation where the {110}, {200}, and {211} reflections contributed (Figs. 5.12f and 5.12g). More importantly, the absence of {1–11} and {1–1–1} demonstrated that the space group was Im3m and not Pn3m (Sagalowicz et al., 2006). The lattice parameter was found to be approximately 14 nm, a value in relatively good agreement with that obtained from the SAXS analysis. In this sample, we also observed a very limited number of particles with a lattice parameter in the range of 10–11 nm. These particles likely had an internal Pn3m phase.

It was important to directly compare the confined nanoscaled internal structure of the aqueous dispersions with those of the corresponding fully hydrated nondispersed bulk systems. This was helpful in elucidating the influence of DGMO on the internal nanostructures of the MLO-based aqueous dispersions, and it was thus possible to better understand the observed structural changes in the dispersed systems. As an example, we compared the scattering curve from the MLO-based dispersion (α = 0, β = 25, gray line) with those from the corresponding nondispersed MLO–DGMO bulk samples at β = 25 with varying water content at 25°C (Fig. 5.13).

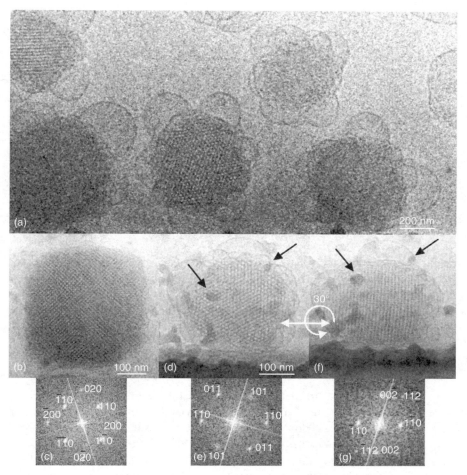

Figure 5.12 Cryo-TEM micrographs of the oil-free cubosomes based on MLO–DGMO ($\alpha = 0$, $\beta = 25$). (a) Micrograph of dispersed cubosomes, (b) micrograph of a cubosome when the electron beam was parallel to [100], and (c) the FFT corresponding to (b). (d) Micrograph obtained when the electron beam was oriented along the [111] direction, and (e) the FFT corresponding to (d). (f) The same particle as in (d) but after tilting it 30° such that the electron beam was parallel to [110], and (g) the Fourier transform of (f). The arrows in (d) and (f) indicate features that are likely to have come from surface contamination. [Reprinted from Yaghmur et al. 2006b).]

The following points summarize our findings with regard to the dispersed and nondispersed oil-free systems:

1. *Enhancement of the Water Solubilization Capacity.* The inclusion of DGMO significantly increased the water solubilization capacity for dispersed and nondispersed bulk MLO–DGMO–water phases. For instance,

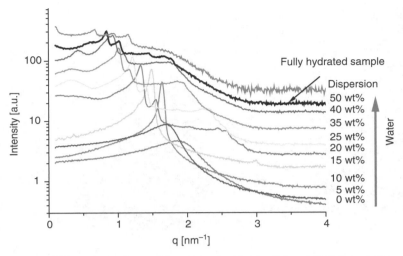

Figure 5.13   Comparison of scattering curve from the oil-free MLO-based dispersion ($\alpha = 0, \beta = 25$, gray line) with those from the oil-free nondispersed MLO–DGMO bulk samples at $\beta = 25$ with varying water content at 25°C. The intensities were shifted by a constant arbitrary factor for the sake of visibility. The thick black line denotes the scattering curve of the fully hydrated bulk sample. [Reprinted from Yaghmur et al. (2006b).]

the maximum solubilized amount of water increased from approximately 32 wt % (de Campo et al., 2004) to approximately 41 wt % water (Yaghmur et al., 2006b) at 25°C when the β value was varied from 0 (MLO–water mixture) to 25 (MLO–DGMO–water system) (Fig. 5.13).

2. *Functionalization of the Self-Assembled Nanostructure.* Our results show that DGMO significantly enlarged the water channels in the fully hydrated bicontinuous cubic phase. For example, the mean lattice parameter, *a*, increased from 8.4 to 10.6 nm with an increasing β value from 0 to 25 (Yaghmur et al., 2006b). This modification was attractive for a number of purposes; in particular, for enhancing the solubilization of active molecules in the internal nanostructures. The functionalization of the V_2 phases has also been described in recent studies (Angelov et al., 2007; Yaghmur et al., 2007).

3. *Dispersed vs. Nondispersed Systems.* We found a disagreement between the nanostructure present in the dispersed particles and that of the fully hydrated nondispersed bulk systems at equivalent β values. A typical example is shown in Figure 5.13: The addition of DGMO promoted the formation of a fully hydrated Pn3m cubic phase in the bulk while the internal structure of the dispersed particles was of the Im3m type. Thus, the behavior differed from that of the MLO-based system described in

the previous sections in which the scattering curves of the dispersions corresponded to those of the nondispersed fully hydrated bulk phases.

4. *Impact of the Stabilizer on the Nanostructure.* The formation of a cubic Im3m phase in the dispersed MLO–DGMO particles instead of a fully hydrated nondispersed cubic Pn3m phase was related to the important role of the stabilizer F127 in the presence of DGMO on the internal structure after the dispersing process. The same structural transition was also observed for the aqueous dispersions of a monoolein (MO)–water system (Larsson, 2000). To illustrate this effect, we explored how the addition of a small amount of F127 affected the MLO–DGMO bulk phases. We found that F127 induced the structural transition from the Pn3m to the Im3m phase (Yaghmur et al., 2006b). In the dispersions, our results indicated that part of F127 was inserted in the confined interior of the dispersed particles, thus changing the symmetry of the cubic phase from Pn3m to Im3m, while the other part performed its job as stabilizer by adhering to the surface of the dispersed particles. This result differed from our findings described in the previous sections. A possible explanation for this discrepancy might simply be the size of the hydrophilic channels in these systems: The hydrophilic channels in the binary MLO–water system were smaller than those in the MLO–DGMO–water or MO–water system (Yaghmur et al., 2006b). This might be the reason why the polymer could not be incorporated as easily in the former case.

In order to modify the MLO-based phases, we also investigated partially replacing MLO by the balanced surfactant PC (Yaghmur et al., 2006b). The MLO–PC system displayed a different behavior as compared to that of the MLO–DGMO mixture at the same β ratio: As described in detail in Yaghmur et al. (2006b), the internal structure contained approximately the same water content as that of MLO ($\beta = 0$). Furthermore, it had a cubic structure of the Pn3m type, possibly a mixture with Im3m (cubosomes), corresponding well with the structure of the nondispersed fully hydrated bulk phase with the same β value (data not shown). The result shows that F127 was very effective as a stabilizer and did not significantly affect the internal nanostructure. However, the fragmentation of the cubic phases into submicron-sized particles only led to a lower degree of order (Yaghmur et al., 2006b). Our findings also imply that the incorporation of the polymer into the lipid bilayers was easier in the MLO–DGMO mixture than into that of an MLO–PC system.

5.6.2 Tuning Back the Curvature of Oil-Loaded MLO-Based Aqueous Dispersions

In order to tune back the oil-loaded aqueous dispersions from hexosomes or EMEs to cubosomes, it is important to replace MLO by a surfactant that has a countereffect with respect to that of TC. This idea is illustrated in the schematic description presented in Figure 5.14. In Figure 5.15, it can be seen that

Oil solubilization:

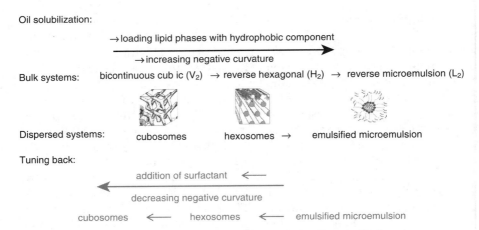

Figure 5.14   Tuning back the curvature of oil-loaded self-assembled nanostructures.

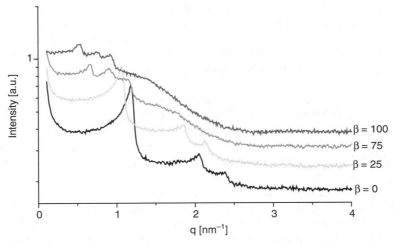

Figure 5.15   Effect of DGMO content on the SAXS curves of TC-loaded aqueous dispersions at 25°C for $\alpha = 6$, and with β values ranging from 0 to 100. The intensities have been shifted by a constant arbitrary factor for the sake of visibility. [Reprinted from Yaghmur et al. (2006b).]

DGMO, at $\alpha = 6$, promoted a reversed transition (from hexosomes at $\beta = 0$ and 25 back to cubosomes at $\beta = 50$ and 100), thus rendering the spontaneous curvature of the oil-loaded interior less negative. However, an amount of DGMO ($\beta = 50$) higher than that of oil was needed to overcome the effect of TC, which indicates that TC had a larger impact on the internal structure than DGMO. At a higher content of solubilized TC [at $\alpha = 75$ as presented in

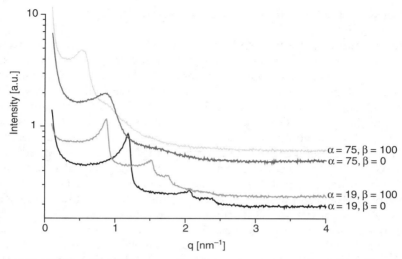

Figure 5.16   Effect of DGMO content on TC-loaded aqueous MLO-based dispersions at 25°C with α ratios of 19 and 75. The scattering curves of the DGMO-free dispersions ($\beta = 0$) are compared to those containing DGMO ($\beta = 100$). The intensities have been shifted by a constant arbitrary factor for the sake of visibility. [Reprinted from Yaghmur et al. (2006b).]

Fig. 5.16), the replacement of a relatively high amount of MLO by DGMO ($\beta = 100$, for $\alpha = 19$)] significantly enlarged the H_2 structure as indicated by the peaks in the scattering curve being shifted to lower angles. However, the DGMO concentration was still not efficient enough to tune back the internal structure to the V_2 phase. The enhanced water solubilization capacity was the underlying reason for this significant change in the internal nanostructures of these oil-loaded dispersions in the presence of DGMO with its bulky hydrophilic group. For instance, increasing the DGMO content in the oil-loaded nondispersed phase (also with $\alpha = 19$ from $\beta = 0$ to 100 significantly increased the content of solubilized water in the H_2 phase from approximately 25 to 35 wt % without any structural transition. Furthermore, the corresponding structure parameter, a, was much larger than that of the ternary MLO–TC–water system at the same water content (Yaghmur et al., 2006b). We also found a good structural agreement between this fully hydrated bulk phase and the dispersed H_2 structure (Yaghmur et al., 2006b). This result was similar to those from previous studies with respect to the stabilization of hexosomes (Gustafsson et al., 1996; Larsson, 2000) indicating that F127 did not influence the internal H_2 structure.

It is also noteworthy that DGMO significantly increased the temperature for all structural transitions (Yaghmur et al., 2006b). For instance, the $H_2 \rightarrow L_2$ transition temperature of the dispersion with $\alpha = 6$ increased from approximately 86 to 92°C when MLO was replaced by MLO–DGMO mixture.

5.7 DIRECT LIPOSOMES–CUBOSOMES TRANSITION

In a recent study, synchrotron SAXS and Cryo-TEM were used to characterize the temperature-induced structural transitions of ME-based aqueous dispersion in the presence of the polymeric stabilizer F127 (Yaghmur et al., 2008). ME is neutral monoacylglycerol having the same molecular weight as its congener MO but with different molecular shape. ME has trans double bonds located at the 9,10 position in its straight acyl chain (C18:1t9), whereas MO has a different configuration (cis double bond in the carbon atom backbone, C18:1c9), which causes a "kink" in the middle of the molecule and reduces its effective length (Fig. 5.17a). This explains the different thermotropic behavior of the ME–water system from the observed temperature dependence of its counterpart MO–water system. One interesting feature of the nondispersed ME–water system is the unique direct transformation under full hydration conditions from L_α to V_2. Therefore, it is well suited to serve as a model system mimicking the first steps of membrane fusion involving lamellar–nonlamellar transitions. Our main focus was to check the possibility of inducing analogous temperature-dependent structural order in the interior of ME-based dispersions. This means to prove that the direct transition from vesicles to cubosomes by heating this dispersion is possible. Our study revealed indeed a direct

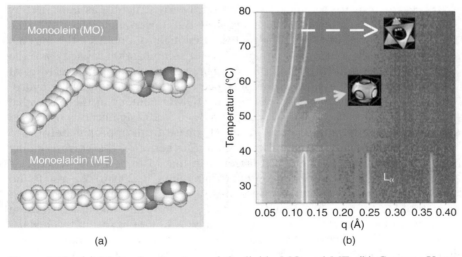

(a)   (b)

Figure 5.17 (a) Molecular structure of the lipids: MO and ME. (b) Contour X-ray diffraction plot covering the regime of the L_α, Im3m, and Pn3m phases as a function of q and temperature. In a temperature scan from 25 to 80°C with a rate of 1°C/min, the ME aqueous dispersion underwent the internal L_α via Im3m to Pn3m structural transition. [Reprinted from Yaghmur et al. (2008).]

liposomes–cubosomes transition in the ME-based aqueous dispersion (Fig. 5.17b). We found that the direct L_α–V_2 transition in the dispersion does not reveal the same mechanism as that of its fully hydrated bulk system. The obtained results suggest that the polymeric stabilizer F127 especially plays a significant role in the membrane fusion processes. It incorporates in considerable amount into the internal nanostructure and leads to the formation of a highly swollen Im3m phase.

5.8 FOOD-GRADE ISAsomes

For the formation of food-grade ISAsomes, it is important to ensure that all components meet the requirements of the food industry. It was, therefore, important to replace the solubilized oil (TC) by food-grade oils such as triglycerides and to stabilize the aqueous dispersions by emulsifiers acceptable in the food industry. We were thus able to form various food-grade ISAsomes that were stabilized by different food-grade surfactants in which TC was replaced by triglycerides such as triolein, or an essential oil [R(+)-limonene], and other hydrophobic food-active materials (data not shown).

5.9 SUMMARY

This chapter has discussed ISAsomes, which are unique emulsion droplets consisting of well-defined interiors with hierarchical nanostructures, and has also focused on how these systems were influenced when the temperature and the solubilizing hydrophobic component were varied and their lipid composition controlled. We have presented findings proving that the formation of these soft nanoparticulate systems was driven simply by the principles of self-assembly. In general, the following major conclusions could be drawn from our investigations:

1. *Formation of MLO-Based ELPs and EMEs.* We showed that the formation of stable ELPs (oil-free dispersions) at high temperatures as well as EMEs (oil-loaded dispersions) at room temperature was possible by tuning the curvature of the inner particle nanostructures. This could be done by varying the temperature and/or altering the system's composition (Fig. 5.9). For more information on the formation and the characterization of other nanostructured aqueous dispersions, the interested reader is directed to two recent reviews on this topic (Boyd et al., 2009; Yaghmur & Glatter, 2009).

2. *Dispersions with Internal Reversible Nanostructures.* The confined nanostructures in the emulsified particles are reversible structures, that is, they exist in thermodynamic equilibrium with the surrounding aqueous phase, and they depend only on the actual temperature and oil content. At each investigated composition and temperature, the MLO-based internal structure was

well preserved during the dispersion procedure: It was identical to the corresponding system of the fully hydrated nondispersed phase. As a consequence, it displayed an interesting temperature-dependent behavior (the *breathing mode*) during heating–cooling cycles, that is, it demonstrated a reversible water deswelling–swelling behavior during heating–cooling cycles.

3. *Control of the Internal Nanostructure.*   Our results revealed that the internal nanostructure of the kinetically stabilized aqueous dispersions based on MLO–water binary and ternary MLO–TC–water mixtures could be modulated by varying the lipid composition: The replacement of MLO by binary MLO–DGMO mixtures significantly affected the internal nanostructures.

4. *Concentrated Nanostructured Dispersions with a Controllable Droplet Size.*   In a recent report (Salentinig et al., 2008), concentrated ISAsomes with various internal structures were also prepared with a laboratory-built shear device. This is achieved within a very short time interval (a few seconds) by using a similar approach to that proposed by Bibette's method (Mason and Bibette, 1996, 1997), which is well-known for the preparation of monodispersed emulsions. The interested reader can find more details on this technique in the contribution of Glatter and Kulkarni in this book.

5. *Potential Applications.*   The formation and characterization of ISAsomes and the functionalization of their confined nanostructures is an important topic in nanotechnology. These dispersions can be utilized in various applications in food and pharmaceutical industries (Boyd, 2005; Drummond and Fong, 2000; Larsson, 2009; Leser et al., 2006; Malmsten, 2006; Mezzenga et al., 2005; Siekmann et al., 2002; Yaghmur and Glatter, 2009). They are particularly interesting for the formulation of nanoparticulate delivery systems of drugs and bioactive materials. It is also interesting to check the potential applications of the concentrated nanostructured aqueous dispersions as reservoirs for enhanced solubilization of bioactive ingredients including proteins and peptides and for the generation of active molecules.

REFERENCES

Almsherqi, Z. A., Kohlwein, S. D., and Deng, Y. (2006). Cubic membranes: A legend beyond the Flatland of cell membrane organization. *Journal of Cell Biology, 173,* 839–844.

Angelova, A., Angelov, B., Papahadjopoulos-Sternberg, B., Bourgaux, C., and Couvreur, P. (2005). Protein driven patterning of self-assembled cubosomic nanostructures: Long oriented nanoridges. *Journal of Physical Chemistry B, 109,* 3089–3093.

Angelov, B., Angelova, A., Garamus, V. M., Lebas, G., Lesieur, S., Ollivon, M., Funari, S. S., Willumeit, R., and Couvreur, P. (2007). Small-angle neutron and x-ray scattering from amphiphilic stimuli-responsive diamond-type bicontinuous cubic phase. *Journal of the American Chemical Society, 129,* 13474–13479.

Barauskas, J., and Landh, T. (2003). Phase behavior of the phytantriol/water system. *Langmuir, 19,* 9562–9565.

Barauskas, J., Johnsson, M., Joabsson, F., and Tiberg, F. (2005). Cubic phase nanoparticles (cubosome): Principles for controlling size, structure, and stability. *Langmuir*, *21*, 2569–2577.

Boyd, B. J. (2005). Controlled release from cubic liquid-crystalline particles. In M. L. Lynch, and P. T. Spicer (Eds.), *Bicontinuous Liquid Crystals*. Surfactant Science Series, Vol. 127, CRC Press, New York, pp. 285–300.

Boyd, B. J., Dong, Y. D., and Rades, T. (2009). Nonlamellar liquid crystalline nanostructured particles: Advances in materials and structure determination. *Journal of Liposome Research*, *19*, 12–28.

Colotto, A., and Epand, R. M. (1997). Structural study of the relationship between the rate of membrane fusion and the ability of the fusion peptide of influenza virus to perturb bilayers. *Biochemistry*, *36*, 7644–7651.

de Campo, L., Yaghmur, A., Sagalowicz, L., Leser, M. E., Watzke, H., and Glatter, O. (2004). Reversible phase transition in emulsified nano-structured lipid systems. *Langmuir*, *20*, 5254–5261.

de Kruijff, B. (1997). Lipids beyond the bilayer. *Nature*, *386*, 129–130.

de Kruijff, B., Killian, J. A., Rietveld, A. G., and Kusters, R. (1997). Phospholipid structure and *Escherichia coli* membranes. In R. M. Epand (Ed.), *Lipid Polymorphism and Membrane Properties*. Current Topics in Membrane, Vol. 44, Academic, San Diego, pp. 477–515.

Deng, Y., Kohlwein, S. D., and Mannella, C. A. (2002). Fasting induces cyanide-resistant respiration and oxidative stress in the amoeba *Chaos carolinensis*: Implications for the cubic structural transition in mitochondrial membranes. *Protoplasma*, *219*, 160–167.

Drummond, C., and Fong, C. (2000). Surfactant self-assembly objects as novel drug delivery vehicles. *Current Opinion in Colloid & Interface Science*, *4*, 449–456.

Ellens, H., Siegel, D. P., Alford, D., Yeagle, P. L., Boni, L., Lis, L. J., Quinn, P. J., and Bentz, J. (1989). Membrane fusion and inverted phases. *Biochemistry*, *28*, 3692–3703.

Guillot, S., Moitzi, C., Salentinig, S., Sagalowicz, L., Leser, M. E., and Glatter O. (2006). Direct and indirect thermal transitions from hexosomes to emulsified microemulsions in oil-loaded monoglyceride-based particles. *Colloids and Surfaces A: Physicochemical and Engineering Aspects*, *291*, 78–84.

Gustafsson, J., Ljusberg-Wahren, H., Almgren, M., and Larsson, K. (1996). Cubic lipid-water phase dispersed into submicron particles. *Langmuir*, *12*, 4611–4613.

Hyde, S. T. (1996). Bicontinuous structures in lyotropic liquid crystals and crystalline hyperbolic surfaces. *Current Opinion in Solid State & Materials Science*, *1*, 653–662.

Hyde, S., Andersson, S., Larsson, K., Blum, Z., Landh, T., Lidin, S., and Ninham, B. W. (1997). *The Language of Shape. The Role of Curvature in Condensed Matter: Physics, Chemistry, Biology*, Elsevier, Amsterdam.

Larsson, K. (1983). Two cubic phases in monoolein–water system. *Nature*, *304*, 664.

Larsson, K. (2000). Aqueous dispersions of cubic lipid-water phases. *Current Opinion in Colloid & Interface Science*, *4*, 64–69.

Larsson, K. (2009). Lyotropic liquid crystals and their dispersions relevant in foods. *Current Opinion in Colloid & Interface Science*, *14*, 16–20.

Leser, M. E., Sagalowicz, L., Michel, M., and Watzke, H. J. (2006). Self-assembly of polar food lipids. *Advances in Colloid and Interface Science*, *123–126*, 125–136.

Lewis, R. N. A. H., Mannock, D. A., and McElhaney, R. N. (1998). Membrane lipid molecular structure and polymorphism. In R. M. Epand (Ed.), *Lipid Polymorphism and Membrane Properties*. Current Topics in Membrane, Vol. 44, Academic, San Diego, pp. 25–102.

Lindblom, G., and Rilfors, L. (1989). Cubic phases and isotropic structures formed by membrane lipids—possible biological relevance. *Biochimica et Biophysica Acta (BBA)—Reviews on Biomembranes*, *988*, 221–256.

Lindblom, G., Larsson, K., Johansson, L., Fontell, K., and Forsén, S. (1979). The cubic phase of monoglyceride-water systems. Arguments for a structure based upon lamellar bilayer units. *Journal of the American Chemical Society*, *101*, 5465–5470.

Lutton, E. S. (1965). Phase behavior of aqueous systems of monoglycerides. *Journal of the American Oil Chemists' Society*, *42*, 1068–1070.

Luzzati, V. (1997). Biological significance of lipid polymorphism: The cubic phases. *Current Opinion in Structural Biology*, *7*, 661–668.

Malmsten, M. (2006). Soft drug delivery systems. *Soft Matter*, *2*, 760–769.

Mason, T. G., and Bibette, J. (1996). Emulsification in viscoelastic media. *Physical Review Letters*, *77*, 3481–3484.

Mason, T. G., and Bibette, J. (1997). Shear rupturing of droplets in complex fluids. *Langmuir*, *13*, 4600–4613.

Mezzenga, R., Schurtenberger, P., Burbidge, A., and Michel, M. (2005). Understanding foods as soft materials. *Natural Materials*, *4*, 729–740.

Moitzi, C., Guillot, S., Fritz, G., Salentinig, S., and Glatter, O. (2007). Phase reorganization in self-assembled systems through interparticle material transfer. *Advanced Materials*, *19*, 1352–1358.

Ortiz, A., Killian, J. A., Verkleij, A. J., and Wilschut, J. (1999). Membrane fusion and the lamellar-to-inverted-hexagonal phase transition in cardiolipin vesicle systems induced by divalent cations. *Biophysical Journal*, *77*, 2003–2014.

Patton, J. S., and Carey, M. C. (1979). Watching fat digestion. *Science*, *204*, 145–148.

Qiu, H., and Caffrey, M. (1998). Lyotropic and thermotropic phase behavior of hydrated monoacylglycerols: Structure characterization of monovaccenin. *Journal of Physical Chemistry B*, *102*, 4819–4829.

Qiu, H., and Caffrey, M. (1999). Phase behavior of the monoerucin water system. *Chemistry and Physics of Lipids*, *100*, 55–79.

Rizwan, S. B., Dong, Y.-D., Boyd, B. J., Rades, T., and Hook, S. (2007). Characterisation of bicontinuous cubic liquid crystalline systems of phytantriol and water using cryo field emission scanning electron microscopy (cryo FESEM). *Micron*, *38*, 478–485.

Sagalowicz, L., Michel, M., Adrian, M., Frossard, P., Rouvet, M., Watzke, H. J., Yaghmur, A., de Campo, L., Glatter, O., and Leser, M. E. (2006). Crystallography of dispersed liquid crystalline phases studied by cryo-transmission on electron microscopy. *Journal of Microscopy—Oxford*, *221*, 110–121.

Sagalowicz, L., Acquistapace, S., Watzke, H. J., and Michel, M. (2007). Study of liquid crystal space groups using controlled tilting with cryogenic transmission electron microscopy. *Langmuir*, *23*, 12003–12009.

Salentinig, S., Yaghmur, A., Guillot, S., and Glatter, O. (2008). Preparation of highly concentrated nanostructured dispersions of controlled size. *Journal of Colloid and Interface Science, 326*, 211–220.

Salonen, A., Guillot, S., and Glatter, O. (2007). Determination of water content in internally self-assembled monoglyceride-based dispersions from the bulk phase. *Langmuir, 23*, 9151–9154.

Siegel, D. P. (1999). The modified stalk mechanism of lamellar/inverted phase transitions and its implications for membrane fusion. *Biophysical Journal, 76*, 291–313.

Siekmann, B., Bunjes, H., Koch, M. H. J., and Kirsten, W. (2002). Preparation and structural investigations of colloidal dispersions prepared from cubic monoglyceride-water. *International Journal of Pharmaceutics, 244*, 33–43.

Spicer, P. T. (2004). Cubosomes: Bicontinuous liquid crystalline nanoparticles. In J. A. Schwarz, C. Contescu, and K. Putyera (Eds.), *Dekker Encyclopaedia of Nanoscience and Nanotechnology*, Marcel Dekker, New York, pp. 881–892.

Yaghmur, A., and Glatter, O. (2009). Characterization and potential applications of nanostructured aqueous dispersions. *Advances in Colloid and Interface Science, 147–148*, 333–342.

Yaghmur, A., de Campo, L., Sagalowicz, L., Leser, M. E., and Glatter, O. (2005). Emulsified microemulsions and oil-containing liquid crystalline phases. *Langmuir, 21*, 569–577.

Yaghmur, A., de Campo, L., Salentinig, S., Sagalowicz, L., Leser, M. E., and Glatter, O. (2006a). Oil-loaded monolinolein-based particles with confined inverse discontinuous cubic structure (Fd3m). *Langmuir, 22*, 517–521.

Yaghmur, A., de Campo, L., Sagalowicz, L., Leser, M. E., and Glatter, O. (2006b). Control of the internal structure of MLO-based isasomes by the addition of diglycerol monooleate (DGMO) and soybean phosphatidylcholine (PC). *Langmuir, 22*, 9919–9927.

Yaghmur, A., Laggner, P., Zhang, S., and Rappolt, M. (2007). Tuning curvature and stability of monoolein bilayers by short surfactant-like designer peptides. *PLoS ONE, 2*, e479.

Yaghmur, A., Laggner, P., Almgren, M., and Rappolt, M. (2008). Self-assembly in aqueous dispersions: Direct vesicles to cubosomes transition. *PLoS ONE, 3*, e3747.

CHAPTER 6

Hierarchically Organized Systems Based on Liquid Crystalline Phases

CHANDRASHEKHAR V. KULKARNI and OTTO GLATTER

Department of Chemistry, University of Graz, Graz, Austria

Abstract

In this chapter, we present hierarchically organized nanostructures formed from lyotropic liquid crystalline (LC) phases. The nano-, micro-, and macroscopic structural hierarchy arises from the kinetic stability of various lyotropic phases dispersed in oil-in-water (O/W) or water-in-oil (W/O) emulsions. When an O/W emulsion consists of a dispersion of LC nanoparticles stabilized by certain stabilizers, it is called an ISAsome, that is, an internally self-assembled particle. In contrast, when the water droplets are dispersed in a continuous film of LC nanostructures, they are called W/O-nanostructured emulsions, which do not require a stabilizing agent. Both emulsions exhibit fascinating properties that can be tuned to a great extent. Such tunability proliferates their performance in various applications. Herein, we discuss the formation, multiscale structure, properties, and their modulation for the aforementioned superstructures formed from LC phases. Focusing further on ISAsomes we present Pickering emulsions stabilized by using various nanoparticles, including synthetic clay Laponite and silica nanoparticles. The transfer of hydrophobic components among several differently nanostructured ISAsomes was studied by time-resolved X-ray scattering; the effects of Isasome-forming components are also illustrated. The continuous aqueous region of ISAsome dispersions can be loaded with water-soluble polymers that form thermoreversible hydrogels. This enables the entrapment of ISAsome systems into such hydrogel networks. Subsequent drying of these loaded systems facilitates immobilization of ISAsomes, which can be easily restored by rehydration of the loaded dry films. The formation of hydrogels in the aqueous reservoirs of W/O-nanostructured emulsions also proved advantageous in terms of tuning their viscosity and, in some cases, enhancing their stability. The current contribution covers systems with diverse structural hierarchy, ranging from equilibrium liquid crystalline nanostructures to the systems with multiple orders of length scales in their structure.

Self-Assembled Supramolecular Architectures: Lyotropic Liquid Crystals, First Edition.
Edited by Nissim Garti, Ponisseril Somasundaran, and Raffaele Mezzenga.
© 2012 John Wiley & Sons, Inc. Published 2012 by John Wiley & Sons, Inc.

6.1   INTRODUCTION—BASIC LYOTROPICS TO HIERARCHICAL STRUCTURES

Amphiphilic molecules, such as lipids, surfactants, and block copolymers, show fascinating polymorphism when mixed with water or other solvents (Hyde et al., 1997; Seddon, 1990). These polymorphs range from simple micelles to complex, yet well-ordered, one-, two-, and three-dimensional assemblies (Fig. 6.1). The lamellar phase is quite common, as it forms an indispensable structural constituent of biological membranes; the latter also consists of other amphiphiles including membrane proteins and fatty acids. Biological lipids self-assemble into some additional shapes, such as tubules, liposomes, and convoluted membranes, which form an integral part of the human body.

The type[1] of fundamental assembly that forms depends on the molecular shape of the amphiphile, as shown in Figure 6.1. Molecules with larger head groups form type 1 (normal) micelles, which are typically observed for surfactants/detergents that contain a single hydrophobic chain; molecules of cylindrical shape with one or two chains form type 0, which is the most common assembly found in nature in the form of biological membranes. Lipids or block copolymers with more hydrophobic character usually form type 2 (inverse) micelles, which are the major focus of the present work. Lipid mesophases can also form, or interchange, with external triggers such as temperature and/or pressure (Kulkarni, 2011; Kulkarni et al., 2011b; Salentinig et al., 2010; Seddon et al., 2006; Yaghmur et al., 2010). The molecular shape and, in turn, the mean curvature[2] of a phase is then altered. For example, chain splay increases with rising temperature (Shearman et al., 2010), and this increases

---

[1]Sign convention of the curvature adopted by different authors may vary, but the current work adopts the convention as: positive curvature toward the chain region (type 1), whereas curvature toward the water region as negative (type 2).
[2]Mean curvature: A mean of two principle curvatures, c_1 and c_2.

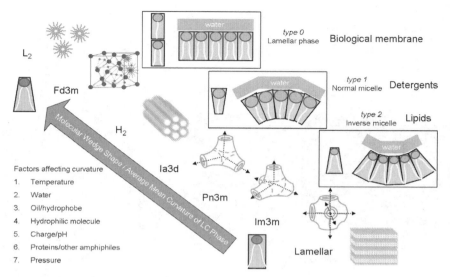

Figure 6.1 Diversity and tuning of lipid polymorphism. Amphiphilic molecules such as lipids have a hydrophilic head group and a long hydrophobic alkyl chain. These molecules self-assemble in aqueous media; according to their molecular shape they form type 0, 1, or 2 micelles, typical examples of which are biological membrane, detergent, and lipid assemblies, respectively (top right). The original molecular shape can be modulated by certain parameters, such as temperature, pressure, water, and composition (see bottom left). An increase in the inverse conical shape of a molecule, typical for type 2 micelles, increases the average negative mean curvature of the lipid mesophases from lamellar toward the L_2 phase via the cubic Im3m, Pn3m, and Ia3d phases, hexagonal, and micellar cubic Fd3m phases, as shown here. Schematic drawings of these (most common) phases show the 1, 2, and 3 dimensionality of the lipid phases (diagonal).

the inverse conical shape of the molecule, which results in more negatively curved mesophases (Fig. 6.1). The lipid molecular shape can also be engineered by altering the unsaturation (number and location), which thus provides an opportunity to finely control the lipid phase behavior (Kulkarni et al., 2010c). More lipid phases can be formed by adding other chemical components (Faham and Bowie, 2002; Guillot et al., 2006; Kulkarni et al., 2011b; Misquitta and Caffrey, 2001; Yaghmur et al., 2006, 2007). Thus, with the varying physico-chemical nature of the amphiphiles, and with varied parameters such as temperature and composition, an assortment of polymorphs can be observed, as shown in Figure 6.1 (see also Fig. 6.2). Due to their dimensional range (from ~2.5 to 25 nm) and their solid–liquid mediation, they are also called liquid crystalline (LC) nanostructures, which can be further developed into fascinating superstructures.

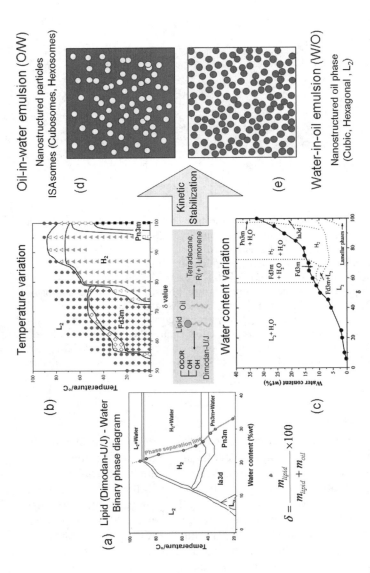

Figure 6.2 Hierarchically ordered structures of lipids: (a) Commonly observed lipid phases for the lipid–water binary phase diagram are indicated on the temperature–water content plot. DU shows two cubic phases, Ia3d and Pn3m, one hexagonal, H₂, and a lamellar phase, L_α along with a molten micellar, L₂, phase. Most of them are also stable in the excess water region, as shown by the phase separation line. The classical phase behavior of such lipids can be further modified by certain additives, such as oils, which, at constant (usually excess) water content, as shown in (b) and at constant temperature as in (c), form additional phases such as the micellar cubic, Fd3m, phase. The range of formations can be tuned by changing δ and/or the temperature. The aforementioned equilibrium bulk nanostructures can then be transformed into hierarchically ordered structures by external energy input with (or sometimes without) additional stabilizer. They can take forms of O/W, as in (d) or W/O emulsions, as in (e). [Figures (a), (b), and (c) were obtained from de Campo et al. (2004), Guillot et al. (2006), and Salonen et al. (2007), respectively, while the remaining part was modified from (Kulkarni et al., 2010b).]

The aforementioned lyotropic phases are mostly equilibrium structures and are typically observed for lipid–water binary mixtures (Fig. 6.2a). Commonly used lipids are monoglycerides; for example, monolinolein and monoolein, which are used in their commercial grades as Dimodan-U/J[3] (DU) and glycerol monooleate (GMO),[4] respectively. Two advantages of these products (consisting of a mixture of saturated and unsaturated lipid chains) are the accessibility of their nonlamellar phases at or below room temperature and their lower degradation rates under ambient conditions, in contrast to their pure counterparts. These advantages, along with their inexpensiveness, increase their applicability in diverse fields (Leser et al., 2006; Sagalowicz et al., 2006a). The addition of extra components to this mixture, for example, hydrophobic additive oil, generates more space for present structures but can also give rise to new phases, such as Fd3m (Yaghmur et al., 2006b), which is absent in the binary phase diagram. This is achieved, specifically, by varying either temperature (Fig. 6.2b) or water content (Fig. 6.2c) while altering the lipid–oil ratio (δ value), as shown by their ternary phase diagrams (Fig. 6.2).

Lyotropic LC phases have many applications in various biotechnological fields, ranging from drug delivery to protein crystallization (Caffrey, 2003; Clogston and Caffrey, 2005; Landau and Rosenbusch, 1996; Sennoga et al., 2003; Shah et al., 2001). However, they face severe problems in the food and pharmaceutical industries, where their rather high viscosity and variable domain consistency causes difficulties in their handling. To facilitate their applicability in various fields, they can be modulated into other forms and/or hierarchical structures. LC nanostructures are spontaneous self-assemblies driven by van der Waals interactions, hydrogen bonding, hydrophobic effects, and interfacial tension. However, with high external energy input it is possible to break these assemblies into hierarchically ordered forms to give partial equilibrium structures. These superstructures usually need to be kinetically stabilized by an external stabilizer.

Depending on the method of preparation and amount of dispersed phase, LC nanostructures reorganize into oil-in-water (O/W) or water-in-oil (W/O) emulsions (Figs 6.2d and 6.2e). The oil phase (lipid or lipid + oil) itself displays a wide range of nanostructures; when dispersed in water in the form of particles, they are termed ISAsomes (Yaghmur et al., 2005). When the oil phase forms a continuous wall-like architecture enclosing water droplets, this is called a W/O nanostructured emulsion (Kulkarni et al., 2010b).

---

[3]Dimodan U/J is a commercial product from Danisco, Brabrand, Denmark, containing more than 98 wt % monoglycerides. The major components are linoleates, forming 62% of the total mixture. The hydrocarbon tails largely consist of C18 chains (91%), the composition of which is as follows: C18:2 (61.9%), C18:1 (24.9%), and C18:0 (4.2%), where the number after the colon indicates the number of unsaturations; additional lipids are C16:0 chains (7.4%) and a residual amount of diglycerides (1.6%).

[4]GMO: Glycerol monooleate (1-monooleoyl-glycerol) contains 98.1 wt % monoolein, which was also obtained from Danisco, Braband, Denmark.

There are other types of hierarchically ordered forms of LC nanostructures not directly linked to the current discussion, such as poly (high internal phase emulsions) (Sergienko et al., 2002), liquid crystalline elastomers (Ohm et al., 2010), bicontinuous interfacially jammed emulsion gels (Cates and Clegg, 2008), polymerized LC phases (Armitage et al., 1996; Forney and Guymon, 2010; Srisiri et al., 1998; Yang et al., 2002), lipid nanocapsules (Huynh et al., 2009), and layer-by-layer assemblies involving LC nanostructures. Structured emulsions, formed primarily from simple monoglycerides, are the focus of the current contribution.

6.2 ISAsomes: FORMATION AND CHARACTERIZATION

Dispersing the LC nanostructures in excess water (typically 80–95%) in the form of submicron-sized (typically 200–400 nm) particles without loss of the original structure has been a requirement for most industrial applications for the last two decades. Cubosomes and hexosomes, which consist of cubic and hexagonal nanostructures, respectively, were originally developed by Gustafsson et al. (1996, 1997). Similar systems were later explored for further internal nanostructures by Yaghmur et al. (2005) who gave them the general term ISAsomes, from internally self-assembled somes (or particles). Subsequent developments on ISAsomes have shown that their internal structure is not only limited to the cubic (Pn3m, Im3m, or Fd3m) and hexagonal but can also include microemulsion (Yaghmur et al., 2005), lamellar (Yaghmur et al., 2008) or even sponge phases (Angelov et al., 2011). Due to the tunable nanostructure of the oil phase, these fascinating superstructures find applications in various scientific, technological, and medical disciplines.

ISAsomes are simply O/W emulsions that have been kinetically stabilized. The primary components of ISAsomes are lipids, stabilizer, and water. Until now the lipids monoolein (MO) and monolinolein (ML) have been utilized directly or in their commercial forms (DU and GMO), although Phytantriol (PT) has also been used in recent years to target specific applications (Lee et al., 2009; Nguyen et al., 2010; Salonen et al., 2010b). Monoelaidin (ME) has the same chemical formula as MO but a different molecular orientation, with a *trans* rather than *cis* conformation of MO. ME is currently being investigated (Yaghmur et al., 2008) as it forms the Im3m cubic phase along with two other phases, Ia3d and Pn3m (Kulkarni, 2009, 2011; Kulkarni et al., 2007, 2009, 2010c).

A range of surfactants, polymers, polysaccharides, and food hydrocolloids are utilized as emulsion stabilizers. In fact, the amphiphilic triblock copolymer, Pluronic F-127 (PEO_{99}-PPO_{67}-PEO_{99}), hereafter simply F-127, has been most widely used for structured emulsions such as ISAsomes (Guillot et al., 2010; Gustafsson et al., 1996, 1997); nevertheless, polysaccharide- and polyethylene-glycol-based stabilizers (Huynh et al., 2009) are also used with varying levels of success. Emulsions are also known to be stabilized by proteins such as casein and lactoglobulin (Dickinson, 2009; Salentinig et al., 2011). Solid particles, such as inorganic nanoparticles (silica nanocolloids) (Muller et al., 2011; Salonen et al., 2010b) or clay platelets (Laponite) (Guillot et al., 2009a; Muller et al., 2010a,b), also stabilize the emulsions, which are then generally called Ramsden–Pickering or Pickering emulsions (Pickering, 1907; Ramsden, 1903). These are discussed in more detail in Section 6.3 of this chapter. Comprehensive literature reviews concerning ISAsomes (Yaghmur and Glatter, 2009) and various emulsifying systems, including surfactants, proteins, hydrocolloids, and nanoparticles, along with their underlying principles (Dickinson, 2009) have recently been published.

Water, the third basic component for ISAsomes, is used in pure form or loaded with certain additives, which can range from drug molecules and perfumes to food and proteins. The amount of water may vary depending on the application; 80–95% water is used for most applications, although it can be as low as 30% for concentrated emulsions (Salentinig et al., 2008).

There are several methods for the preparation of stable ISAsomes (Garg et al., 2007; Salentinig et al., 2008; Spicer et al., 2001), but the two main approaches are "top-down" and "bottom-up." The top-down approach starts with the formation of the bulk lyotropic phase, which is subsequently dispersed in the water by using a high-energy source, such as ultrasonication, high-pressure homogenization, or strong shear forces. In the bottom-up approach, ISAsome formation is initiated by a precursor at the molecular length scale. Progressive hydration and nucleation then lead to structured nanoparticles. The ISAsome precursors can be prepared by at least three different techniques; namely, hydrotrope (Spicer et al., 2001), spray–drying (Spicer et al., 2002), or freeze–drying.

For lower concentrations of dispersed phase (~10–30% ISAsomes in the total dispersion), ultrasonication is very convenient; however, for concentrations above ~30%, the shearing method produces better results. Shearing can also produce highly concentrated ISAsome dispersions (up to ca. 70% of dispersed hydrophobic material) (Salentinig et al., 2008). This technique can be easily scaled up for the continuous production of ISAsome systems. We have also shown recently that by simply optimizing shear conditions one can create W/O nanostructured emulsions that, most interestingly, do not require any type of stabilizer (Kulkarni et al., 2010a). These are discussed further in the last section of this chapter.

As mentioned earlier, the internal structure of ISAsomes can be tuned to create various liquid crystalline phases. It is known that the addition of diglycerol monooleate, or F-127, converts Pn3m into Im3m (Guillot et al., 2010; Yaghmur et al., 2006a), the latter also controls the size of the ISAsomes. The particle size can be also tuned by varying the concentration of the dispersed phase and the shear rate of the Couette cell (Salentinig et al., 2008). The lattice parameters of the nanostructures can also be modulated by varying the lipid components; for example, GMO and PT are used for many applications similar to DU-based systems (Dong et al., 2010; Lee et al., 2009; Nguyen et al., 2010). Structural tuning can also be induced by addition of various component/s such as oil [tetradecane (TC), R(+)limonene (LM), etc.], glycerol, or other surfactants (Engelskirchen et al., 2011; Mezzenga et al., 2005; Sagalowicz et al., 2006b; Ubbink et al., 2008; Yaghmur et al., 2006a, 2009). Further details on the nanostructural tuning of ISAsomes are provided elsewhere in this book.

Temperature (Guillot et al., 2006; Yaghmur et al., 2010) or pressure (Yaghmur et al., 2010) can be used to interconvert ISAsome nanostructures; for instance, from H_2 to L_2. The internal structure of ISAsomes can be also triggered by changes in effective charge (Muller et al., 2010b; Salonen et al., 2008, 2010b) or pH (Salentinig et al., 2010). It is possible to mix two different types of ISAsomes and—due to the interparticle transfer of hydrophobic molecules—one can create new types of ISAsomes different from either of the starting ISAsomes (Moitzi et al., 2007). The rate of transfer can be modulated by changing the ISAsomes' components; this feature is explained in more detail in Section 6.4.

Liquid crystalline materials and their hierarchical structures have similarities at the nanostructural level; however, the hierarchical structures have certain advantages over the bulk systems: The consistency of the bulk phase is broken in ISAsomes (by shear or ultrasonication), which eases their handling, mainly due to reduced viscosity, and also reduces the effective cost by utilizing less bulk material. The bulk phase is usually oil continuous; in contrast, ISAsome systems are water continuous and contain a variable volume of water. These nanostructured assemblies are a well-known vehicle for drug delivery as well as for the controlled release and uptake of functional molecules (Garg et al., 2007; Hirlekar et al., 2010; Lee et al., 2009; Nguyen et al., 2010; Rizwan et al., 2010; Sagalowicz et al., 2006a; Yaghmur and Glatter, 2009). They find further applications in the fields of pharmaceuticals and foods since they can be made easily from food-grade or biocompatible components (Mezzenga et al., 2005; Sagalowicz et al., 2006a; Ubbink et al., 2008; Yaghmur and Glatter, 2009). ISAsomes act as bioadhesives (Geraghty et al., 1997) in some applications, as they are (or can be made) biodegradable by simple enzyme action.

The release and uptake rates of ISAsomes can be controlled by adding certain gelling agents to the dispersion; this arrests the dynamics of the ISAsomes by creating a hydrogel in the continuous water phase. ISAsomes entrapped in thermoreversible hydrogels (Guillot et al., 2009b; Tomsic et al., 2009) and their properties are discussed in Section 6.5. Recently, we have shown that it is also possible to dry the ISAsome-arrested polymer matrix and obtain a solid film that can be later resolubilized to restore the ISAsomes (Kulkarni et al., 2011a), as elucidated further in Section 6.5.

6.3 ISAsomes STABILIZED BY NANOPARTICLES: PICKERING EMULSIONS

Although surfactant stabilizers are widely used for stabilizing structured emulsions, including ISAsomes, they have certain disadvantages; for instance, they affect the nanostructure at higher surfactant concentrations (Guillot et al., 2010; Muller et al., 2010a; Sagalowicz et al., 2006c). Moreover, the most commonly used triblock copolymer, F-127, can actually penetrate the cubosomes to some extent and thereby influence (disrupt) the structure of at least their outer shell (Guillot et al., 2006, 2010; Gustafsson et al., 1997). Surfactant stabilizers can be avoided by using solid particles (Cui et al., 2010; Muller et al., 2010b; Salonen et al., 2008, 2010b); such emulsions are sometimes called surfactant-free emulsions. The solid particles adsorb onto the surface of so-called dispersed phase, creating a steric barrier at the interface and stabilizing the system (Hunter et al., 2008). Such stabilization is well known from the pioneering work of Ramsden (1903) and Pickering (1907), and the term Pickering emulsions has been established over the years to describe emulsions stabilized in this manner. In principle, the overall emulsion stability varies inversely with particle size, as the smaller particles pack more efficiently, forming a homogeneous interfacial layer (Hunter et al., 2008). The particle size ranges from nanometer to micrometer, stabilizing up to millimeter-sized droplets, which is not commonly observed for surfactant-based systems (Aveyard et al., 2003). Protein-stabilized food emulsions can also be termed Pickering emulsions, even though most proteins used for this purpose are also amphiphilic.

A variety of solid particles are utilized for Pickering emulsions, including latex particles, metal oxides, carbon and sulfate particles (Dai et al., 2008), polymeric rods (Noble et al., 2004), grafted polymer brushes (Saigal et al., 2010), silica (Binks and Lumsdon, 2000), and clay. The latter two are best suited for stabilizing ISAsomes (Fig. 6.3) because they allow the emulsion stability to be controlled by changing one or more of the following parameters: charge,

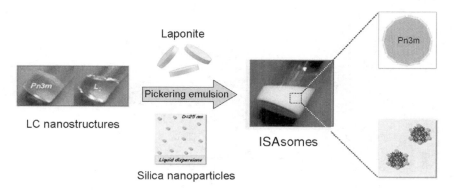

Figure 6.3   Pickering-type nanostructured emulsions. LC nanostructures/phases can be stabilized as submicron-sized particles by using inorganic solids such as the clay platelets Laponite and silica nanoparticles. In other words, ISAsomes retain their original LC nanostructure (from the bulk phase) when stabilized by nanoparticles. [Figure modified from images in Muller et al. (2010b) and Salonen et al. (2010b).]

pH, and temperature; although it is also possible to alter conventional parameters, such as concentration, particle size, and shape, for this purpose. More importantly, it is possible to adjust the surface characteristics (with respect to interaction with emulsion medium, in other words, their wettability and interparticle interactions) (Muller et al., 2010a,b; Salonen et al., 2008, 2010b).

Natural and synthetic clay-based systems are well known for forming nanocomposites with inorganic and organic (especially polymeric) materials (Agrawal et al., 2004; Gupta and Bhattacharya, 2008). Clays have a platelike structure in which an octahedral metal hydroxide layer is sandwiched between two tetrahedral silica layers. Due to the characteristic high-aspect ratio (1 nm thick and ~30 nm mean diameter), large surface area, and surface functionalization, they are ideal for enhancing these materials' properties and for stabilizing Pickering emulsions. Ashby and Binks (2000) were the first to focus on specific effects of such clay nanoparticles as stabilizers. However, Salonen et al. (2008) and Muller et al. (2010b) have also explored similar clay pellets for stabilizing structured emulsions such as ISAsomes. They have demonstrated that the ISAsomes can be stabilized by using Laponite (Laponite XLG from Southern Clay Products, Gonzales, Texas) particles regardless of the internal LC structure (Muller et al., 2010a; Salonen et al., 2008). The natural clay Montmorillonite (SPV Wyoming provided from Comptoir des Minéraux, France), having larger-diameter pellets, can also stabilize ISAsomes to a similar extent as that of Laponite (Guillot et al., 2009a). Figure 6.4 shows the small-

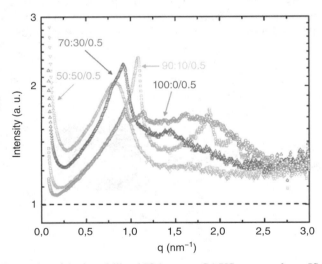

Figure 6.4 Laponite- (clay) stabilized ISAsomes. SAXS pattern from ISAsomes with varying proportions of DU and TC. Each curve is labeled with its DU:TC ratio together with the Laponite concentration: 100:0/0.5 (Pn3m, open circles), 90:10/0.5 (H₂, open squares), 70:30/0.5 (Fd3m, up triangles), and 50:50/0.5 (EME, down triangles). [From Salonen et al. (2008).]

TABLE 6.1   Effect of Stabilizers F-127 and Laponite on Internal Nanostructure (Type of LC Phase) and Radius of ISAsomes Formed from DU:TC Mixtures

DU:TC/ Stabilizer	Internal Nanostructure of ISAsomes			Average Radius of ISAsomes $\langle R \rangle$ (nm)		
	F-27	Laponite	Laponite— After 3 Weeks	F-127	Laponite	Laponite— After 3 Weeks
100:0/0.5	Pn3m + Im3m	Pn3m	H_2	116	91	88
90:10/0.5	H_2	H_2	H_2	114	93	89
70:30/0.5	Fd3m	Fd3m	Fd3m	110	92	88
50:50/0.5	L_2	L_2	L_2	99	83	81

Source: Data taken from Salonen et al. (2008).

angle X-ray scattering (SAXS) pattern of ISAsomes stabilized by 0.5 wt % of Laponite and with varying internal nanostructure as controlled by the ratio between lipid (DU) and oil (TC). The stabilization of ISAsomes was found to differ from that of O/W Pickering emulsions, where stabilization was only achieved with the addition of salt (Ashby and Binks, 2000). In contrast, ISAsome stabilization was assisted by the presence of DU and thus did not require salt addition. This also prevents the development of repulsive interactions between the clay pellets (Salonen et al., 2008).

Laponite clay as a stabilizer for ISAsomes has certain advantages over F-127. For instance, Laponite does not penetrate the ISAsomes as much as F-127; in other words, Laponite particles do not change the internal (Pn3m) nanostructure of cubosomes, in contrast with F127-stabilized ISAsomes (Table 6.1), which have Im3m nanostructure coexisting with the Pn3m. Also, Laponite-stabilized ISAsomes were smaller than those stabilized by F-127, which indicates that the effective surface area covered per gram of Laponite was higher than for F-127.

Laponite has long-term (~3 weeks) effects on the internal structure of cubosomes made from DU. The Pn3m phase is converted into hexagonal (H_2) due to Laponite-induced high pH (pH of 9–10 for <2 wt % Laponite pellets in water) (Fig. 6.5a). To separate the effect of pH from that of stabilizer type, a high pH (pH 10) was introduced into the F-127-stabilized ISAsomes. Initially coexisting Pn3m and Im3m nanostructures (Fig. 6.5c) were transformed into hexagonal structure within the same period (~3 weeks) as shown in Figure 6.5e. The phase transition can therefore be attributed to the hydrolysis of ester bonds from the monoglyceride (DU) at high pH (Salonen et al., 2008). On the other hand, if the H_2 phase transition is desired, this can be achieved without the use of an external oil additive, as is usually required (Guillot et al., 2006).

Phytantriol, however, was found to be quite stable against such phase transitions (Fig. 6.5b). Interestingly, the lattice parameters of the PT cubic phase

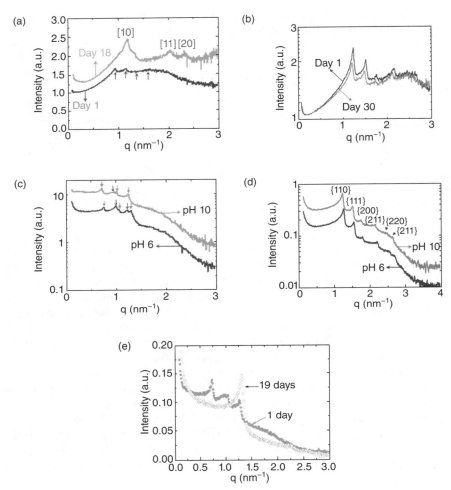

Figure 6.5   Laponite clay effect on the stability of ISAsomes. (a) Laponite-stabilized DU cubosomes after 1 day (cubic, Pn3m + Im3m) and 18 days (H₂). (b) Laponite-stabilized PT cubosomes after 1 and 30 days. The figure clearly demonstrates the stabilization by Laponite of these PT-based Pn3m cubosomes. (c) F-127-stabilized DU cubosomes at neutral pH and immediately at pH 10. (d) F-127-stabilized PT-based cubosomes; lower curve in water, upper curve at pH 10. Graphs have been shifted for the sake of clarity. Peak indices correspond to Pn3m nanostructure. (e) DU : TC 100 : 0 sample stabilized with F-127 and mixed with a buffer at pH 10. After one day, the particles display Pn3m + Im3m internal structure (●), which transforms into H₂ within 19 days (◇). Small arrows above or below the peaks indicate cubic nanostructure. [Graphs are taken or modified from Muller et al. (2010a,b) and Salonen et al. (2008).]

TABLE 6.2   Effect of Stabilizer Concentration[a]

PT:TC	Concentration of Silica Nanocolloids (wt %)	Internal Nanostructure of ISAsomes	Average Radius of ISAsomes $\langle R \rangle$ (nm)
50:50	0.5	L_2	140
50:50	1.0	L_2	286
50:50	1.75	L_2	203
100:0	0.5	Pn3m	140
100:0	1.0	Pn3m	142

[a]Dynamic light scattering (DLS) results for 100:0 and 50:50 mixtures of PT:TC as a function of silica nanocolloid concentration.

(Pn3m) were seen to increase in both Laponite-stabilized ISAsomes (Fig. 6.5b) and F-127-stabilized samples at high pH (Fig. 6.5d) (Muller et al., 2010b). In conclusion, increasing the pH can change the structure and hydration of the LC phases and release trapped molecules. For many potential applications there might be substantial benefits in stabilizing cubosomes using clay-disk-like particles instead of F-127.

In recent work, Salonen et al. (2010b) also used spherical silica nanocolloids for stabilizing ISAsomes. These colloids were 25 nm (Ludox TM 50, Du Pont) and carrying negative charge that was partially screened by sodium counterions. The pH of a 50-wt % stock solution was measured to be 8.85 at 25°C. As observed for Laponite (Table 6.1), the silica particles also affected the stability of monoglyceride-based cubosomes with Pn3m nanostructure in case of DU/TC. However, these nanocolloids were successfully utilized for stabilizing the ISAsomes in the PT/TC system (Table 6.2).

The influence of the concentration of stabilizing silica nanocolloids on the LC nanostructure was also investigated (Salonen et al., 2010b) in order to understand the fundamental difference between the stabilization pathways of different internal structures of ISAsomes and to quantify this effect (Table 6.2). Far fewer silica particles were required to stabilize cubic phases than L_2 phases of similar size. Moreover, the concentration of silica particles was also found to control the size of L_2-based ISAsomes but not that of cubosomes. Further, free nanoparticles coexisted with the cubosomes.

Very recent observations show that the addition of sodium dodecyl sulfate increases the stability of ISAsomes prepared from silica particles (diameter 18 nm) (Dulle et al., in preparation). Subsequent addition of $C_{12}E_5$ showed even more interesting results with enhanced charge and stability as well as a decrease in the free silica particles in the dispersion. ISAsomes with hexagonal nanostructure were observed under cryo–transmission electron microscopy (TEM) (Fig. 6.6) and were seen to be covered and stabilized with silica particles on the surface. Residual nonadsorbed stabilizer particles were also observed as monomers or as aggregates.

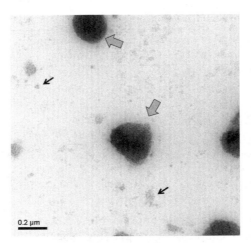

Figure 6.6   Stabilization by silica nanoparticles. Cryo-TEM image of ISAsomes with hexagonal structure stabilized by silica nanoparticles. Block arrows indicate ISAsomes covered with particulate silica, and the headed arrows show the silica particles in free or aggregated form in the solution.

6.4   TRANSFER KINETICS OF HYDROPHOBIC MOLECULES IN MIXED-ISAsome SYSTEMS

Two different ISAsome dispersions—with different nanostructures—can easily be mixed and equilibrated quite rapidly. In order to understand more about these interesting equilibration kinetics, various aspects of mixed-ISAsome systems were studied by Moitzi et al. (2007). Two differently structured ISAsomes were mixed for this purpose, and the kinetics of structural transitions was followed using a time-resolved experimental setup on a lab-based SAXS machine. The nanostructures of the ISAsomes were tuned by changing the composition of their components [TC and monolinolein (MLO)]. They were followed by using the published phase diagram (Guillot et al., 2006), which was defined by the δ value, as given by the formula shown in (Fig. 6.2).

In the simplest example, ISAsomes with H_2 (hexagonal, $\delta = 84$) and L_2 ($\delta = 57$) nanostructures were mixed in 1:1 proportions; L_2 is also called an emulsified microemulsion (EME) (Yaghmur et al., 2005). The resulting equilibrated ISAsomes should have an average composition of $\delta = 70.5$, which falls in the micellar cubic Fd3m region of the phase diagram (Guillot et al., 2006). This is exactly what was found in the time-resolved SAXS studies, where the initial phases (H_2 and EME) disappear and Fd3m forms gradually, as shown in Figure 6.7a. The SAXS signatures for corresponding phases were then fitted by a weighted sum of phase fractions (Fig. 6.7b) and were used to quantify the

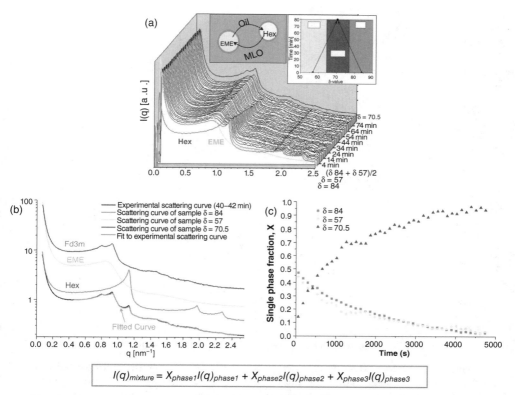

$$I(q)_{mixture} = X_{phase1}I(q)_{phase1} + X_{phase2}I(q)_{phase2} + X_{phase3}I(q)_{phase3}$$

Figure 6.7 Transfer kinetics among mixed ISAsomes: (a) Time-resolved three-dimensional stack plot of SAXS scattering curves during material exchange of H_2- and EME-based ISAsomes at 25°C. The decrease in intensities of the H_2 and EME phase signatures indicates their disappearance, while the gradual increase in intensity of the Fd3m signature indicates its formation. The compositional change (δ value) is shown schematically as an inset on the right. (b) SAXS scattering curves of various nanostructures are indicated. The experimental data were fitted to the master curve, which was calculated by using the formula shown below the graphs for $I(q)_{mixture}$. It was fitted as a weighted sum of the scattering curves of the single phases $[I(q)_{phase1}, I(q)_{phase2}, \ldots]$ using a least-squares algorithm (Moitzi et al., 2007). The weights of the best fit ($X_{phase1}, X_{phase2}, \ldots$) can be interpreted as fractions of the corresponding phases in the investigated samples. By this procedure one can follow the time-resolved changes in concentration of the different phases shown in (c).

kinetic events of disappearing and newly forming ISAsome nanostructures (Fig. 6.7c).

The hexagonal phase exhibits the sharpest peak in its SAXS scattering curves (Fig. 6.7b); the transfer kinetics were followed by tracking the changes in this peak intensity. The corresponding rate constant was then calculated by fitting the curve to a single exponential decay, which is characteristic for a

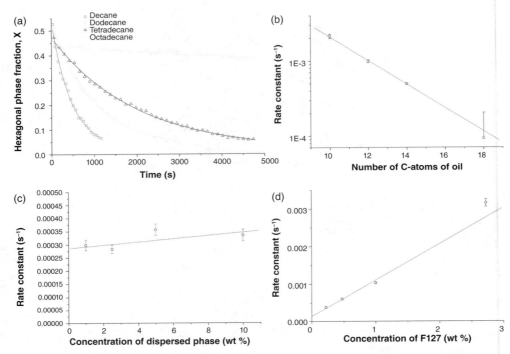

Figure 6.8   Factors affecting the transfer kinetics of ISAsomes. (a) Disappearance of the H_2 phase for various oils. The curves are used to determine the corresponding rate constants plotted in (b) against the number of C atoms in the oil used. Rate constants determined independently for varying amounts of dispersed phase (lipid or lipid + oil) and F-127 concentrations are shown in (c) and (d), respectively. [Figures (b), (c), and (d) are modified from Moitzi et al. (2007).]

first-order kinetics process. The decay rate was found to be 4.94×10^{-4} s^{-1} for the curve shown in Figure 6.7c.

Here, the principal questions were: What is the process actually leading to equilibration of the final nanostructure? Does it occur via fusion/coalescence of nanostructured droplets (ISAsomes) or via transfer of oil molecules through the aqueous medium? Studies in this direction (e.g., Moitzi et al., 2007) show that the latter process is much more likely and is supported by the following observations.

In similar experiment to that shown in Figure 6.7, various rate constants were obtained by changing the type of oil from decane to octadecane, as shown in Figure 6.8a. The rate constant decreased linearly with C-chain length of the oil molecules (Fig. 6.8b). This indicates that the transfer of oil between ISAsomes depends on the solubility of the oil in water, which decreases with increasing C-chain length of the oil molecule. Thus, the material transfer process is rate limited by the transfer of oil through the water phase.

More importantly, coalescence was expected to lead to the growth of dispersion droplets, which was not observed in the measurement of ISAsome size by DLS; on the contrary, ISAsome size remained practically constant. These experiments were performed with a three-dimensional DLS setup in order to suppress multiple scattering in the turbid dispersions. Another important observation regarding the stability of ISAsomes to coalescence was that there was no significant influence of the concentration of the dispersed phase (hydrophobic material containing lipid or lipid + oil) on the transfer rates, as shown in Figure 6.8c. The transport of oil and/or MLO among the ISAsomes is presumed to be supported by surfactant (F-127) micelles. The linear increase in rate constant with increasing concentration of added F-127 (Fig. 6.8d) is direct evidence for this fact.

The mechanism of material transfer in ISAsome systems appears to be very similar to Ostwald ripening (Kabalnov, 2001; Taylor, 2003). However, due to comparable sizes of individual ISAsomes and relatively low surface tensions, the entropy of mixing becomes the main driving force. This is a characteristic driving force for so-called compositional ripening and is usually two to three orders of magnitude larger than that typically observed for Ostwald ripening (Taisne et al., 1996).

The question that now arises is: What is being transferred, oil or MLO, or both? Mixtures of two different systems—one with no oil (i.e., plain cubosomes; $\delta = 100$) and the other with no MLO (i.e., alkane emulsion; $\delta = 0$)—were used (Salonen et al., 2010a). The resulting phases thus have the average composition of an EME ($\delta = 50$). The kinetics of formation of the EME were measured with the established protocol discussed above. However, in this case, it was more important to monitor (using DLS) the changes in respective droplet size. This was done for two different sets of experiments as follows:

In the first set of experiments, the radius of initial cubosomes was kept constant at ~110 nm, while the initial emulsion-droplet radius of the different oils was varied. The radius of the resulting EMEs (after the transfer process) was plotted against the droplet radius of the initial oil emulsions, as shown in Figure 6.9a.

In the second set of experiments, the size of the initial cubosome was varied, while the initial emulsion droplet size was kept constant for the different alkanes (110 nm for decane and 140 nm for tetradecane and octadecane). The radius of the resulting EMEs was plotted against cubosome radius, as shown in Figure 6.9b.

The rates of increase in the radius of EME droplets (after equilibration) for three different oils were obtained from a linear fit to the data, as shown in Figures 6.9a and 6.9b. These rates were plotted as a function of alkane chain length, as shown in Figure 6.10, which sheds light on the transfer mechanism. In the case of shorter C-chain oils, such as decane, the transfer of oil to cubosomes was found to be much faster than the transfer of monoglycerides to the emulsion droplets, but for less soluble oils the trend was reversed. These rates are equal at the crossover point (see Fig. 6.10) where the monoglyceride (18

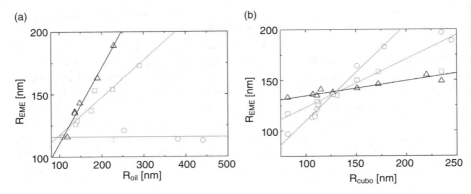

Figure 6.9   Size changes due to material transfer. Radius (R_{EME}) of the resulting EMEs as a function of (a) initial emulsion oil droplet radius for three different *n*-alkanes and (b) initial cubosome size (for a 50:50 mixture of cubosomes and oil emulsion). Data is shown for *n*-decane (circles), *n*-tetradecane (squares), and *n*-octadecane (triangles). The lines are linear fits to the data.

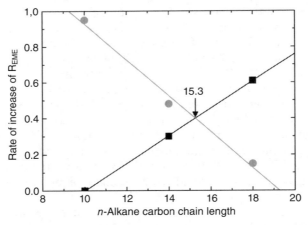

Figure 6.10   Transfer rate is affected by the length of the hydrophobic component. Rate of increase of radius (R_{EME}) as a function of *n*-alkane carbon-chain length for cubosomes (squares) and for oil emulsions (circles). [Taken from Salonen et al. (2010a).] Further interesting examples of material transfer can be found in the original publications (Moitzi et al., 2007; Salonen et al., 2010a).

C-chain lipid) behaves like an alkane with about 15.3 carbons in such a compositional ripening. However, MLO is amphiphilic due to its glycerol moiety; also, the molecule has extremely large inverse conical shape due to its two (cis) double bonds. Thus its effective chain length is reduced (14.23 Å for pure ML, according to unpublished results)(Kulkarni et al., 2010c).

There are several key aspects of the transfer kinetics between ISAsomes:

1. New ISAsomes have been created with new structures, without the aid of additional energy, such as ultrasonication, shearing, or vortexing, which is generally required for the (initial) production of ISAsomes. In contrast, the energy involved here arises simply from entropy-driven events, albeit without destroying the ISAsome signatures.

2. No coalescence or uncontrolled aggregation was observed during the mixing of two different LC nanostructured ISAsomes, despite the fact that they are all involved in transferring their internal material to each other and in equilibrating to the new type of LC-nanostructured ISAsomes. Strong evidence for this is that the sizes of the initial and final ISAsomes were practically the same (±2 nm).

3. The transfer—and thus the equilibration—of the systems was relatively fast (<2 h). This is astonishing given that the equilibration of lyotropic phases, especially viscous cubic phases (e.g., Pn3m), takes much longer, from hours to days.

4. The transfer behavior is universal with regard to the type of internal nanostructure of ISAsomes.

5. Rather than random mixing of components, the transfer events occur quite monotonically, as indicated by the first-order kinetics of the transfer (Fig. 6.7), and the dependence of the kinetics on the type of component (Fig. 6.8). Thus, it was possible to control the transfer rates by changing: (a) the carbon-chain length of oils (n-alkanes), (b) the concentration of additional stabilizer (F-127,) and (c) the type of primary ISAsome nanostructure (including EME). Currently, we are studying the effects of arrested dynamics of ISAsomes in hydrogels and the effects of stabilizer type on the rate of material transfer in ISAsomes.

6. Transfer of the hydrophobic material, that is, oil or lipid molecules (or both), occurs through the aqueous medium. We speculate that these hydrophobic molecules are transported preferentially via free F-127 micelles, which act as shuttles between the two different nanostructured ISAsomes until they equilibrate into the average nanostructure, as established by the temperature–composition phase diagram of the lipid and the oil. We are currently investigating the transfer kinetics of surfactant-free systems (i.e., Pickering emulsions) to provide evidence for this hypothesis.

The overall transfer kinetics among the ISAsome systems could function as a potential microlaboratory in which it is possible to perform controlled reactions that are otherwise too complex to follow. For instance, components A and B could be loaded into two different ISAsomes, which would allow slow, controlled, molecular-level transfer, and subsequent contact between them for further reaction. This may prove advantageous for exothermic reactions, where direct mixing of the reactants may cause rapid reaction or even explosion.

6.5   ISAsomes EMBEDDED IN THERMOREVERSIBLE POLYMERIC GELS AND FILMS

The basis for all our systems is the nanostructured liquid crystalline phase, which is oil continuous and has a high viscosity that is fixed for a given LC phase and temperature. LC phases converted into ISAsomes are water-continuous dispersions. It is possible to increase the viscosity of these dispersions by adding a certain type of hydrogelling agent. In this section we describe the embedding of ISAsomes in hydrogel networks and their further immobilization by dehydration of the hydrogels to give dry films.

Two different hydrogelling agents, namely, κ-carrageenan (KC) and methylcellulose (MC), (Tomsic et al., 2008) have been exploited for this purpose. KC is a linear sulfated polysaccharide extracted from red seaweed; whereas MC is a hydrophobically modified form of cellulose that does not occur in nature. The structural units of MC and KC are shown in Figure 6.11. Both of these polymers form thermoreversible hydrogels in aqueous medium, with or without the addition of salts. KC forms a viscous gel at room temperature, while MC undergoes a sol–gel transition at elevated temperature. Due to the fact that the two polymers have independent gelling mechanisms, it was also possible to obtain a mixed (double) gel system containing a 1:1 mixture of these gelling agents. The resulting system is a gel at lower and higher temperature with a narrow window consisting of the sol state (Tomsic et al., 2008). This narrow window facilitates the homogeneous incorporation of colloidal ISAsomes. This property could be potentially utilized to release gel-embedded functional material at specific sites under fluid (sol) conditions.

Guillot et al. (2009b) and Tomsic et al. (2009) have performed systematic studies on aqueous systems of the two hydrogelators (KC and MC) and their mixtures for incorporating various ISAsomes. Investigations focused on the changes as a function of hydrogelator concentration, type of ISAsome nano-

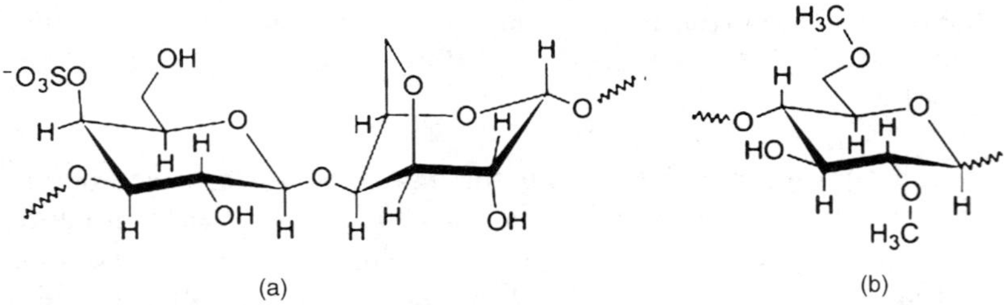

(a)   (b)

Figure 6.11   Hydrogelling agents. Structural representation of (a) KC and (b) MC building blocks.

structure (LC phase), and temperature recycling (up and down scans) (Guillot et al., 2009b; Tomsic et al., 2008, 2009). Various characterization tools, including SAXS, DLS, differential scanning calorimetry (DSC), and oscillatory rheological experiments, were utilized to demonstrate the incorporation of ISAsomes into the hydrogels.

A mere 2% gelling agent is sufficient to form quite a strong network (storage modulus, $G'\sim10^4-10^6$ Pa). The ISAsomes stay practically intact, having the same size before and after gelation and dilution (Fig. 6.12a), as confirmed by DLS (Tomsic et al., 2009). However, the phase behavior of ISAsomes was, in general, only slightly affected by the gel interactions, although it

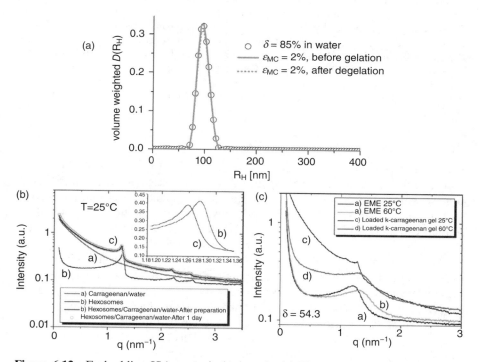

Figure 6.12 Embedding ISAsomes in hydrogels: (a) Thermoreversibility test of the integrity of hexosomes (DU/TC system, $\delta = 85$) as particles during gelation–degelation of 2% MC by DLS. The data were obtained via inverse Laplace transformation for the individual steps of the temperature cycle 25°C (sol)–70°C (gel)–25°C (sol) on 4000× diluted samples. (b) SAXS data from hexosomes (DU/LM system, $\delta = 83.3$). Inset: the scattering curve of the KC–water system was subtracted from the hexosomes–KC–water scattering curve. There was no change observed even after one day in the hexosome–KC system. The hexagonal phase remained intact; only the peak position was shifted. (c) SAXS data from EME (DU/LM system, $\delta = 54.3$) at 25 and 60°C. The EME converts into Fd3m micellar phase due to interference from KC. [Figures (a) and (b) were taken from Tomsic et al. (2009) and (c) from Guillot et al. (2009b).]

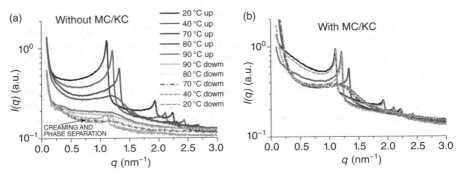

Figure 6.13 Enhanced ISAsome stability due to being embedded in hydrogels. SAXS studies with a stepwise temperature scan of 20–90–20°C on the DU/TC system with $\delta = 85$ and $\phi = 5\%$ of the dispersed phase. (a) Free ISAsomes (without gelator) show phase separation due to creaming at higher temperatures; during the downward scan, the curves do not show the same appearance as during the upward scan (see lower set of broken curves). (b) With 2% gelator (MC:KC 1:1), the absence of creaming can be clearly observed from the downward scan, thus revealing the high-temperature stabilizing effect of the MC–KC (1:1) polymer mixture on the ISAsomes.

occasionally showed more significant effects (Guillot et al., 2009b). Hexosomes (DU/LM system, $\delta = 83.3$) in KC were investigated using SAXS (Fig. 6.12b), which showed that the lattice parameter of the hexagonal nanostructure increased from 5.65 to 5.75 nm. However, in a further study, the KC-loaded EME (DU/LM system, $\delta = 54.3$) was converted into the micellar cubic Fd3m phase at 25°C, as shown by curve c in Figure 6.12c.

An important advantage of the incorporation of ISAsomes into such gelled polymer systems is the increase in their stability. This is particularly true at higher temperature, when free-ISAsome systems undergo rather rapid creaming (Fig. 6.13a). The addition of hydrogelators reduces the temperature-induced phase separation of ISAsomes, as shown by the temperature scans on ISAsome-loaded gel systems in Figure 6.13b. This can be of particular importance during formulation of more complex systems.

ISAsomes in polymer hydrogels are a fascinating example of structural hierarchy: LC nanostructures form the internal architecture of submicron-sized ISAsome particles, which are in turn entrapped in a macromolecular (polymer) hydrogel network. These systems have some important differences from the bulk lyotropic mesophases: First, they are still continuous in the aqueous phase, with a water content of up to 90% or more, but at the same time they contain a stabilized LC phase inside the ISAsomes. In contrast, bulk phases cannot hold such high amounts of water. Second, the viscoelastic properties of bulk phases can only be controlled by changing the temperature and/or the composition of the LC phase itself, whereas ISAsome–gel systems can be controlled by changing the concentration and type of hydrogelator, the concentration of the dispersed phase, and/or the temperature. This allows a wide range of tuning parameters, resulting in a rather broad regime of viscoelastic properties. In most of the observations related to entrapment of ISAsomes in hydrogels, the nanostructures and sizes of ISAsomes are quite thermoreversible, and they equilibrate rapidly; on the contrary, bulk LC phases undergo slow transitions that are more often metastable, especially in the cooling direction.

The preparation of ISAsomes by the shearing technique can be difficult, primarily due to the large difference in viscosities of the components (oil phase and water phase). It has been observed that the addition of hydrogelator MC to the aqueous phase affords more comparable viscosities, which facilitates mixing at high temperature. This, in turn, leads to stable ISAsome solutions at room temperature (unpublished results). A further advantage of hydrogel formation in ISAsome systems is the possibility of controlling (more specifically, reducing) the rate of interparticle material transfer, as shown by our results. In addition, we have investigated the arrested dynamics of ISAsomes in such hydrogel networks by multispeckle DLS (data not shown) (Iglesias et al., in preparation).

ISAsome–hydrogel systems can be dried into thin films of variable characteristics (Fig. 6.14a) (Kulkarni et al., 2011a). Films were formed from ISAsome-loaded KC and MC gels upon drying at low or high temperature, respectively. The film thickness was found to increase with the amount of ISAsome loading. Due to dehydration, the dried hydrogel network becomes compact and thereby immobilizes the ISAsomes (Fig. 6.14b). The ISAsomes can, however, be remobilized upon rehydration of such films. The ISAsome-loaded films were transparent (less transparent for increased loading) and became turbid upon rehydration, as shown in Figure 6.14c. Details of rehydration and ISAsome restoration are explained in the following discussion.

During the drying of the loaded film, the ISAsomes also lose their water, and the original H_2 nanostructure is converted into L_2 (Fig. 6.15a). Nevertheless, the original nanostructure (H_2) is restored upon rehydration of the film. Using time-resolved SAXS, we were able to track the nanostructural changes in the ISAsome system upon partial hydration (equal amount of water added to the dried material) of a loaded KC film. The ISAsomes rapidly (<300 s) take up the water and regain their original H_2 nanostructure, which is then stabilized (900 s) (Fig. 6.15b). Partially hydrated film is essentially thermoreversible, as shown in Figure 6.15c. The size of the ISAsomes remains practically the same upon loading into the hydrogel network and subsequent drying into film form, as indicated by DLS measurements performed on the diluted (1000×) mixtures of ISAsomes, resolubilized loaded gel and resolubilized loaded dry film (Fig. 6.15d). Solid films containing ISAsomes are promising systems for the handling and storage of ISAsomes, as well as for potential materials science applications that require functional surfaces.

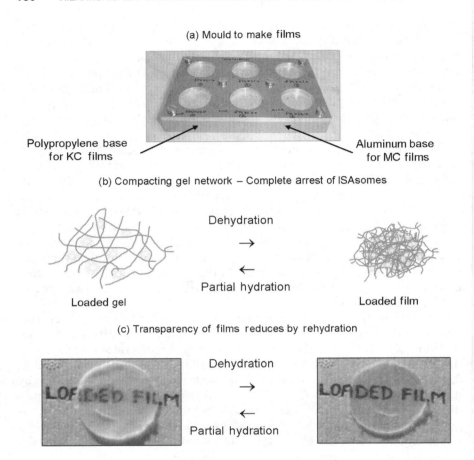

Figure 6.14   Formation of solid films from ISAsome-loaded hydrogels: (a) A 6-well plate mold that generates hydrogel films; each well has 30 mm diameter and 5 mm depth. About 3 ml hydrogel mixture was placed in each well. When dried, this afforded a thin film of ca. <0.5 mm thickness. Polypropylene and aluminum bases were used for drying KC and MC films, respectively, at lower and higher temperature. (b) Illustration of the compacting of the hydrogel network, which immobilizes the ISAsomes. (c) Rehydration restores the ISAsomes, as indicated by the increased turbidity of the film.

6.6   WATER-IN-OIL NANOSTRUCTURED EMULSIONS WITH LIQUID CRYSTALLINE OIL PHASES

As discussed above, ISAsomes are kinetically stabilized O/W emulsions that require some kind of stabilizer extra to the primary lipid–surfactant component. ISAsomes are primarily dispersed LC phases in water that take

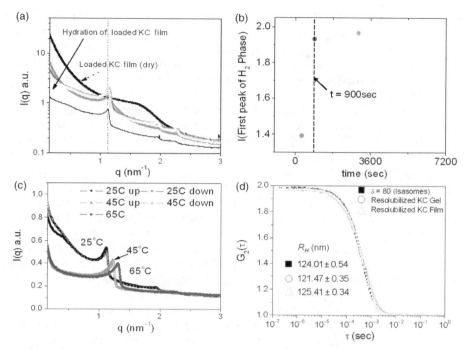

Figure 6.15 Rehydration of hydrogel films to restore ISAsomes. (a) Thick solid curve shows a broad peak from the L_2 nanostructure formed due to water loss from the H_2-nanostructured ISAsomes. Rehydration of the KC film leads to restoration of H_2 nanostructure, indicated by the sharp peak. The H_2 nanostructure remains essentially the same size upon hydration (as shown by the dotted line; lattice parameter 6.49 nm) as it is in the loaded gel (thin curve with lower intensity). (b) The intensity of the first peak of the H_2 nanostructure [from (a)] plotted against film hydration time. The absorption of water to restore H_2 nanostructure is very fast (300 s; first point) and is complete in about 900 s, after which there is very little change in the peak intensity. (c) The upward and downward temperature scans on partially hydrated KC film show thermoreversibility with respect to the nanostructural dimensions of the ISAsomes (6.49 nm at 25°C, 5.93 nm at 45°C, and 5.53 nm at 65°C). (d) Resolubilization and dilution studies by DLS indicate that the ISAsomes stay practically intact (with respect to their size) during their entrapment and immobilization in gel and film.

particulate (colloidal) form. However, it is also possible to form reverse (W/O) emulsions, where the dispersed phase is water and the continuous phase can be various LC phases (Caldero et al., 2009; Solans and Esquena, 2008; Solans et al., 2001, 2003). These are usually concentrated emulsions,[5] the stability of which is also usually governed by an external stabilizing agent. However, we

---

[5]Concentrated emulsions are either stable due to emulsion stabilizer or due to a very high concentration of internal phase or both. When the dispersed phase is above 74%, they are called HIPRE (high internal phase ratio emulsions) or HIPE (high internal phase emulsions), or alternatively gel emulsions or bi-liquid foams.

have in fact prepared W/O emulsions without any external stabilizer (Kulkarni et al., 2010a,b). In addition, it was possible to vary the amount of dispersed water (from 50 to ~90%) as well as the type of LC nanostructure (e.g., cubic—Pn3m or Fd3m and H_2). The formation and characterization of these W/O nanostructured emulsions (Kulkarni et al., 2010b) are discussed in the following.

Water-in-oil nanostructured emulsions, hereafter nanostructured emulsions, were prepared using a custom-made shearing device based on the Couette cell, details of which have been published elsewhere (Salentinig et al., 2008). The oil phase (lipid or lipid+oil) and aqueous phase (water or water+hydrogelator) were injected simultaneously from the bottom into a prechamber and mixed using a propeller-type device. The resultant raw emulsion was then pushed into the main Couette shear cell with vertical axis, where the raw emulsion was converted into a nanostructured emulsion. The whole ensemble was thermally controlled using a circulating water bath. The temperature was set such that the corresponding LC phase was in the molten state (L_2 phase). Shear rates were in the range of 31,000–78,500 s^{-1}; the higher values were required for emulsions with high water content (e.g., 90% water) (Kulkarni et al., 2010b).

These nanostructured emulsions also present a high level of structural hierarchy and have been characterized using a wide range of techniques. The type and dimensions of LC nanostructures were determined from SAXS studies (data not shown). Polarizing optical microscopy and confocal microscopy were utilized to examine microstructures involving various sizes of water droplets (Fig. 6.16). Confocal microscopy was also capable of showing the lipid network, which was stained by Nile Red dye (Sigma Aldrich, St. Louis, MO).

There is a very strong link between the nano/microstructure and the properties of nanostructured emulsions, which facilitates their wider tunability. The rheological properties of W/O-nanostructured emulsions can be modulated by changing the composition of the components (δ and ϕ), temperature (T), and the concentration of hydrogelling agent (ε) (Fig. 6.17). The viscoelastic properties of these emulsions were investigated by using a strain-controlled rheometer (Anton Paar Physics UDS 200, Graz, Austria).

It was also possible to form a hydrogel in the water reservoirs of these nanostructured emulsions. This characteristic also allowed us to increase the stability against water separation (particularly for the bicontinuous Pn3m nanostructure) and also increase the viscosity of emulsions. In general, concentrated emulsions are stable if the volume fraction of an internal phase is above the critical value[6] of 0.74; otherwise, they need an external stabilizer.

---

[6] The number 0.74 is a volume fraction of monodispersed spherical droplets or particles resulting into the most compact arrangement, which governs the stability. This number may vary for deformed or polydisperse spheres. It is also called a critical volume fraction.

Nanostructured emulsion based on H_2 ($\phi = 80$, $\delta = 80$)

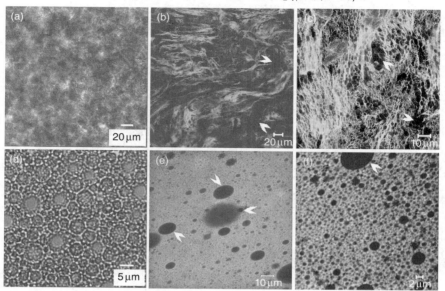

Nanostructured emulsion based on L_2 ($\phi = 80$, $\delta = 50$)

Figure 6.16 Microstructural investigations of nanostructured emulsions. Optical micrographs (a) and (d) show birefringence from the lipid phase. Confocal micrographs clearly show the lipid network (in light gray) in H_2 [(b) and (c)], and EME [(e) and (f) nanostructured emulsions. In images (e) and (f) the network appears denser than in (d) because of hydrogel formation in the aqueous regions of the EME-based emulsion. Arrows indicate droplet inhomogeneities, which represent large water droplets (~20 μm). The water droplets responsible for emulsion stability [as in (d)] are in the range of 2–8 μm in size.

In our case, the nanostructured emulsions were stable above 74% water, but for lower values of the dispersed phase, the stability is presumed to be given by the LC phase itself (being rather viscous). Only in the case of EME as the structure-forming unit was the stability below 80% water too low, leaving the nanostructured emulsions more prone to phase—especially oil—separation.

The most novel things about these W/O-nanostructured emulsions are: (1) they do not require an external stabilizing agent, and (2) their structural hierarchy (Fig. 6.18) can be tuned on a very broad scale. Thus, these emulsions have very good potential for applications in materials science and biotechnology.

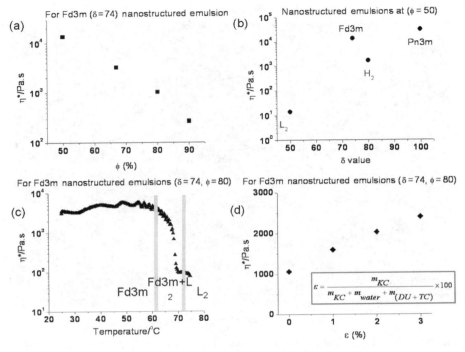

Figure 6.17   Modulating the rheological properties of nanostructured emulsions. The complex viscosity (η^*) was determined from the strain-controlled rheometer. It could be modulated by changing (a) the lipid-phase composition (δ), (b) the concentration of the dispersed phase, that is, water (ϕ), (c) the temperature (which resulted in phase transition), and (d) concentration of the hydrogelling agent (ε).

6.7 SUMMARY

This chapter starts with a discussion on the diversity of LC phases driven by certain parameters; for instance, temperature variation or addition of oil can increase the inverse conical shape of a molecule, which stabilizes highly curved LC phases. In order to increase their applicability it is customary to reorganize such LC phases into some type of hierarchically ordered structures. The O/W or W/O type of emulsions are good options because they enable very broad tunability, which can enhance their utility for specific applications. These emulsions are simply kinetically stabilized LC phases that, in the particulate form, are called ISAsomes and in the continuous-film form (holding large amount of water) are called nanostructured emulsions.

The second section deals with the preparation and characterization of ISAsomes systems. These systems show a high degree of structural tunability based on (1) the composition of the hydrophobic phase (δ) (Guillot et al., 2006); (2)

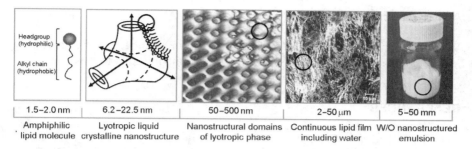

1.5–2.0 nm	6.2–22.5 nm	50–500 nm	2–50 μm	5–50 mm
Amphiphilic lipid molecule	Lyotropic liquid crystalline nanostructure	Nanostructural domains of lyotropic phase	Continuous lipid film including water	W/O nanostructured emulsion

Figure 6.18 Structural hierarchy of W/O-nanostructured emulsions. Represented by various length scales.

the amount of dispersed phase (ϕ); (3) temperature-induced nanostructural interconversions (Guillot et al., 2006; Muller et al., 2010a); (4) particle size (given by R_H) (Guillot et al., 2010; Salentinig et al., 2008); (5) the amount and type of stabilizer (Guillot et al., 2009a, 2010; Muller et al., 2010a; Salonen et al., 2010b); (6) the type of lipid (Muller et al., 2010b; Nguyen et al., 2010); (7) pH (Muller et al., 2010a; Salentinig et al., 2010); (8) the amount and type of additive [e.g., R (+) limonene (Guillot et al., 2006) or oleic acid (Salentinig et al., 2010)]; (9) pressure-induced behavioral changes (Yaghmur et al., 2010); (10) the method of preparation (Salentinig et al., 2008), that is, ultrasonication or shearing; and (11) intermixing of two differently structured ISAsomes (Moitzi et al., 2007; Salonen et al., 2010a). Some of these aspects are described in the second and third sections. These properties (among others) facilitate the modulation of ISAsome systems for their potential applications in a wide range of fields.

Interparticle transfer among mixed ISAsomes renders smart systems possible where the material transfer could be performed in a controlled manner and the rate of transfer systematically modulated, as presented in Section 6.4. Initial addition of hydrogelling agents reduces the diffusion dynamics of ISAsomes, while further addition enables ISAsome entrapment followed by immobilization upon drying, which is discussed in Section 6.5. ISAsome–polymer matrix systems are good candidates for housing various functional molecules for transport or storage under dry conditions in the form of loaded hydrogel films.

In the final section of the chapter, we presented inverse (W/O) emulsions prepared from LC phases (in contrast to ISAsome dispersions, which are O/W emulsions). Both types presented here are good examples of hierarchically ordered structures that originate from lyotropic LC phases. Hydrogelled aqueous regions of these emulsions demonstrate further possibilities in terms of tuning their properties and hence increasing their performance in applied science and technology.

ACKNOWLEDGMENTS

We would like to thank all the current and former members of the Scattering Methods Group, University of Graz, for their contributions to this work.

REFERENCES

Agrawal, V., Kulkarni, C. V., Deshmukh, Y., and Guruswamy, K. (2004). Synthesis of "designer" nanofillers with tailored surface functionalities. In MACRO 2004, International Conference on Polymers for Advanced Technologies. Thiruvananthapuram, India.

Angelov, B., Angelova, A., Mutafchieva, R., Lesieur, S., Vainio, U., Garamus, V. M., Jensen, G. V., and Pedersen, J. S. (2011). SAXS investigation of a cubic to a sponge (L3) phase transition in self-assembled lipid nanocarriers. *Physical Chemistry Chemical Physics, 13*, 3073–3081.

Armitage, B., Bennett, D., Lamparski, H., and O'Brien, D. (1996). Polymerization and domain formation in lipid assemblies. In *Biopolymers Liquid Crystalline Polymers Phase Emulsion*. Springer, Berlin/Heidelberg, Vol. 126, pp. 53–84.

Ashby, N. P., and Binks, B. P. (2000). Pickering emulsions stabilised by Laponite clay particles. *Physical Chemistry Chemical Physics, 2*(24), 5640–5646.

Aveyard, R., Binks, B. P., and Clint, J. H. (2003). Emulsions stabilised solely by colloidal particles. *Advances in Colloid and Interface Science, 100–102*, 503–546.

Binks, B. P., and Lumsdon, S. O. (2000). Influence of particle wettability on the type and stability of surfactant-free emulsions. *Langmuir, 16*(23), 8622–8631.

Caffrey, M. (2003). Membrane protein crystallization. *Journal of Structural Biology, 142*(1), 108–132.

Caldero, G., Llinas, M., Garcia-Celma, M. J., and Solans, C. (2009). Studies on controlled release of hydrophilic drugs from W/O high internal phase ratio emulsions. *Journal of Pharmaceutical Sciences, 99*(2), 701–711.

Cates, M. E., and Clegg, P. S. (2008). Bijels: A new class of soft materials. *Soft Matter, 4*(11), 2132–2138.

Clogston, J., and Caffrey, M. (2005). Controlling release from the lipidic cubic phase. Amino acids, peptides, proteins and nucleic acids. *Journal of Controlled Release, 107*(1), 97–111.

Cui, Z. G., Yang, L. L., Cui, Y. Z., and Binks, B. P. (2010). Effects of surfactant structure on the phase inversion of emulsions stabilized by mixtures of silica nanoparticles and cationic surfactant. *Langmuir, 26*(7), 4717–4724.

Dai, L. L., Tarimala, S., Wu, C. Y., Guttula, S., and Wu, J. (2008). The structure and dynamics of microparticles at Pickering emulsion interfaces. *Scanning, 30*(2), 87–95.

de Campo, L., Yaghmur, A., Sagalowicz, L., Leser, M. E., Watzke, H., and Glatter, O. (2004). Reversible phase transitions in emulsified nanostructured lipid systems. *Langmuir, 20*(13), 5254–5261.

Dickinson, E. (2009). Hydrocolloids as emulsifiers and emulsion stabilizers. *Food Hydrocolloids, 23*(6), 1473–1482.

Dong, Y. D., Tilley, A. J., Larson, I., Lawrence, M. J., Amenitsch, H., Rappolt, M., Hanley, T., and Boyd, B. J. (2010). Nonequilibrium effects in self-assembled mesophase materials: Unexpected supercooling effects for cubosomes and hexosomes. *Langmuir, 26*(11), 9000–9010.

Dulle, M., Ahualli, S., Iglesias, G. R., Pirolt, F., and Glatter, O. (in preparation).

Engelskirchen, S., Maurer, R., and Glatter, O. (2011). Effect of glycerol addition on the internal structure and thermal stability of hexosomes prepared from phytantriol. *Colloids and Surfaces A: Physicochemical and Engineering Aspects, 391*(1–3), 95–100.

Faham, S., and Bowie, J. U. (2002). Bicelle crystallization: A new method for crystallizing membrane proteins yields a monomeric bacteriorhodopsin structure. *Journal of Molecular Biology, 316*(1), 1–6.

Forney, B. S., and Guymon, C. A. (2010). Nanostructure evolution during photopolymerization in lyotropic liquid crystal templates. *Macromolecules, 43*(20), 8502–8510.

Garg, G., Saraf, S., and Saraf, S. (2007). Cubosomes: An overview. *Biological & Pharmaceutical Bulletin, 30*(2), 350–353.

Geraghty, P. B., Attwood, D., Collett, J. H., Sharma, H., and Dandiker, Y. (1997). An investigation of the parameters influencing the bioadhesive properties of Myverol 18–99/water gels. *Biomaterials, 18*, 63–67.

Guillot, S., Moitzi, C., Salentinig, S., Sagalowicz, L., Leser, M. E., and Glatter, O. (2006). Direct and indirect thermal transitions from hexosomes to emulsified microemulsions in oil-loaded monoglyceride-based particles. *Colloids and Surfaces A: Physicochemical and Engineering Aspects, 291*(1–3), 78–84.

Guillot, S., Bergaya, F., de Azevedo, C., Warmont, F., and Tranchant, J. F. (2009a). Internally structured pickering emulsions stabilized by clay mineral particles. *Journal of Colloid Interface Science, 333*(2), 563–569.

Guillot, S., Tomsic, M., Sagalowicz, L., Leser, M. E., and Glatter, O. (2009b). Internally self-assembled particles entrapped in thermoreversible hydrogels. *Journal of Colloid and Interface Science, 330*(1), 175–179.

Guillot, S., Salentinig, S., Chemelli, A., Sagalowicz, L., Leser, M. E., and Glatter, O. (2010). Influence of the stabilizer concentration on the internal liquid crystalline order and the size of oil-loaded monolinolein-based dispersions. *Langmuir, 26*(9), 6222–6229.

Gupta, R., and Bhattacharya, S. (2008). Polymer-clay nanocomposites: Current status and challenges. *Indian Chemical Engineer, 50*(3), 242–267.

Gustafsson, J., Ljusberg-Wahren, H., Almgren, M., and Larsson, K. (1996). Cubic lipid–water phase dispersed into submicron particles. *Langmuir, 12*(20), 4611–4613.

Gustafsson, J., Ljusberg-Wahren, H., Almgren, M., and Larsson, K. (1997). Submicron particles of reversed lipid phases in water stabilized by a nonionic amphiphilic polymer. *Langmuir, 13*(26), 6964–6971.

Hirlekar, R., Jain, S., Patel, M., Garse, H., and Kadam, V. (2010). Hexosomes: A novel drug delivery system. *Current Drug Delivery, 7*, 28–35.

Hunter, T. N., Pugh, R. J., Franks, G. V., and Jameson, G. J. (2008). The role of particles in stabilising foams and emulsions. *Advances in Colloid Interface Science, 137*(2), 57–81.

Huynh, N. T., Passirani, C., Saulnier, P., and Benoit, J. P. (2009). Lipid nanocapsules: A new platform for nanomedicine. *International Journal of Pharmaceutics, 379*(2), 201–209.

Hyde, S. T., Andersson, S., Larsson, K., Blum, Z., Landh, T., Lidin, S., and Ninham, B. W. (1997). *The Language of Shape: The Role of Curvature in Condensed Matter: Physics, Chemistry and Biology*. Elsevier Science, Amsterdam.

Iglesias, G. R., Pirolt, F., Kulkarni, C. V., and Glatter, O. (in preparation). Controlling the rate of material transfer among internally nanostructures lipid particles.

Kabalnov, A. (2001). Ostwald ripening and related phenomena. *Journal of Dispersion Science and Technology, 22*(1), 1–12.

Kulkarni, C. V. (2009). *In Cubo Crystallization of Membrane Proteins*. University of London, London.

Kulkarni, C. V. (2011). Nanostructural studies on monoelaidin-water systems at low temperatures. *Langmuir, 27*(19), 11790–11800.

Kulkarni, C. V., Ces, O., and Templer, R. H. (2007). Monoelaidin-water phase behaviour with temperature. In European Biophysics Congress, 14–18th July, 2007, London, United Kingdom.

Kulkarni, C. V., Ces, O., and Templer, R. H. (2009). Mesosphere behavior of a "trans" conformational lipid molecule. In 23rd Conference of the European Colloid and Interface Society, 6–11 Sept. 2009, Antalya, Turkey.

Kulkarni, C. V., Iglesias, G. R., and Glatter, O. (2010a). Nanostructured water in oil emulsions with high water content formed without stabilizer. In International Soft Matter Conference July 5–8, 2010. Granada, Spain.

Kulkarni, C. V., Mezzenga, R., and Glatter, O. (2010b). Water-in-oil nanostructured emulsions: Towards the structural hierarchy of liquid crystalline materials. *Soft Matter, 6*(21), 5615–5624.

Kulkarni, C. V., Tang, T. Y., Seddon, A. M., Seddon, J. M., Ces, O., and Templer, R. H. (2010c). Engineering bicontinuous cubic structures at the nanoscale—the role of chain splay. *Soft Matter, 6*, 3191–3194.

Kulkarni, C. V., Tomsic, M., and Glatter, O. (2011a). Immobilization of nanostructured lipid particles in polysaccharide films. *Langmuir, 27*(15), 9541–9550.

Kulkarni, C. V., Wachter, W., Iglesias, G. R., Engelskirchen, S., and Ahualli, S. (2011b). Monoolein: A magic lipid? *Physical Chemistry Chemical Physics (PCCP), 13*, 3004–3021.

Landau, E. M., and Rosenbusch, J. P. (1996). Lipidic cubic phases: A novel concept for the crystallization of membrane proteins. *PNAS, 93*(25), 14532–14535.

Lee, K. W., Nguyen, T. H., Hanley, T., and Boyd, B. J. (2009). Nanostructure of liquid crystalline matrix determines in vitro sustained release and in vivo oral absorption kinetics for hydrophilic model drugs. *International Journal of Pharmaceutical, 365*(1–2), 190–199.

Leser, M. E., Sagalowicz, L., Michel, M., and Watzke, H. J. (2006). Self-assembly of polar food lipids. *Advances in Colloid and Interface Science, 123–126*, 125–136.

Mezzenga, R., Schurtenberger, P., Burbidge, A., and Michel, M. (2005). Understanding foods as soft materials. *Nature Materials, 4*(10), 729–740.

Misquitta, Y., and Caffrey, M. (2001). Rational design of lipid molecular structure: A case study involving the C19:1c10 monoacylglycerol. *Biophysical Journal, 81*(2), 1047–1058.

Moitzi, C., Guillot, S., Fritz, G., Salentinig, S., and Glatter, O. (2007). Phase reorganization in self-assembled systems through interparticle material transfer. *Advanced Materials, 19*(10), 1352–1358.

Muller, F., Salonen, A., and Glatter, O. (2010a). Monoglyceride-based cubosomes stabilized by Laponite: Separating the effects of stabilizer, pH and temperature. *Colloids and Surfaces A: Physicochemical and Engineering Aspects, 358*(1–3), 50–56.

Muller, F., Salonen, A., and Glatter, O. (2010b). Phase behavior of Phytantriol/water bicontinuous cubic Pn3m cubosomes stabilized by Laponite disc-like particles. *Journal of Colloid and Interface Science, 342*(2), 392–398.

Muller, F., Salonen, A., Dulle, M., and Glatter, O. (2011). Salt-induced behavior of internally self-assembled nanodrops: Understanding stabilization by charged colloids. In V. Starov (Ed.), *Progress in Colloid and Polymer Science, Vol. 138: Trends in Colloid and Interface Science* (Vol. XXIV): Springer, Berlin.

Nguyen, T. H., Hanley, T., Porter, C. J., Larson, I., and Boyd, B. J. (2010). Phytantriol and glyceryl monooleate cubic liquid crystalline phases as sustained-release oral drug delivery systems for poorly water soluble drugs I. Phase behaviour in physiologically-relevant media. *Journal of Pharmacy and Pharmacology, 62*(7), 844–855.

Noble, P. F., Cayre, O. J., Alargova, R. G., Velev, O. D., and Paunov, V. N. (2004). Fabrication of "hairy" colloidosomes with shells of polymeric microrods. *Journal of the American Chemical Society, 126*(26), 8092–8093.

Ohm, C., Brehmer, M., and Zentel, R. (2010). Liquid crystalline elastomers as actuators and sensors. *Advanced Materials, 22*(31), 3366–3387.

Pickering, S. U. (1907). Emulsions. *Journal of Chemical Society, 91*, 2001.

Ramsden, W. (1903). Separation of solids in the surface-layers of solutions and "suspensions" (observations on surface-membranes, bubbles, emulsions, and mechanical coagulation)—Preliminary account. *Proceedings of the Royal Society of London, 72*(477–486), 156–164.

Rizwan, S. B., Boyd, B. J., Rades, T., and Hook, S. (2010). Bicontinuous cubic liquid crystals as sustained delivery systems for peptides and proteins. *Expert Opinion on Drug Delivery, 7*, 1133–1144.

Sagalowicz, L., Leser, M. E., Watzke, H. J., and Michel, M. (2006a). Monoglyceride self-assembly structures as delivery vehicles. *Trends in Food Science & Technology, 17*(5), 204–214.

Sagalowicz, L., Mezzenga, R., and Leser, M. E. (2006b). Investigating reversed liquid crystalline mesophases. *Current Opinion in Colloid & Interface Science, 11*(4), 224–229.

Sagalowicz, L., Michel, M., Adrian, M., Frossard, P., Rouvet, M., Watzke, H. J., Yaghmur, A., de Campo, L., Glatter, O., and Leser, M. E. (2006c). Crystallography of dispersed liquid crystalline phases studied by cryo-transmission electron microscopy. *Journal of Microscopy, 221*(Pt 2), 110–121.

Saigal, T., Dong, H., Matyjaszewski, K., and Tilton, R. D. (2010). Pickering emulsions stabilized by nanoparticles with thermally responsive grafted polymer brushes. *Langmuir, 26*(19), 15200–15209.

Salentinig, S., Yaghmur, A., Guillot, S., and Glatter, O. (2008). Preparation of highly concentrated nanostructured dispersions of controlled size. *Journal of Colloid and Interface Science, 326*(1), 211–220.

Salentinig, S., Sagalowicz, L., and Glatter, O. (2010). Self-assembled structures and pKa value of oleic acid in systems of biological relevance. *Langmuir*, *26*(14), 11670–11679.

Salentinig, S., Sagalowicz, L., Leser, M. E., Tedeschi, C., and Glatter, O. (2011). Transitions in the internal structure of lipid droplets during fat digestion. *Soft Matter*, *7*(2), 650–661.

Salonen, A., Guillot, S., and Glatter, O. (2007). Determination of water content in internally self-assembled monoglyceride-based dispersions from the bulk phase. *Langmuir*, *23*(18), 9151–9154.

Salonen, A., Muller, F. O., and Glatter, O. (2008). Dispersions of internally liquid crystalline systems stabilized by charged disklike particles as Pickering emulsions: Basic properties and time-resolved behavior. *Langmuir*, *24*(10), 5306–5314.

Salonen, A., Moitzi, C., Salentinig, S., and Glatter, O. (2010a). Material transfer in cubosome-emulsion mixtures: Effect of alkane chain length. *Langmuir*, *26*(13), 10670–10676.

Salonen, A., Muller, F. O., and Glatter, O. (2010b). Internally self-assembled submicrometer emulsions stabilized by spherical nanocolloids: Finding the free nanoparticles in the aqueous continuous phase. *Langmuir*, *26*(11), 7981–7987.

Seddon, J. M. (1990). Structure of the inverted hexagonal (HII) phase, and non-lamellar phase transitions of lipids. *Biochimica et Biophysica Acta (BBA)—Reviews on Biomembranes*, *1031*(1), 1–69.

Seddon, J., Squires, A., Conn, C., Ces, O., Heron, A., Mulet, X., Shearman, G., and Templer, R. (2006). Pressure-jump X-ray studies of liquid crystal transitions in lipids. *Philosophical Transactions of the Royal Society A: Mathematical, Physical and Engineering Sciences*, *364*(1847), 2635–2655.

Sennoga, C., Heron, A., Seddon, J. M., Templer, R. H., and Hankamer, B. (2003). Membrane-protein crystallization in cubo: Temperature-dependent phase behaviour of monoolein-detergent mixtures. *Acta Crystallographica Section D—Biological Crystallography*, *59*, 239–246.

Sergienko, A. Y., Tai, H., Narkis, M., and Silverstein, M. S. (2002). Polymerized high internal-phase emulsions: Properties and interaction with water. *Journal of Applied Polymer Science*, *84*(11), 2018–2027.

Shah, J., Sadhale, Y., and Chilukuri, D. M. (2001). Cubic phase gels as drug delivery systems. *Advanced Drug Delivery Reviews*, *47*(2–3), 229–250.

Shearman, G. C., Ces, O., and Templer, R. H. (2010). Towards an understanding of phase transitions between inverse bicontinuous cubic lyotropic liquid crystalline phases. *Soft Matter*, *6*(2), 256–262.

Solans, C., and Esquena, J. (2008). Highly concentrated (Gel) emulsions as reaction media for the preparation of advanced materials. In P. D. E. Platikanov (Ed.), *Highlights in Colloid Science*, Wiley-VCH, Weinheim, Germany, pp. 291–297.

Solans, C., Pinazo, A., Calderó, G., and Infante, M. R. (2001). Highly concentrated emulsions as novel reaction media. *Colloids and Surfaces A: Physicochemical and Engineering Aspects*, *176*(1), 101–108.

Solans, C., Esquena, J., and Azemar, N. (2003). Highly concentrated (gel) emulsions, versatile reaction media. *Current Opinion in Colloid & Interface Science*, *8*(2), 156–163.

Spicer, P. T., Hayden, K. L., Lynch, M. L., Ofori-Boateng, A., and Burns, J. L. (2001). Novel process for producing cubic liquid crystalline nanoparticles (Cubosomes). *Langmuir*, *17*(19), 5748–5756.

Spicer, P. T., Small, W. B., Lynch, M. L., and Burns, J. L. (2002). Dry powder precursors of cubic liquid crystalline nanoparticles (cubosomes). *Journal of Nanoparticle Research*, *4*, 297–311.

Srisiri, W., Benedicto, A., O'Brien, D. F., Trouard, T. P., Orädd, G., Persson, S., and Lindblom, G. (1998). Stabilization of a bicontinuous cubic phase from polymerizable monoacylglycerol and diacylglycerol. *Langmuir*, *14*(7), 1921–1926.

Taisne, L., Walstra, P., and Cabane, B. (1996). Transfer of oil between emulsion droplets. *Journal of Colloid and Interface Science*, *184*(2), 378–390.

Taylor, P. (2003). Ostwald ripening in emulsions: estimation of solution thermodynamics of the disperse phase. *Advances in Colloid and Interface Science*, *106*, 261–285.

Tomsic, M., Guillot, S., Sagalowicz, L., Leser, M. E., and Glatter, O. (2009). Internally self-assembled thermoreversible gelling emulsions: ISAsomes in methylcellulose, kappa-carrageenan, and mixed hydrogels. *Langmuir*, *25*(16), 9525–9534.

Tomsic, M., Prossnigg, F., and Glatter, O. (2008). A thermoreversible double gel: Characterization of a methylcellulose and [kappa]-carrageenan mixed system in water by SAXS, DSC and rheology. *Journal of Colloid and Interface Science*, *322*(1), 41–50.

Ubbink, J., Burbidge, A., and Mezzenga, R. (2008). Food structure and functionality—a soft matter perspective. *Soft Matter*, *4*(8), 1569–1581.

Yaghmur, A., de Campo, L., Sagalowicz, L., Leser, M. E., and Glatter, O. (2005). Emulsified microemulsions and oil-containing liquid crystalline Phases. *Langmuir*, *21*(2), 569–577.

Yaghmur, A., de Campo, L., Sagalowicz, L., Leser, M. E., and Glatter, O. (2006a). Control of the internal structure of MLO-based isasomes by the addition of diglycerol monooleate and soybean phosphatidylcholine. *Langmuir*, *22*(24), 9919–9927.

Yaghmur, A., deCampo, L., Salentinig, S., Sagalowicz, L., Leser, M. E., and Glatter, O. (2006b). Oil-loaded monolinolein-based particles with confined inverse discontinuous cubic structure (Fd3m). *Langmuir*, *22*(2), 517–521.

Yaghmur, A., and Glatter, O. (2009). Characterization and potential applications of nanostructured aqueous dispersions. *Adv Colloid Interface Sci*, *147–148*, 333–342.

Yaghmur, A., Kriechbaum, M., Amenitsch, H., Steinhart, M., Laggner, P., and Rappolt, M. (2010). Effects of pressure and temperature on the self-assembled fully hydrated nanostructures of monoolein-oil systems. *Langmuir*, *26*(2), 1177–1185.

Yaghmur, A., Laggner, P., Almgren, M., and Rappolt, M. (2008). Self-assembly in monoelaidin aqueous dispersions: direct vesicles to cubosomes transition. *PLoS One*, *3*(11), e3747.

Yaghmur, A., Laggner, P., Zhang, S., and Rappolt, M. (2007). Tuning curvature and stability of monoolein bilayers by designer lipid-like peptide surfactants. *PLoS ONE*, *2*(5), e479.

Yang, D., O'Brien, D. F., and Marder, S. R. (2002). Polymerized bicontinuous cubic nanoparticles (cubosomes) from a reactive monoacylglycerol. *Journal of the American Chemical Society*, *124*(45), 13388–13389.

Synthesis and Alignment of Nanostructured Materials Using Liquid Crystals

IDIT AMAR-YULI, ABRAHAM ASERIN, and NISSIM GARTI

Casali Institute of Applied Chemistry, The Institute of Chemistry, The Hebrew University of Jerusalem, Jerusalem, Israel

Abstract

Liquid crystal (LC) science and technology have made important contributions to nanoscience and nanotechnology in areas such as medical diagnostics, drug delivery, and in high-tech devices. The major nanotechnology contribution is their capability to provide new synthetic procedures, self-assembly, and alignment of nanoscale materials with controlled uniform size, shape, and dimensionality. The self-assembly of the liquid crystalline systems, particularly the lyotropic LC type, enables the integration of lipophilic and hydrophilic reagents that meet to react at the interface. Therefore the reaction and the products properties are governed by the host liquid crystalline medium and by the external conditions (e.g., light, heat, chemical environment, and electric and magnetic field), which directly affect the LC hosting system.

This chapter attempts to give an overview of current research in the fields of synthesis, self-assembly, and alignment of nanomaterials using mainly lyotropic LCs and partially thermotropic LCs as direct and reverse templates. In the near future, liquid crystals are expected to play a prevalent role in nanoscience and nanotechnology and can potentially be used as new functional materials for electron, ion, molecular transporting, sensory, catalytic, optical, and bioactive materials.

Self-Assembled Supramolecular Architectures: Lyotropic Liquid Crystals, First Edition.
Edited by Nissim Garti, Ponisseril Somasundaran, and Raffaele Mezzenga.
© 2012 John Wiley & Sons, Inc. Published 2012 by John Wiley & Sons, Inc.

7.1   INTRODUCTION

Liquid crystalline materials, combining order (crystal) and mobility (liquid) at the molecular-nanoscale level, appear as leading candidates for the synthesis, self-assembly, and alignment of nanoscale materials with controlled size, shape, dimensionality, and structure (Brad et al., 2005; Wang et al., 2009). In addition, the dynamic functional liquid crystals (LCs) are known to respond to external fields (such as electric and magnetic fields) and interact with surfaces, thereby influencing their structure and properties (Hegmann et al., 2007). As a result of the LC intrinsic orientational order together with the molecular motion, they are recognized in bioscience (as model systems for cell membranes) and have been widely used in displays, sensors, optical elements (controllable lenses and lasers), and as drug delivery vehicles (Hegmann et al., 2007; Wang et al., 2009).

Liquid crystals are commonly divided into two general classes: *thermotropic* (TLCs) and *lyotropic* liquid crystals (LLCs) according to the composition and formation conditions. In general, molecular shape, microsegregation of incompatible parts, specific molecular interactions, and self-assembly are important factors that derive the formation of both thermotropic and lyotropic liquid phases (Bisoyi and Kumar, 2011). TLCs are defined as *ordered fluid phases* (based on pure compound) and exhibit diverse phases that differ in their degree of order as a function of temperature (Douglas et al., 2008; Hyde, 2001; Hyde et al., 1997). The molecules that tend to form TLC structures have anisotropic shapes (e.g., typically rod or disk shaped) and contain a relatively rigid core and a number of flexible alkyl tails (Fig. 7.1). In general, the rigid cores of these molecules promote ordered packing, whereas the flexible tails tend to disorder the system until a compromise is met to produce an *ordered fluid state* (Hegmann et al., 2007). The mesophase formation in this kind of thermotropic LC is driven by the segregation of chemically incompatible subunits from one another, such as the segregation of rigid aromatic cores from flexible alkyl tails within a molecule (Fig. 7.1).

For example, in the nematic phase, the least ordered TLC phase, the aggregates solely possess orientational ordering and no positional ordering. It is generally formed when the volume fraction of an incompatible core unit exceeds that of the incompatible wings and may result for both rod- and disk-shaped molecules (Fig. 7.1). In this phase, the aggregates are aligned along a direction that is termed the director n (Fig. 7.1). The director n is oriented either parallel to the long molecular axis for rodlike aggregates (N_u, the uniaxial nematic phase), parallel to the column axis for columnar aggregates formed by amphiphilic or disklike molecules (N_{Col}, the columnar nematic phase), or parallel to the short molecular axis for disklike aggregates of amphiphiles (i.e., platelike micelles), molecules, or particles (N_D, the discotic nematic phase) (Goodby, 2002).

In contrast to the thermotropic LC phases, the lyotropic systems contain at least two chemical components, an organic molecule (amphiphilic molecule)

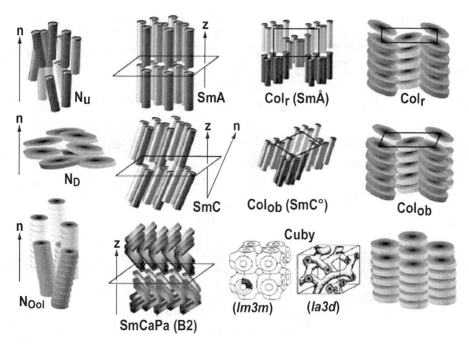

Figure 7.1   Presentation of main types of nematic and positional ordered thermotropic LC phases: N_u = uniaxial nematic phase, N_D = discotic nematic phase, N_{Col} = columnar nematic phase, L_a = lamellar or smectic A phase, SmA = smectic A phase, SmC = smectic C phase, SmC_aP_a = anticlinic antiferroelectric polar smectic C phase, Col_r = rectangular columnar phase, Col_{ob} = oblique columnar phase, Col_h = hexagonal columnar phase, Cub_V = bicontinuous cubic phase, and Cub_I = micellar cubic phase (space groups are in italics). [Adapted from Hegmann et al., (2007).]

and a solvent (usually water) (Hyde, 2001; Hyde et al., 1997). The addition of a solvent such as water to the amphiphilic molecules selectively hydrates the hydrophilic moiety of each molecule (ionic or neutral), avoiding the hydrophobic regions. This hydrogen bonding between the solvent (e.g., water) and the polar moiety of the solute, and the interplay (van der Waals interactions) between the hydrophobic moieties drives the molecules to self-assemble, thereby segregating the hydrophobic chain away from contact with the water. As a result, the strongly hydrogen-bonded polar moiety is frozen, while the apolar, hydrophobic part is molten (Larsson, 1989; Larsson et al., 1980).

In this case, the degree of order of the different phases depends on the components' composition; however, they are also influenced by other external parameters such as temperature and pressure (Seddon, 1990; Tiddy, 1980).

The LLC systems may exhibit a rich polymorphism of structures depending on the molecular shape/packing preferences of the LLC molecules (Fig. 7.2;

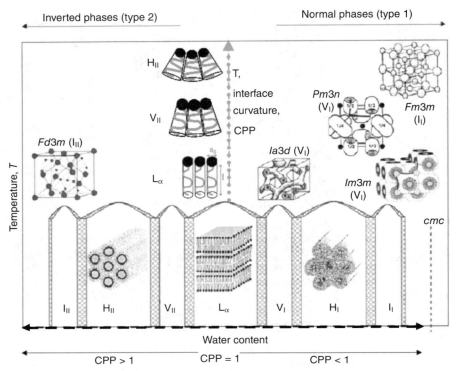

Figure 7.2   Hypothetical lyotropic binary phase diagram where phase transitions can be induced by varying water content or temperature. The indicated mesophases are: L_α = lamellar, Ia3d, Pn3m, or Im3m = direct and inverted bicontinuous cubic (V_I and V_{II}, respectively), Fd3m = direct and inverted micellar/discontinuous cubic (I_I and I_{II}, respectively), and H_I and H_{II} = direct and inverted hexagonal phase. The cmc (far right in the figure) represents the critical micellar concentration of surfactant from which it creates micelles.

Israelachvili et al., 1976; Larsson et al., 1980) and interfacial curvature energy considerations (Duesing et al., 1997; Templer et al., 1998; Vacklin et al., 2000). The major mesophases of the LLCs are lamellar (L_α); normal and reverse hexagonal (H_I and H_{II}); and cubic bicontinuous and discontinuous structures (V_I, V_{II} and I_I, I_{II}, respectively) (Fig. 7.2).

Thermotropic and lyotropic LC phases have various properties that make them ideal for use as advanced materials for potential nanomaterial applications (Miller et al., 1999). Since LCs combine order and mobility at the molecular, nanoscale level, their ordered structures can respond to external fields or to chemical changes as desired for specific applications such as controlled release (Douglas et al., 2008).

Their highly uniform, porous nanoscale architectures can be used also to incorporate, crystallize, or synthesize (growing nanostructured) hydrophobic as well as hydrophilic reagents or materials in separate sections. Thus, it is possible to engineer the environments in those regions for specific applications by wise selection of the surfactant head group and/or tail and solvent (Douglas et al., 2008). In addition, the microscopic structure of the LC systems (i.e., symmetry and structural parameter such as lattice parameter) can be controlled on the 1 to 5 nm scale via molecular design. Moreover, this ability can be extended to the macroscopic scale through appropriate alignment by external conditions.

The main advantage of synthesis in LC systems is expressed by the ability to produce the desired materials, whose nanostructures are determined by the rich thermotropic and lyotropic polymorphism, while their shape, size, and dimensionality can be manipulated by selecting the templating phases of LC (will be elaborated further).

The aim of this chapter is to provide an overview of the main research fields obtainable by diverse LC phases during the last several decades. The role of liquid crystalline systems, or more specific their interface abilities, in various fields will be highlighted.

7.2 NANOPARTICLE SYNTHESIS—LIQUID CRYSTALLINE PHASES PERFORM AS TEMPLATES

7.2.1 LC Surfactant Assembly as Directing Agent

The main concern in nanoparticle synthesis is the ability to form stable particles of controlled size, shape, and dispersity. Additional aim might be also to gain nanoparticles that can be redissolved in solvents without obtaining aggregation or decomposition.

The use of "hard" templates on the basis of silica (Giersig et al., 1997), anodic aluminum oxide (Li et al., 2008; Orikasa et al., 2006; Zhou et al., 2002), and mesoporous carbon (Dong et al., 2003; Lee et al., 2004; Xia and Mokaya, 2005) to synthesis nanoparticles has been known and used for decades. However, the "soft" templates gained more attention over the last decade and were found to be useful for designing novel nanomaterials such as spherical, hollow, one-dimensional (1D) nanowires, nanorods and nanotubes, and two-dimensional (2D) ordered arrays (Van Bommel et al., 2003). This is due to their diversity in structures, and structure parameters with lattice constants range from several nanometers to tens of nanometers. The soft templates can be composed of soft compounds such as biomolecules (Zhou et al., 2009), polymers forming gels (Shchukin et al., 2005), and amphiphilic molecules forming emulsions or liquid crystals (Li and Coppens, 2005; Zhang et al., 2007).

The resultant nanostructured materials have potential applications in various fields, such as materials science, biomedical science, electronics, optics, magnetism, energy storage, electrochemistry, and the like (Bruchez et al., 1998; Caruso et al., 1998; Gleiter, 2000).

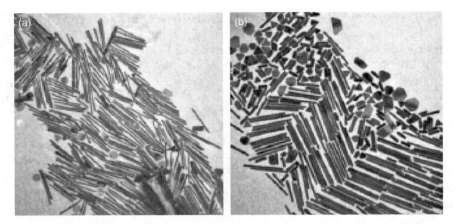

Figure 7.3   TEM micrographs of gold nanorods synthesized from (a) CTAB-stabilized 8 nm seeds and (b) CTAB-stabilized 16 nm seeds (scale bars = 500 nm). [Adapted from Cole and Murphy (2004).]

The Brust–Schiffrin reaction has been successfully used to produce spherical gold and silver nanoparticles (Ahmadi et al., 1996; Bradley et al., 2000; Petroski et al., 1998). Recently, the use of this method has been expanded to the formation of gold and silver nanorods by a seed-mediated growth approach (Jana et al., 2001a; Landskron and Ozin, 2004). In the first step, spherical gold and silver nanoparticles (diameter of ~3.5 nm) were produced by hydride reduction of hydrogen tetrachloroaurate or silver nitrate, respectively, in the presence of sodium citrate (Cole and Murphy, 2004; Jana et al., 2001a,b). Then, these spherical nanoparticles were transferred to a mixture of cetyltrimethylammonium bromide (CTAB), hydrogen tetrachloroaurate, or silver nitrate to produce rod-shaped nanoparticles (Fig. 7.3; Cole and Murphy, 2004). The CTAB, an ionic surfactant capable of forming LLC phases, acts as a directing agent, thereby forming a bilayer on the nanorods. Its stronger binding to the side edges than to the ends of the rods allows growth only in one direction. Moreover, it was further found that using a mixture of CTAB, sodium hydroxide, and ascorbic acid in the second step achieved longer and monodispersed nanorods (Jana et al., 2001a; Sertova et al., 2006). The ascorbic acid, being a weak reducing agent, was incapable of reducing the hydrogen tetrachloroaurate in the absence of the gold seeds, therefore minimizing additional nucleation during particle growth (Jana et al., 2001a).

A similar approach, for example, utilizing an amphiphilic molecule as a directing agent, has been used to form silicate nanorods (Ying et al., 1999). The inorganic aluminosilicate mesoporous material (belonging to the M41S family of molecular sieves) was synthesized by the addition of an aluminum source to a mixture of a silica source (e.g., tetraethylorthosilicate), CTAB, a base (e.g., sodium hydroxide), and water. Initially the mixture was aged at

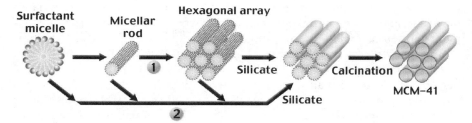

Figure 7.4 Two possible pathways for the LC templating mechanism. [Adapted from Ying et al., (1999).]

elevated temperatures (≥100°C) for long periods (24–144 h), thereby reaching solid precipitation. The inorganic components (which were negatively charged at the high pH used) mostly interact with the positively charged surfactant ammonium head group and condense into solid organic–inorganic nanorods. Then, after the product is filtered, washed, and dried, it is calcined at ~500°C to burn off the surfactant and to produce the mesoporous material (Ying et al., 1999).

The mechanistic pathway raised by the authors is that the inorganic materials mediated the ordering of the surfactants into the hexagonal arrangement, especially since the surfactant concentration was lower than the amount required for the formation of a hexagonal LC. Chen et al. (1993) used in situ nitrogen-14 nuclear magnetic resonance (^{14}N-NMR) spectroscopy to evaluate a similar product, MCM-41 silicate mesoporous material arranged in a hexagonal symmetry. According to their postulation, in the first step, two to three silicate precursor monolayers are deposited onto isolated surfactant micellar rods (Fig. 7.4; Ying et al., 1999). The randomly oriented silicate-encapsulated rods were further packed into the hexagonal structure. Finally, heating and aging completed the condensation of the silicate nanorods. It was also demonstrated that the structures and pore dimensions of mesoporous materials can be controlled by the properties of the surfactants, such as surfactant chain length and the solution chemical properties (Kresge et al., 1992).

Additional reports, using surfactant as a phase transfer agent, that is, as an assistant in 1D growth that is not based on silicate materials, are also available such as selenium and tellurium nanowires and nanorods, respectively (Gates et al., 2000; Liu et al., 2003). Generally, by using this approach, the resulting pore size is barely predictable, although it is highly reproducible. This is due to the fact that the final structural properties, including pore size, are developing during the final condensation process.

7.2.2 LC Structures as Direct Templates and Reverse Templates

Nanoscale particles usually do not induce significant distortions in the LC phases. Therefore, various nanomaterials have been dispersed and studied in

LC medium, while the hosting systems act as tunable solvents for the nanomaterials, and, being anisotropic media, they can provide a controlled alignment and self-assembly for the nanomaterials into larger organized structures in multiple dimensions (Hegmann et al., 2007; Kumar, 2007).

Furthermore, since the LC medium responds to external stimuli, the dispersed nanomaterials can be forced to track the order of the host medium, which develops the functionality of the host system. Diverse LC phases have been successfully used as templates for the synthesis of nanostructure materials, either by producing porous inorganic replicates of LLC origin (termed as direct templating) or by using LC mesophases as nanoreactors to produce the corresponding morphologies (reverse templating) (Attard et al., 1995, 1997; Ding et al., 2000).

7.2.2.1 *Direct Templating Method (Nanocasting)* The direct templating procedure enables the formation of porous materials exhibiting uniform pore size, morphology, and three-dimensional (3D) distribution. Additionally, these structural characteristics as well as their macroscopic shape can be adjusted by modifying the composition of the liquid crystalline systems. However, it should be noted that this process is restricted to the formation of frameworks that can remain stable even after surfactant removal and when exposed to air (Schüth 2001).

This technique is termed *true LC templating* and *nanocasting* and is widely used in the production of porous materials used in catalysis or adsorption techniques (Antonietti and Ozin, 2004; Attard et al., 1995, 1997; Landskron and Ozin, 2004).

Generally, in this approach, the continuous phase of the direct liquid crystalline structure (Fig. 7.5; Antonietti, 2006) may either be composed of a siliceous material or an aqueous metal salt. In the next step, the metal salt will be reduced, thus resulting in an ordered network of the surrounding host medium.

Figure 7.5 Illustration of a typical templating process using lyotropic liquid crystalline systems (in this case with a hexagonal array). The LLC phase is surrounded with a high concentration of inorganic precursor (left), and the continuous phase is solidified (middle). After solidification of the inorganic phase, the template is removed, and a solid negative replica of the original phase is obtained (right). [Adapted from Antonietti (2006).]

Finally, upon calcination a porous nanostructure is obtained retaining the structure of the original liquid crystal. Taken together, this technique allows rational prediction of the pore properties since it starts from an organized, LC phase, in contrast to the former approach that utilized LC surfactant assembly as a directing agent.

The production of silica-based mesoporous material was first introduced by Attard and co-workers in 1995. They synthesized a silica mesostructure based on the H_I LC templating process using polyethylene oxide surfactant (as was generally described in Fig. 7.5). The tetramethylorthosilicate (TMOS) in the continuous phase was gelled in the H_I system, and then the LC was destabilized and removed by methanol to produce the final hexagonal mesoporous material. It was further shown that the formation of $Ia\bar{3}d$ and lamellar silica-based materials was also possible by changing the tail chain length of the surfactant (Antonietti, 2006; Attard et al., 1995).

Many types of silica-based (Doshi et al., 2003; Kresge et al., 1992; Thomas et al., 2003) as well as nonsiliceous mesoporous nanomaterials (Kresge et al., 1992; Lyons et al., 2002)—for example, metal oxides, CdS and CdSe composites, and Pt/Ru alloys—have been successfully synthesized through this method using polymer or oligomer surfactant systems.

It has been shown that this technique offers the possibility to create an exact 1:1 negative copy of the origin LC phase. It has been suggested to replace a majority of the solvent by a metal or metal oxide precursor with similar polarity and condensing this precursor around the aggregates. For instance, the precursor hydrated silicic acid shows a very similar polarity and proton-bridging behavior as water, and therefore the final structure is very similar to that found in water (Antonietti, 2006). X-ray measurements performed during the process showed that the final structure of the mesoporous material preserved all structural features during the solidification process, shrinking only slightly on condensation (Attard et al., 1995, 1997).

Additionally, comparing transmission electron microscopy (TEM) images of original branched lamellar LC phase with its replicate (Figs. 7.6a and 7.6b, respectively; Hentze et al., 1999) reveals that one is the "negative" of the other, while it is almost impossible to differentiate them without a priori knowledge.

To conclude, it has been demonstrated that in this approach, the pores size and shape are determined solely by the size and shape of the LC template that preexists in solution prior to the solidification process.

It was further demonstrated that upon addition of salt to the synthesis mixture, the mesoporous silicates can be modified from the original LC phase (McGrath et al., 1997). McGrath et al. revealed that after the addition of sodium chloride the mesoporous material deviated from the ideal hexagonal structure to form a sponge phase (L_3, randon bicontinuous cubic phase) of mesoporous silicate (McGrath et al., 1997). Ethylenediaminetetraacetic acid (EDTA) was also found to induce the formation of L_3 at the expense of hexagonal structures (McGrath et al., 2000).

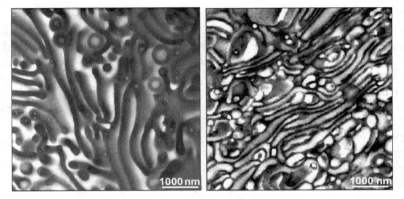

Figure 7.6 Illustration to show that the final porous structures are an almost perfect 1:1 to the original block copolymer mesophase (a negative image), where on the left there is the lyotropic phase and on the right the corresponding silica. [Adapted from Hentze et al., (1999).]

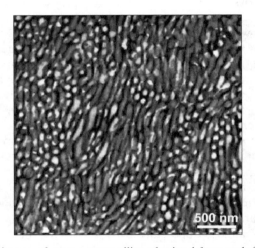

Figure 7.7 TEM image of mesoporous silica obtained from poly(butadiene-*b*-ethylene oxide) (PB–PEO) block copolymer templates illustrating the unusual bicontinuous spongelike L₃ phase (Förster et al., 2001).

Figure 7.7 presents the replica of the L_3 spongelike phase, which is usually not found in the traditional phase diagrams of block copolymers (Förster et al., 2001). The sponge phase exhibits a multiply connected, bicontinuous disordered phase without long-range order, yet with a well-defined characteristic length of 160 nm, which is related to an average unit cell size.

7.2.2.2 *Reverse Templating* The inner structure (channels) of the reverse LC phases can also be used to produce an ordered array of nanoparticles, performing as a nanoreactor. Therefore, by controlling the type of liquid crystalline phase, one can manipulate the size and shape of the nanoparticles grown within, which are known to determine the electronic, optical, magnetic, and chemical characteristics of the nanoparticles (Bisoyi and Kumar, 2011; Ding and Gin, 2000). The resultant nanomaterials can be applicable in catalysis, batteries, fuel cells, capacitors, and sensors, for example.

The alignment of β-lactoglobulin (βlg) fibrils due to their confinement within three different lyotropic liquid crystals—lamellar, cubic, and hexagonal symmetries—was studied by Amar-Yuli et al. (2011) using atomic force microscopy (AFM).

Two-dimensional order parameter (S) of the amyloid fibrils, representing their orientation distribution, has been calculated based on AFM analysis and according to the following formula, $S = \langle 2\cos^2\theta - 1 \rangle$, where θ is the angle between an individual βlg fibril and the common director vector existing in the case of a nonrandom distributed fibrils orientation. The order parameter values range between 0 and 1, where the former corresponds to an isotropic structure and the latter represents a perfect alignment along the director (Amar-Yuli et al., 2011). Figures 7.8a–7.8c depict AFM images of βlg fibrils that were embedded in the three mesophases—lamellar, cubic, and hexagonal (L_α, Pn3m, and H_{II}, respectively). The fibrils that were confined both within the water-filled lamellae, in the lamellar structure (Fig. 7.8a) and in the water channels of the isotropic cubic (Fig. 7.8b), demonstrated no specific orientation and yielded very low order parameter values (0.05 and <0.2, respectively).

In contrast, incorporation of the fibrils in the anisotropic, unidimentional, reverse hexagonal phase, revealed a strong templating effect of the anisotropic mesophase on the fibrils alignment (Fig. 7.8c). This was witnessed by the nearly perfect orientation of the fibrils in correspondence to the hexagonally packed water channels (Fig. 7.8c). The order parameter values reached a value as high as 0.97, which indicates a nearly perfect alignment (Amar-Yuli et al., 2011).

The authors also showed that this specific anisotropy coupling mechanism can be used to affect the order of the mesophase by manipulating that of the fibrils. For this objective, amyloid fibrils mixed iron(III) salts and reduced with $NaBH_4$, producing βlg fibrils coated with magnetic nanoparticles that were used as stimuli-responsive triggers tuned by an external field (Amar-Yuli et al., 2011).

The 2D scattering pattern of the magnetic nanoparticle-coated amyloid fibrils confined within H_{II}, L_α, and Pn3m LLC is shown in Figure 7.9a–i based on small-angle neutron-scattering measurements. Prior to magnetic field exposure, the first Bragg reflection of the mesophases (in H_{II}, Fig. 7.9a, L_α, Fig. 7.9d, and Pn3m, Fig. 7.9g) was azimuthally symmetric. The authors showed that by applying external magnetic field of 1.1T while cooling from a temperature above the isotropic micellar state, they can tune the alignment of both anisotropic LLCs. As a result, a pronounced anisotropic pattern of the H_{II} (Fig. 7.9b) and L_α (Fig. 7.9e) host mesophase, with maximum intensities perpendicular to the magnetic field, that is, with the mesophase interfaces parallel to the field direction, was obtained. Furthermore, by rotating the

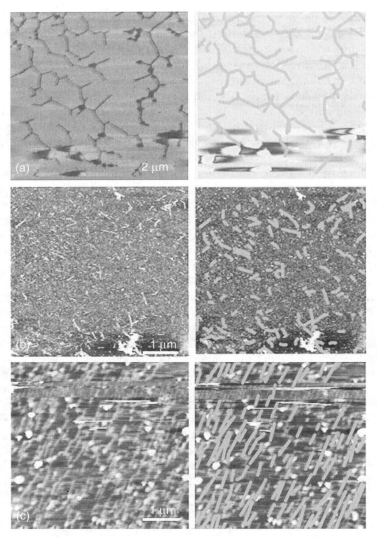

Figure 7.8   AFM images of β-lactoglobulin fibrils assembly, after 2-h sonication treatment, when confined in the (a) lamellar, the (b) cubic, and the (c) hexagonal phase. The right column depicts the fibrils tracking needed for the calculation of the 2D order parameter (Amar-Yuli et al., 2011).

magnetic field by 90° and clearing the alignment by raising the temperature up to the micellar isotropic fluid and cooling down, they realigned the hexagonal (Fig. 7.9c) and lamellar (Fig. 7.9f) domains in the same direction of the applied field. On the contrary, it was impossible to affect the alignment of the isotropic cubic phase. Solely isotropic nonoriented 2D scattering patterns

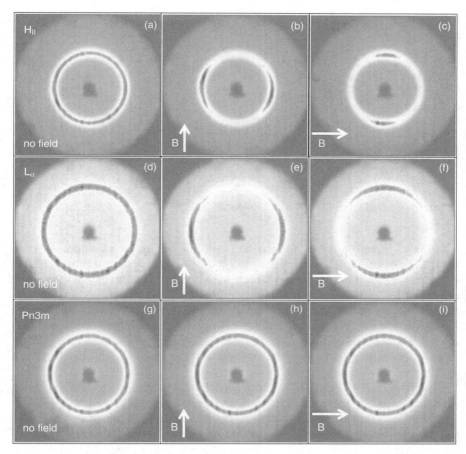

Figure 7.9 2D small-angle neutron scattering (SANS) patterns of: (upper row) H_{II}, (middle row) L_α, and (lower row) Pn3m mesophases containing amyloid fibrils coated with iron-oxide magnetic nanoparticles. Labels (a), (d), and (g) represent the SANS measurements in the absence of the external magnetic field. Labels (b), (c), (e), (f), (h), and (i) indicate the measurements performed under exposure to 1.1T external magnetic field applied in two different directions tilted by 90° (Amar-Yuli et al., 2011).

were detected both prior (Fig. 7.9g) and after (Figs. 7.9h and 7.9i) exposure to the same magnetic field.

The mechanism for this magnetically assisted orientation, introduced by the authors, was that the columnar hexagonal and lamellar phases nucleating from the isotropic fluid upon cooling and in the presence of aligned fibrils, used these latter as heterogeneous nucleating surfaces, leading to a preferential orientation of the resulting mesophase.

One of the common explored classes of nanomaterials is metal nanoparticles. In general, in the first step of metal nanoparticle synthesis, a metal salt

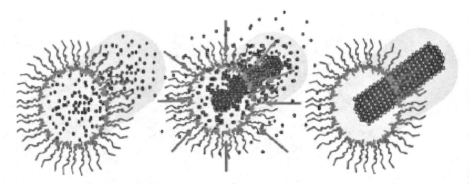

Figure 7.10 Schematic representation of nanoparticle synthesis utilizing LLCs as nanoreactors. [Adapted from Karanikolos et al., (2006).]

solution is mixed with an appropriate amount of liquid crystalline host to form the desired LLC phase. Then, the nanoparticles aggregate into clusters to form a single nanostructure, which ideally will take on the shape of the original LLC nanodomain (as illustrated in Fig. 7.10). When successful templating effect occurs, the columnar phases will form tube, wire, or rodlike nanostructures, whereas cubic and lamellar phases usually will result in spherical or disklike nanostructures (Ding and Gin, 2000; Karanikolos et al., 2006). Finally, the produced nanoparticles can be collected through centrifugation or filtering.

Dellinger and Braun (2004) have used LLCs to synthesize nanoparticles. They exploited the restricted dimensions of the hydrophilic and hydrophobic domains (2–10 nm) in order to selectively soluble the reagents in only one of the nanoscopic dimentions. The authors new approach included synthesis of Bi and PbS (lead sulfide, a semiconductor compound) nanoparticles by bringing together reactants via the shear mixing of two liquid crystals, while the mixture containing dissolved reactants that are selectively soluble in one of the two phases (Dellinger and Braun, 2004).

In the case of BiOCl synthesis both lamellar and hexagonal LLC hosts were used (Dellinger and Braun, 2004). The LLC was added to aqueous $BiCl_3$ while simultaneously heating and stirring, and upon cooling to room temperature BiOCl was precipitated through the diffusion of ammonia into the LLC lattice. It was found that the lamellar phase produced spherical 5-nm particles, whereas the hexagonal LLC phase formed monodisperse 100-nm-wide and 250-nm-long rods. Furthermore, an isotropic fluid host yielded disk-shaped particles with 50- to 250-nm diameters (Dellinger and Braun, 2004).

PbS nanoparticles were prepared by shear mixing of a liquid crystal containing solutions of $Pb(NO_3)_2$ or $Pb(CH_3COO)_2$ (~0.005–0.1 M) with additional LCs containing saturated solutions of the extrinsic reducing agent H_2S (~0.1 M) produced by bubbling H_2S through water. As a result an LC that contained PbS nanoparticles was formed.

Figure 7.11 depicts micrographs of PbS nanoparticles produced from lead nitrate via shear mixing in the hexagonal, lamellar, and inverse hexagonal phases based on the nonionic amphiphile oligo(ethylene oxide)10 oleyl ether (Dellinger and Braun, 2004). By altering the liquid crystalline phase from hexagonal to lamellar to inverse hexagonal, the geometry of the aqueous domains has been modified from *3D continuous nanoreactors* to sheetlike *2D nanoreactors* to rodlike *1D nanoreactors*. The size of the nanoparticles was clearly dependent on the morphology of the phases. When the hexagonal phase was used, with a 3D continuous medium, the particles produced were somewhat polydispersed and exhibited the cubic morphology common for PbS. Changing to lamellar and then to inverse hexagonal phases resulted in the production of spherical morphology particles of reduced size.

Qi et al. (1999) have been demonstrated that ribbons of silver nanoparticles can be synthesized in lamellar LCs based on tetraethylene glycol monododecyl ether ($C_{12}E_4$) and aqueous silver ion solution by the reduction of the silver ions using the surfactant itself as reductant.

The ribbons were found to consist of close-packed silver nanoparticles about 2–3 nm and a few larger silver nanoparticles distributed or attached on the ribbons. The authors revealed that altering the surfactant concentration (e.g., decreasing $C_{12}E_4$ content, which increases the water layer thickness) did not affect the size of the smaller particles forming the ribbons; however, it had a direct impact on the proportion and/or size of the larger particles (Qi et al., 1999). After 20 days, the ribbons of silver nanoparticles were formed (~ 300 nm wide and 7 µm long) twisted and folded, thereby forming complex 3D configuration.

In other research, reverse hexagonal LCs were used as template to synthesize nanowires (e.g., ZnS) by γ-ray irradiation (Jiang et al., 1993). However, due to the random orientation of the hexagonal cylinders, the resultant nanowires were relatively short (<2 µm). An additional method for the formation of nanowires was demonstrated by Huang et al. (2002). The authors showed that cuprite (Cu_2O) semiconductor nanowires can be fabricated by electrodeposition within H_{II} LLC based on sodium bis(2-ethylhexyl) sulfosuccinate (AOT) (Huang et al., 2002). By changing the time of electrodeposition, they exhibited the ability to control the growth of the nanowires up to tens of micrometers. Upon relatively short electrodeposition time (1 h) a large number of nanowires of 3–5 µm length were obtained (Fig. 7.12a) while longer deposition time (2 h) resulted in longer and higher diameter nanowires (10 µm length and 100 nm diameter; Figs. 7.12b and 7.12c; Huang et al., 2002). The diameter values of the nanowires were also revealed to be controlled by adjusting the water-to-surfactant ratio. It should be noted that in the absence of LLC performing as a template for this synthesis only micrometer-sized copper crystals were produced from bulk $CuCl_2$ aqueous solution at the same conditions. Additionally, a short electrode distance during electrodeposition was shown to improve the alignment of the liquid crystalline phase for producing nanowires with a high aspect ratio (Huang et al., 2002).

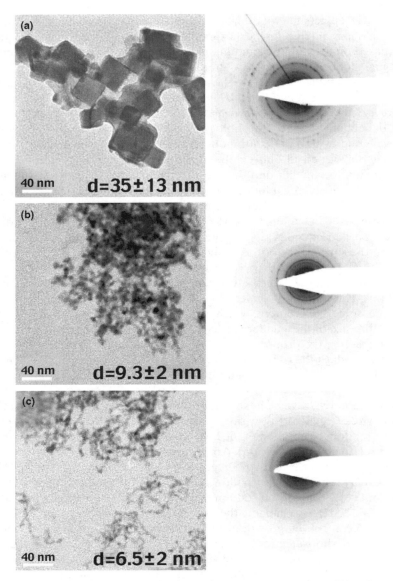

Figure 7.11   TEM micrographs (left) and selected area electron diffraction patterns (right) of PbS nanoparticles produced by shear mixing LLCs containing $Pb(NO_3)_2$ and H_2S in: (a) hexagonal phase, (b) lamellar phase, and (c) inverse hexagonal phase (Dellinger and Braun, 2004).

Figure 7.12 SEM images of cuprite nanowires electrodeposited from reverse hexagonal liquid crystalline phase AOT/p-xylene/CuC$_{12}$ for different deposition times: (a) 1 h, (b) 2 h, and (c) 2 h, at different magnifications (Huang et al., 2002).

Expanded study of utilizing electrophoretic deposition and LLC for nanomaterials fabrication was carried out by Zhang et al. (2006). They reported on the electrophoretic assembly of silver nanoparticles (by using NaBH$_4$ to reduce AgNO$_3$) in lamellar LLC composed of anionic surfactant–sodium bis-(2-ethylhexyl) sulfosuccinate (AOT), isooctane, and water.

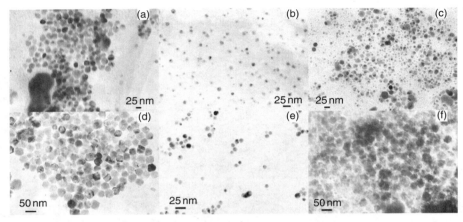

Figure 7.13   TEM images of deposits at different conditions. Electric field strength (V/mm) and deposition time (s) were: (a) 2.0, 4000; (b) 2.0, 4000; (c) 1.0, 4000; (d) 3.3, 4000; (e) 2.0, 1800; and (f) 2.0, 7000. (b) Image of deposit in Ag–SO hydrosol for reference while others are those in hybrids (Zhang et al., 2006).

The electric field strength and the length of electrophoretic time were shown to affect the deposit morphology as depicted in Figure 7.13. Comparing the nanoparticles obtained using lamellar LLC (AOT/Ag–sodium oleate hybrid, Fig. 7.13a) to a common technique as previously reported (Wang et al., 1999), forming sodium oleate-stabilized silver colloids (Ag–SO hydrosol), both under fixed electric field strength and time (2.0 V/mm for 4000 s) demonstrated the advantages of the LLC system. Utilizing the lamellar LLC reservoir causes the fabrication of larger silver nanoparticles with a much denser distribution and ordered arrays (Fig. 7.13a), while those in Figure 7.13b appeared as individual ones and no significant ordering was found. Deposition under different electric field strengths at 1.0, 2.0, and 3.3 V/mm for 4000 s led to the formation of larger nanoparticles of 10 nm (Fig. 7.13c), 15–20 nm (Fig. 7.13a), and 30–50 nm (Fig. 7.13d), respectively, proportional to the field strength. A long electrophoretic time had a strong impact on the morphology of deposits. A short time (less than 1800 s) was not enough to bring the particles together and resulted in the formation of only dispersed particles of about 10 nm (Fig. 7.13e). Longer deposition time (4000 s) enabled assembling ordered arrays with more uniform size of 15–20 nm (Fig. 7.13a). After 7000 s the particle sizes increased and the particles were accompanied by coalescence (Fig. 7.13f).

To elaborate on these results the authors proposed a schematic description of the electrophoretic deposition process in lamellar LLC (Fig. 7.14) (Zhang et al., 2006). As illustrated in Figure 7.14a, with no electric field or low electric field strength, the oleates form a capping layer along the LLC water layer to protect the nanoparticles from contacting by steric repulsion, thus preventing

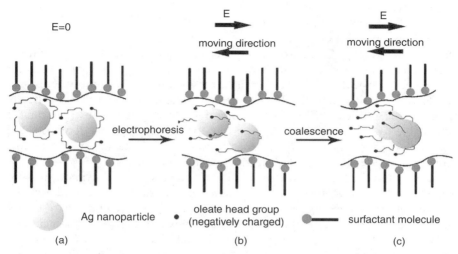

Figure 7.14 Schematic description of electrophoretic deposition (EPD) process in lamellar LLC. (a) Ag nanoparticles capped by oleates in water layer without electric field; (b) rearrangement of capping layer under higher electric field; and (c) larger particle formation by coalescence (Zhang et al., 2006).

nanoparticle aggregation (Zhang et al., 2006). The dissociation of capped oleate molecules allows the nanoparticle surface to be fully negatively charged and therefore movable to the anode under electric field (i.e., electrophoresis). Unlike the particles moving randomly in hydrosol, particles in LLC hybrid are confined in the water layer and move parallel along the surfactant bilayers, directing the particles to be arranged densely and ordered. Additionally, the limited space also enhanced contact and coalescence of nanoparticles, hence greater size of particles and denser arrays were obtained from hybrid compared to those from hydrosol, as shown in Figures 7.13a and 7.13b.

Applying higher field strength, the electrostatic attractions caused the oleates to move toward the anode. The Ag particle surface is then partially exposed due to the rearrangement or deformation of the oleate capping layer (Fig. 7.14b). As a result, particle coalescence and larger particle formation took place to reduce the high surface tension or energy in the nanoscale dimension (Fig. 7.14c).

Higher electric field in a given time period affected the size of the nanoparticles as was shown in Figures 7.13a, 7.13c, and 7.13d (Zhang et al., 2006). The higher the electric field, the more particles contacted and, hence, the larger particles were formed as was seen in Figures 7.13c, 7.13a, and 7.13d. Yet, in the LLC viscous medium, electrophoretic time also exhibited an important effect on the final product. As time increased, more particles reached the anode surface and formed arrays, therefore, larger and heavier aggregation of particles was obtained, as shown in Figure 7.13e, 7.13a, and 7.13f.

In summary, following this study, electrophoretic deposition of nanoparticles templated by LLC can provide a promising approach for obtaining densely arranged and ordered nanoparticles, and their morphology can be in tune with the water layer width, the electric field strength, and deposition time.

Recently, Yoshida et al. (2010) introduced a method that can drastically simplify the fabrication of gold nanoparticle–TLC suspensions and can be applicable to any type of material and LC phase. The authors demonstrated that metal nanoparticle–TLC suspensions can be fabricated by sputter depositing the target material (gold) on the host liquid crystal.

They exploited the fact that certain LC molecules possess a vapor pressure smaller than 1 Pa at room temperature, which is sufficient to undergo the sputter deposition process, which is performed at a pressure of a few to several tens of pascals. Several nematic samples are shown in Figure 7.15a, including pure 4-pentyl-40-cyanobiphenyl (5CB) forming a nematic LC structure, between 24 and 35°C, and after sputter deposition of gold (for 30 s, and 3, 10,

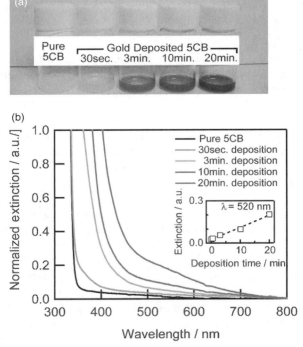

Figure 7.15 (a) Photograph of the nematic LC 5CB deposited with gold for various durations and (b) extinction spectra of 5CB deposited with gold for various periods, as indicated in the figure (Yoshida et al., 2010).

and 20 min, respectively; Yoshida et al., 2010). The vapor pressure of 5CB is reported to be less than 0.23 Pa at room temperature (Deschamps et al., 2008), which enables gold deposition without any noticeable evaporation of the host. The pure 5CB appears white at room temperature because of light scattering, while the gold-deposited samples exhibit a brown-reddish color (deposition time dependent). The extinction spectra of the material shown in Figure 7.15b demonstrates that the color of the LC is attributed to the local surface plasmon resonance of gold, which is observed as a slight shoulder at around 520 nm. A linear relationship between the deposition time and the concentration of the gold nanoparticles was revealed (inset in Fig. 7.15b). It should be noted that the nanoparticles were reported as very stable in the LC phase and did not show any observable changes for at least 3 months (Yoshida et al., 2010).

Microscopy images (data not shown) were evidence for obtaining spherical nanosized particles with a mean diameter of 2.9 nm and a standard deviation of 0.6 nm. It has been emphasized that the maximum diameter to which the nanoparticles grow is limited to a certain value determined by the physical properties of the host LC, such as its wettability and elastic constants. Finally, the electrooptic properties of a twisted nematic cell with the specific gold nanoparticles show remarkable improvement.

7.3   CONCLUSIONS

In the past several decades, many methods have been developed to synthesize nanostructured materials with controlled size, shape, dimensionality, and structure, including lithographic techniques and hard-template methods. However, the use of soft (liquid crystalline) templates for nanomaterial syntheses exhibits many incomparable advantages over other methods. The diversity of LC mesophases, particularly those of lyotropic type, has been utilized as directing agent, direct or reverse templates, and proven to be useful routes to prepare porous organic materials, oxides, metal/alloy, and others. So far, the TLCs have not yet been used to as great (an extent for the synthesis of metallic, magnetic, or semiconducting nanoparticles as the lyotropic LCs have.

One of the most attractive features of these methods is the versatility of the resultant structure of the nanomaterials, transferred from the LC templates to the desired materials. The reactants are confined in the limited dimension of the LC and the structures of long-chain order affect the nucleation and growth processes of the products, which could be applied to control the synthesis of the nanomaterials with desired morphologies.

As a result, rich morphologies of the synthesized nanomaterials are obtained, including spherical particles, hollow nanostructures, 1D nanowires, nanorods and nanotubes, and 2D ordered arrays.

We demonstrated that among those three main methods based on liquid crystalline systems, the direct and reverse templating methods were the superior ones. These preparation techniques were shown to be easily reproducible and more importantly allowed rational prediction of the nanoparticle properties such as size, shape, and dimensionality. On the contrary, nanoparticle characteristics of nanomaterials synthesized using the surfactant assembly as a directing agent were barely predictable, although they showed high reproducibility. This is due to the fact that the final structural properties, including pore size, are developed during the final condensation process.

To conclude, the design of LC structures for nanomaterial production has and will further contribute to the nanotechnology revolution of many high-tech applications in the fields of nanoscale electronics, electrooptics, sensors, and display devices.

REFERENCES

Ahmadi, T. S., Wang, Z. L., Green, T. C., Henglein, A., and El-Sayed, M. A. (1996). Shape-controlled synthesis of colloidal platinum nanoparticles. *Science, 272,* 1924–1926.

Amar-Yuli, I., Adamcik, J., Lara, C., Bolisetty, S., Vallooran, J. J., and Mezzenga, R. (2011). Templating effects of lyotropic liquid crystals in the encapsulation of amyloid fibrils and their stimuli-responsive magnetic behavior. *Soft Matter, 7,* 3348–3357.

Antonietti, M. (2006). Silica nanocasting of lyotropic surfactant phases and organized organic matter: Material science or an analytical tool? *Philosophical Transactions of the Royal Society A, 364,* 2817–2840.

Antonietti, M., and Ozin, G. A. (2004). Promises and problems of mesoscale materials chemistry or why Meso? *Chemistry A European Journal, 10,* 28–41.

Attard, G. S., Glyde, J. C., and Göltner, C. G. (1995). Liquid-crystalline phases as templates for the synthesis of mesoporous silica. *Nature, 378,* 366–368.

Attard, G. S., Corker, J. M., Göltner, C. G., Henke, S., and Templer, R. H. (1997). Liquid-crystal templates for nanostructured metals. *Angewandte Chemie International Edition, 36,* 1315–1317.

Bisoyi H. K., and Kumar, S. (2011). Liquid-crystal nanoscience: An emerging avenue of soft self-assembly. *Chemical Society Reviews, 40,* 306–319.

Brad, C., Chen, X., Narayanan, R., and El-Sayed, M. (2005). Chemistry and properties of nanocrystals of different shapes. *Chemical Reviews, 105,* 1025–1102.

Bradley, J. S., Tesche, B., Busser, W., Maase, M., and Reetz, M. T. (2000). Surface spectroscopic study of the stabilization mechanism for shape-selectively synthesized nanostructured transition metal colloids. *Journal of the American Chemical Society, 122,* 4631–4636.

Bruchez, M., Moronne, M., Gin, P., Weiss, S., and Alivisatos, A. P. (1998). Semiconductor nanocrystals as fluorescent biological labels. *Science, 281,* 2013–2016.

Caruso, F., Caruso, R. A., and Mohwald, H. (1998). Nanoengineering of inorganic and hybrid hollow spheres by colloidal templating. *Science, 282,* 1111–1114.

Chen, C. Y., Burkett, S. L., Li, H. X., and Davis, M. E. (1993). Studies on mesoporous materials. II. Synthesis mechanism of MCM-41. *Microporous Materials, 2,* 27–34.

Cole, A., and Murphy, C. J. (2004). Seed-mediated synthesis of gold nanorods: Role of the size and nature of the seed. *Chemistry of Materials*, *16*, 3633–3640.

Dellinger, T. M., and Braun, P. V. (2004). Lyotropic liquid crystals as nanoreactors for nanoparticle synthesis. *Chemistry of Materials*, *16*, 2201–2207.

Deschamps, J., Martin Trusler, J. P., and Jackson, G. (2008). Vapor pressure and density of thermotropic liquid crystals: MBBA, 5CB, and novel fluorinated mesogens. *Journal of Physical Chemistry B*, *112*, 3918–3926.

Ding, J. H., and Gin, D. L. (2000). Catalytic Pd nanoparticles synthesized using a lyotropic liquid crystal polymer template. *Chemistry of Materials*, *12*, 22–24.

Dong, A., Ren, N., Tang, Y., Wang, Y., Zhang, Y., Hua, W., and Gao, Z. (2003). General synthesis of mesoporous spheres of metal oxides and phosphates. *Journal of the American Chemical Society*, *125*, 4976–4977.

Doshi, D. A., Gibaud, A., Goletto, V., Lu, M. C., Gerung, H., Ocko, B., Han, S. M., and Brinker, C. J. (2003). Peering into the self-assembly of surfactant templated thin-film silica mesophases. *Journal of the American Chemical Society*, *125*, 11646–11655.

Douglas, L. G., Cory, S. P., Jason, E. B., and Robert, L. K. (2008). Functional lyotropic liquid crystal materials. *Structure and Bonding*, *128*, 181–222.

Duesing, P. M., Templer, R. H., and Seddon, J. M. (1997). Quantifying packing frustration energy in inverse lyotropic mesophases. *Langmuir*, *13*, 351–359.

Förster, S., Berton, B., Hentze, H. P., Krämer, E., Antonietti, M., and Lindner, P. (2001). Lyotropic phase morphologies of amphiphilic block copolymers. *Macromolecules*, *34*, 4610–4623.

Gates, B., Yin, Y. D., and Xia, Y. N. (2000). A solution-phase approach to the synthesis of uniform nanowires of crystalline selenium with lateral dimensions in the range of 10–30 nm. *Journal of the American Chemical Society*, *122*, 12582–12583.

Giersig, M., Ung, T., LizMarzan, L. M., and Mulvaney, P. (1997). Direct observation of chemical reactions in silica-coated gold and silver nanoparticles. *Advanced Materials*, *9*, 570–575.

Gleiter, H. (2000). Nanostructured materials: Basic concepts and microstructure. *Acta Materialia*, *48*, 1–29.

Goodby, J. W. (2002). Twist grain boundary and frustrated liquid crystal phases. *Current Opinion in Colloid and Interface Science*, *7*, 326–332.

Hegmann, T., Qi, H., and Marx, V. M. (2007). Nanoparticles in liquid crystals: Synthesis, self-assembly, defect formation and potential applications. *Journal of Inorganic and Organometallic Polymers and Materials*, *17*, 483–508.

Hentze, H. P., Krämer, E., Berton, B., Förster, S., Antonietti, M., and Dreja, M. (1999). Lyotropic mesophases of poly(ethylene oxide)-*b*-poly(butadiene) diblock copolymers and their crosslinking to generate ordered gels. *Macromolecules*, *32*, 5803–5809.

Huang, L., Wang, H., Wang, Z., Mitra, A., Zhao, D., and Yan, Y. (2002). Cuprite nanowires by electrodeposition from lyotropic reverse hexagonal liquid crystalline phase. *Chemistry of Materials*, *14*, 876–880.

Hyde S. T. (2001). *Handbook of Applied Surface and Colloid Chemistry*, Wiley, New York, Chapter 16.

Hyde, S. T., Andersson, S., Larsson, K., Blum, Z., Landh, T., Lidin, S., and Ninham, B. W. (1997). *The Language of Shape. The Role of Curvature in Condensed Matter: Physics, Chemistry and Biology*, Elsevier, Amsterdam.

Israelachvili, J. N., Mitchell, D. J., and Ninham, B. W. (1976). Theory of self-assembly of hydrocarbon amphphiles into micelles and bilayers. *Journal of Chemical Society*, 72, 1525–1568.

Jana, N. R., Gearheart, L., and Murphy, C. J. (2001a). Seed-mediated growth approach for shape-controlled synthesis of spheroidal and rod-like gold nanoparticles using a surfactant template. *Advanced Materials*, 13, 1389–1393.

Jana, N. R., Gearheart, L., and Murphy, C. J. (2001b). Wet chemical synthesis of silver nanorods and nanowires of controllable aspect ratio. *Chemical Communications*, 7, 617–618.

Jiang, X. C., Xie, Y., Lu, J., Zhu, L., He, W., and Qian, Y. (1993). Simultaneous in situ formation of ZnS nanowires in a liquid crystal template by γ-irradiation. *Chemistry of Materials*, 13, 1213–1218.

Karanikolos, G. N., Alexandridis, P., Mallory, R., Petrou, A., and Mountziaris, T. J. (2006). Water-based synthesis of ZnSe nanostructures using amphiphilic block copolymer stabilized lyotropic liquid crystals as templates. *Nanotechnology*, 16, 3121–3128.

Kresge, C. T., Leonowicz, M. E., Roth, W. J., and Vartuli, J. C. (1992). Ordered mesoporous molecular sieves synthesized by a liquid-crystal template mechanism. *Nature*, 359, 710–712.

Kumar, S. (2007). Discotic-functionalized nanomaterials. *Synthesis and Reactivity in Inorganic and Metal-Organic Chemistry*, 37, 327–331.

Landskron, K., and Ozin, G. A. (2004). Periodic mesoporous dendrisilicas. *Science*, 306, 1529–1532.

Larsson, K. (1989). Cubic lipid-water phases: Structures and biomembrane aspects. *Journal of Physical Chemistry*, 93, 7304–7314.

Larsson, K., Fontell, K., and Krog, N. (1980). Structural relationships between lamellar, cubic and hexagonal phases in monoglyceride-water systems—possibility of cubic structures in biological-systems. *Chemistry and Physics of Lipids*, 27, 321–328.

Lee, J., Han, S., and Hyeon, T. (2004). Synthesis of new nanoporous carbon materials using nanostructured silica materials as templates. *Journal of Materials Chemistry*, 14, 478–486.

Li, W., and Coppens, M. (2005). Synthesis and characterization of stable hollow Ti–Silica microspheres with a mesoporous shell. *Chemistry of Materials*, 17, 2241–2246.

Li, D., Thompson, S. R., Bergmann, G., and Lu, J. (2008). Template-based synthesis and magnetic properties of cobalt nanotube arrays. *Advanced Materials*, 20, 4575–4578.

Liu, Z., Hu, Z., Xie, Q., Yang, B., Wu, J., and Qian, Y. (2003). Surfactant-assisted growth of uniform nanorods of crystalline tellurium. *Journal of Materials Chemistry*, 13, 159–162.

Lyons, D. M., Ryan, K. M., and Morris, M. A. (2002). Preparation of ordered mesoporous ceria with enhanced thermal stability. *Journal of Materials Chemistry*, 12, 1207–1212.

McGrath, K. M., Dabbs, D. M., Yao, N., Aksay, I. A., and Gruner, S. M. (1997). Formation of a silicate L_3 phase with continuously adjustable pore sizes. *Science*, 277, 552–556.

McGrath, K. M., Dabbs, D. M., Yao, N., Edler, K. J., Aksay, I. A., and Gruner, S. M. (2000). Silica gels with tunable nanopores through templating of the L_3 phase. *Langmuir, 16,* 398–406.

Miller, S. A., Ding, J. H., and Gin, D. L. (1999). Nanostructured materials based on polymerizable amphiphiles. *Current Opinion in Colloid and Interface Science, 4,* 338–347.

Orikasa, H., Inokuma, N., Okubo, S., Kitakami, O., and Kyotani, T. (2006). Template synthesis of water-dispersible carbon nano "test tubes" without any post treatment. *Chemistry of Materials, 18,* 1036–1040.

Petroski, J. M., Wang, Z. L., Green, T. C., and El-Sayed, M. A. (1998). Kinetically controlled growth and shape formation mechanism of platinum nanoparticles. *Journal of Physical Chemistry B, 102,* 3316–3320.

Qi, L., Gao, Y., and Ma, J. (1999). Synthesis of ribbons of silver nanoparticles in lamellar liquid crystals. *Colloids and Surfaces A, 157,* 285–294.

Schüth, F. (2001). Non-siliceous mesostructured and mesoporous materials. *Chemistry of Materials, 13,* 3184–3195.

Seddon, J. M. (1990). Structure of the inverted hexagonal (H_{II}) phase and non-lamellar phase transitions of lipids. *Biochimica et Biophysica Acta, 1031,* 1–69.

Sertova, N., Toulemonde, M., and Hegmann, T. (2006). Towards recyclable nanoporous polymer membranes for the synthesis of one-dimensional nanoscale gold colloids. *Journal of Inorganic and Organometallic Polymers and Materials, 16,* 91–96.

Shchukin, D. G., Yaremchenko, A., Ferreira, M., and Kharton, V. (2005). Polymer gel templating synthesis of nanocrystalline oxide anodes. *Chemistry of Materials, 17,* 5124–5129.

Templer, R. H., Seddon, J. M., Warrender, N. A., Syrykh, A., Huang, Z., Winter, R., and Erbes, J. (1998). Modeling the phase behavior of the inverse hexagonal and inverse bicontinuous cubic phases in 2:1 fatty acid/phosphatidylcholine mixtures. *Journal of Physical Chemistry B, 102,* 7262–7271.

Thomas, A., Schlaad, H., Smarsly, B., and Antonietti, M. (2003). Replication of lyotropic block copolymer mesophases into porous silica by nanocasting: learning about finer details of polymer self-assembly. *Langmuir, 19,* 4455–4459.

Tiddy, G. J. T. (1980). Surfactant-water liquid crystal phases. *Physics Reports, 57,* 1–46.

Vacklin, H., Khoo, B. J., Madan, K. H., Seddon, J. M., and Templer, R. H. (2000). The bending elasticity of 1-monoolein upon relief of packing stress. *Langmuir, 16,* 4741–4748.

Van Bommel, K. J. C., Friggeri, A., and Shinkai, S. (2003). Organic templates for the generation of inorganic materials. *Angewandte Chemie International Edition, 42,* 980–999.

Wang, W., Chen, X., and Efrima, S. (1999). Silver nanoparticles capped by long-chain unsaturated carboxylates. *Journal of Physical Chemistry B, 103,* 7238–7246.

Wang, C., Chen, D., and Jiao, X. (2009). Lyotropic liquid crystal directed synthesis of nanostructured materials. *Science and Technology of Advanced Materials, 10,* 023001–023012.

Xia, Y., and Mokaya, R. (2005). Hollow spheres of crystalline porous metal oxides: A generalized synthesis route via nanocasting with mesoporous carbon hollow shells. *Journal of Materials Chemistry, 15,* 3126–3131.

Ying, J. Y., Mehnert, C. P., and Wong, M. S. (1999). Synthesis and applications of supramolecular-templated mesoporous materials. *Angewandte Chemie International Edition, 38,* 56–77.

Yoshida, H., Kawamoto, K., Kubo, H., Tsuda, T., Fujii, A., Kuwabata, S., and Ozaki. M. (2010). Nanoparticle-dispersed liquid crystals fabricated by sputter doping. *Advanced Materials, 22,* 622–626.

Zhang, G., Chen, X., Zhao, J., Chai, Y., Zhuang, W., and Wang, L. (2006). Electrophoretic deposition of silver nanoparticles in lamellar lyotropic liquid crystals. *Materials Letters, 60,* 2889–2892.

Zhang, H., Wu, J., Zhou, L., Zhang, D., and Qi, L. (2007). Facile synthesis of monodisperse microspheres and gigantic hollow shells of mesoporous silica in mixed water-ethanol solvents. *Langmuir, 23,* 1107–1113.

Zhou, Y., Huang, J., Shen, C., and Li, H. (2002). Synthesis of highly ordered LiNiO$_2$ nanowire arrays in AAO templates and their structural properties. *Materials Science and Engineering: A, 335,* 260–267.

Zhou, H., Fan, T., Li, X., Ding, J., Zhang, D., Li, X., and Gao, Y. (2009). Bioinspired bottom-up assembly of diatom-templated ordered porous metal chalcogenide meso/nanostructures. *European Journal of Inorganic Chemistry, 2,* 211–215.

CHAPTER 8

Recent Developments in Lyotropic Liquid Crystals as Drug Delivery Vehicles

DIMA LIBSTER, ABRAHAM ASERIN, and NISSIM GARTI

Casali Institute of Applied Chemistry, The Institute of Chemistry, The Hebrew University of Jerusalem, Jerusalem, Israel

Abstract

Recently, self-assembled lyotropic liquid crystals (LLCs) of lipids and water have attracted the attention of both the scientific and the applied research communities due to the remarkable structural complexity and practical potential of these nanostructures in diverse applications.

The phase behavior of mixtures of glycerol monooleate (monoolein, GMO) was particularly well studied due to the potential utilization of these systems in drug delivery systems, food products, and encapsulation and crystallization of proteins. The present chapter summarizes structural features of LLCs and recent systematic efforts to utilize these for solubilization and the potential release of drugs and biomacromolecules. One of the most interesting applications is the implementation of cell-penetrating peptides in the reversed hexagonal mesophase to enhance the skin-penetrating pattern of a model drug (sodium diclofenac).

Liquid crystal vehicles were shown to allow "on demand" targeted release, based on controlling the polymorphism of lyotropic liquid crystalline mesophases. Novel liquid crystalline matrix–gold nanorod hybrid materials were reported to induce light-triggered phase transition of liquid crystalline phases. Hydrophobized gold nanorods (GNRs) have been incorporated within the LLCs, composed of phytantriol and water, to provide remote heating, and trigger the phase transitions on irradiation at close to their resonant wavelength. A new pathway to pH-responsive LLCs, enabling the controlled release of hydrophilic drugs diffusing through the water channels of the mesophases, was also investigated. The system is capable of self-assembling into a reverse bicontinuous cubic phase of Im3m symmetry at pH 7 and transforming into a reverse columnar hexagonal phase at pH 2.

Self-Assembled Supramolecular Architectures: Lyotropic Liquid Crystals, First Edition.
Edited by Nissim Garti, Ponisseril Somasundaran, and Raffaele Mezzenga.
© 2012 John Wiley & Sons, Inc. Published 2012 by John Wiley & Sons, Inc.

Lyotropic liquid crystals were shown to entrap several nucleotides into cubic and lamellar monoolein-based mesophases in order to protect them and enable their release. Deoxyribonucleic acid (DNA) within two types of reverse columnar hexagonal mesophases was studied, one based on pure nonionic lipids and the other decorated by cationic lipids to induce opposite charges at the surfaces of the water channels of the mesophases. This provided new opportunities in the design technologies for DNA transfection and for gene delivery.

The main outcomes of the described research demonstrated that control of the physical properties of hexagonal LLC on different length scales is key for rational design of these systems as delivery vehicles for both low-molecular-weight therapeutics and biomacromolecules.

8.1 INTRODUCTION

Liquid crystals (LCs) are self-assembled organized mesophases with properties intermediate to those of crystalline solids and isotropic liquids (Gin et al., 2008). In LC phases, long-range periodicity exists, although the molecules exhibit a dynamical disorder at atomic distances, as is the case in liquids. Accordingly, these materials can also be considered ordered fluids (Larsson, 1989). Lyotropic LCs (LLCs) are materials that are composed from at least two molecules: an amphiphilic molecule and its solvent. A hydrophilic solvent, such as water, hydrates the polar moieties of the amphiphiles via hydrogen bonding, while the flexible aliphatic tails of the amphiphiles aggregate into fused hydrophobic regions based on van der Waals interactions. In addition to morphologic dependence on the chemical composition, LLC are also sensitive to external parameters, such as temperature and pressure (Amar-Yuli, 2008; Gin et al., 2008; Larsson, 1989). As a function of the molecular shape of

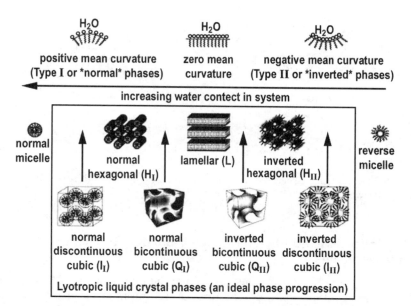

Figure 8.1 Schematic representations of common LLC phases formed by amphiphiles in water (Gin et al., 2008).

the surfactants, packing parameters, and interfacial curvature energy considerations, LLCs can be formed with aqueous domains ranging from planar bilayer lamellae to extended, cylindrical channels, to three-dimensional interconnected channels and manifolds (Gruner, 1989). These mesophases are defined as lamellar (L_α), hexagonal (H), bicontinuous cubic [Q (or V)], and discontinuous cubic (I) phases, based on their symmetry (Seddon, 1990). In addition, most lyotropic mesophases exist as symmetric pairs, a "normal" (type I) oil-in-water (O/W) system, consisting of lipid aggregates in a continuous water matrix, and a topologically "inverted" (type II) water-in-oil (W/O) version. The head-groups hydrated by water are arranged within a continuous nonpolar matrix, which is composed of the fluid hydrocarbon chains (Seddon, 1990). In addition to its biological significance, inverse lipid phases could be useful as host systems for incorporation of food additives (Sagalowicz et al., 2006a,b), the crystallization of membrane proteins for drug delivery (Cherezov et al., 2006; Sagalowicz et al., 2006b), and for inorganic synthesis (Boyd et al., 2007).

The lamellar structure does not possess any intrinsic curvature and is considered as the midpoint of an ideal, symmetrical LLC phase progression (Fig. 8.1) (Gruner, 1989; Seddon, 1990). The current review is mainly focused only on the inverted (W/O) mesophases (cubic and hexagonal), representing an important class of nanostructures for potential applications.

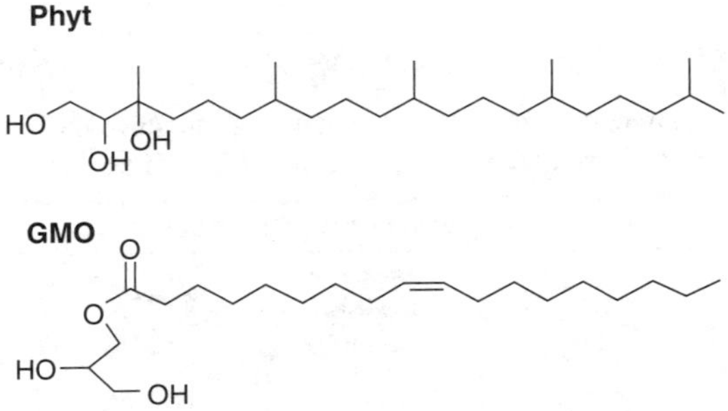

Phyt

GMO

Figure 8.2 Chemical structures for phytantriol (Phyt) and glyceryl monooleate (GMO) (Rizwan et al., 2009).

8.1.1 Monoolein and Phytantriol: Main Building Blocks of Lipid Mesophases

Only a few synthetic amphiphiles can mimic the behavior of biological lipids and form inverted mesophases. The unsaturated monoglycerides (monoolein and monolinolein) belong to this category. Glycerol monooleate (GMO, monoolein) is a polar lipid commonly used as a food emulsifier. Presently, GMO is the preferred amphiphile for formulating LC phases for scientific research and drug delivery. This is a nontoxic, biocompatible, and biodegradable lipid, which possesses low water solubility but swells and forms several LC phases in excess water (Drummond and Fong, 2000).

Another lipid, phytantriol, was recently shown to form cubic LC. The advantage of this lipid, compared to GMO, is that it is not susceptible to esterase-catalyzed hydrolysis and therefore provides additional stability to the mesophases. Phytantriol is commonly used as an ingredient in the cosmetics industry for improving moisture retention. Its phase behavior as a function of water concentration and temperature is very similar to that of GMO, although structurally they are very different (Fig. 8.2) (Rizwan et al., 2009).

8.1.2 Cubic Phases

The bicontinuous cubic phase is the most complicated among all the LLCs. Here the lipids are located in a complex, optically isotropic three-dimensional (3D) lattice. Cubic structures can be either normal (type I, O/W) or inverse (type II, W/O), and their topology is differentiated as bicontinuous or micellar, resulting in seven cubic space groups. Only three space groups are both inverse and bicontinuous structures (Fig. 8.3): the gyroid (G) type (Ia3d, denoted Q^{230}), the diamond (D) type (primitive lattice Pn3m, denoted Q^{229}), and the primitive (P) type (body-centered lattice Im3m, denoted Q^{224}) (Gruner et al., 1988). The aqueous channels of G surface consist of two separate, left-handed and right-handed helical channels. The aqueous channels can extend through the matrix,

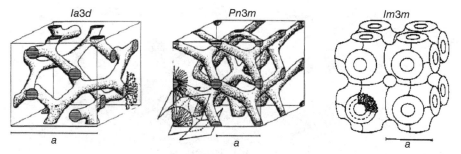

Figure 8.3 Schematic illustrations of the inverse bicontinuous cubic phases Ia3d (gyroid type), Pn3m (diamond type), and Im3m (primitive type) (Seddon et al., 2000).

but the centers of the water channels never intersect. Hence they are connected and the structure adopts a helical arrangement. The structure D is characterized by a bilayer, which separates two interpenetrating aqueous channel systems forming a diamond lattice. In this configuration four aqueous channels of the D surface meet at a tetrahedral angle of 109.5°. Structure P contains two aqueous channel systems that are separated by a bilayer. The unit cell possesses three mutually perpendicular aqueous channels, connected to contiguous unit cells, taking shape of a cubic array.

8.1.3   Reverse Hexagonal Mesophase as Delivery Vehicles

The reverse hexagonal mesophase is of a primitive type (P6mm) and characterized by one cylinder per unit cell corner. These densely packed, straight, water-filled cylinders exhibit two-dimensional ordering. Each cylinder is surrounded by a layer of surfactant molecules that are perpendicular to the cylinder interface such that their hydrophobic moieties point outward from the water rods (Fig. 8.4). There is a growing indication that inverse hexagonal mesophases play structural and dynamic roles in biological systems (Gradzielski, 2004; Seddon, 1990; Yang et al., 2003). These systems are assumed to be active as transient intermediates in biological phenomena that require topological rearrangements of lipid bilayers, such as membrane fusion/fission and the transbilayer transport of lipids and polar solutes (Gradzielski, 2004; Seddon, 1990; Yang et al., 2003). H_{II} mesophases have recently been considered promising drug delivery vehicles, mainly owing to their unique structural features (Boyd et al., 2006, 2007; Lopes et al., 2007).

Another LC structure, based on the close packing of inverse micelles, is a 3D hexagonal inverse micellar phase, of the space group P63/mmc, recently discovered by Shearman et al. (2006). Ternary lipid mixtures comprising two lipids, dioleoylphosphatidylcholine (DOPC), dioleoylglycerol (DOG), and cholesterol, of molar ratios 1:2:1 and 1:2:2, in the excess water-induced formation of 3D hexagonal mesophases over a temperature range of 16–52°C

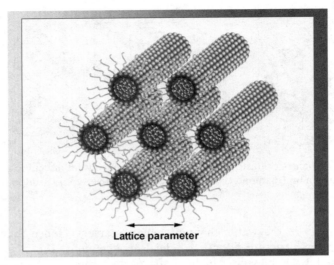

Figure 8.4   Schematic presentation of H_{II} mesophase showing the packing of water-filled rods surrounded by lipid layers.

and a wide range of pressures (1–3000 bars). The new inverse micellar phase consists of an hexagonal close packing (hcp) of identical, spherical inverse micelles, with two lattice parameters, a and c, that were found to be $a = 71.5$ Å and $c = 116.5$ Å (Fig. 8.5). This is the first new inverse LLC phase, which was reported for two decades, and it is the only known phase whose structure consists of a close packing of identical inverse micelles.

8.2   PHASE BEHAVIOR

Most of the common surfactants form direct (type I) phases, where the interface bends away from the polar solvent toward the tail region. However, most biological amphiphiles (such as phospholipids) form type II LLC, where the interface curves toward the polar region. Only a few synthetic amphiphiles can mimic the behavior of biological lipids and form inverted mesophases. The unsaturated monoglycerides (monoolein and monolinolein) belong to this category (Drummond and Fong, 2000). GMO, monoolein, is a polar lipid that is commonly used as a food emulsifier. Presently, GMO is the preferred amphiphile for formulating LC phases for scientific research and drug delivery. This is a nontoxic, biocompatible, and biodegradable lipid, which possesses low water solubility, but in excess water it swells and forms several LC phases.

Monoolein–water is one of the most extensively investigated types of lipid-based mesophases (Drummond and Fong, 2000). The phase diagram of this system (Fig. 8.6) revealed complex structural behavior. At room temperature

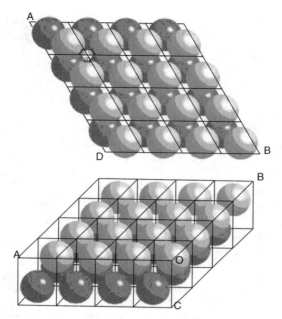

Figure 8.5 Plan (top) and perspective (bottom) views of the schematic structure of the 3-D hexagonal inverse micellar phase. The spheres represent the polar regions (water cores plus lipid head-groups), and the remaining fluid volume is filled by the hydrophobic regions of the lipid molecules. The different shading of the two identical layers of spheres is purely for clarity.

the following phase sequence existed upon increasing hydration: lamellar crystalline phase (L_c) in coexistence with an L_2 phase, lamellar mesophase (L_α), and the inverted bicontinuous cubic mesophases—gyroid Ia3d and diamond Pn3m. Upon heating, at about 85°C, the cubic phase is transformed into the H_{II} mesophase, followed by the micellar phase.

These concentration and temperature-dependent structural transitions can be qualitatively explained in terms of an effective critical packing parameter (CPP), as developed by Israelachvili et al. (1976). According to this theory, amphiphiles possess geometric parameters characterized by the critical packing parameter, CPP = $V_s/a_0 l$, where V_s is the hydrophobic chain volume, a_0 is the polar head-group area, and l is the length of the chain in its molten state. The packing parameter is useful in predicting which phases can be preferentially formed by a given lipid, since it connects the molecular shape and properties to the favored curvature of the polar–apolar interface, and therefore the topology and shape of the aggregate. The main factors responsible for alterations of the mentioned parameters are the molecular shape of the surfactant, the chemical composition, and the temperature. According to the theory developed by Israelachvili and co-workers, inverse mesophases are formed by amphiphiles with CPP > 1, which adopt inversed cone-shaped geometry. Such

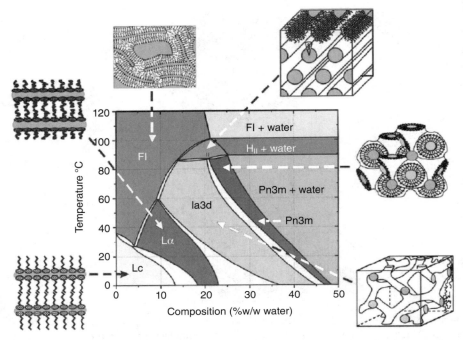

Figure 8.6 Temperature–composition phase diagram of the monoolein–water system (up to 50 wt % water). A cartoon representation of the various phase states is included in which colored zones represent water. The mesophases are as follows: L_c, crystalline lamellar; L_α, lamellar; Ia3d, gyroid inverted bicontinuous cubic; Pn3m, primitive inverted bicontinuous cubic; H_{II}, inverted hexagonal, and F1, reverse micelles fluid phases (Cherezov et al., 2006).

lipids should possess a small head-group area as compared to that of the tail region. In contrast, direct mesophases are preferred when the surfactant head cross section is larger than that of the tail, resulting in CPP < 1. In the case of lamellar mesophase these parameters are equal and hence result in CPP = 1. The size of the polar head-group area is dictated by both molecular shape and the degree of hydration.

Within the boundaries of a given mesophase the head-group area normally increases with increasing hydration and decreases upon temperature rise. The length and the volume of the lipophilic tails are also greatly affected by a temperature increase. Stronger thermal motion of the tails eventually decreases the length of the tails but increases their volume.

It was shown for monoglyceride-based LLC that H_{II} mesophases are formed from amphiphiles with CPP ~ 1.7, cubic phases with CPP ~ 1.3, and lamellar structures with CPP = 1 (Larsson, 1989). The lamellar phase is characterized by zero curvature since the cross sections of the polar heads and the lipophilic tails are similar. Upon increased hydration, cubic phases (Ia3d and Pn3m) with higher curvature and consequently higher CPP are preferred. The effective CPP theory can supply a reasonable explanation for the temperature-induced structural shifts from lamellar through cubic to reverse hexagonal phases, requiring higher curvature than in the lamellar phase. Increasing the thermal motion of both the hydrocarbon chains and the water molecules would increase the CPP values by way of expanding the volume of the lipophilic moiety, but decreasing the chain length and the head-group area. This leads to an increase in curvature and therefore induces the formation of cubic and hexagonal mesophases.

It should be noticed that the CPP concept may only be used as a rule of thumb to predict transitions between various forms of LLC. For example, it was shown that it fails to predict the appearance of the inverse bicontinuous phases and an intermediate phase (Shearman et al., 2006; Shah and Paradkar, 2005). This is the reason a thermodynamic approach considering the total free energy of the LLC systems was applied by Seddon and co-workers (Seddon et al., 1996, 2000; Shearman et al., 2006); however, this is beyond the scope of this chapter.

8.3 H_{II} MESOPHASE COMPOSED OF GMO–TRIGLYCERIDE–WATER AS DRUG DELIVERY SYSTEMS

It was shown that in the monoolein-based system, the cubic phase is transformed into an H_{II} mesophase upon heating at ca. 85°C (Drummond and Fong, 2000). CPP theory can supply a reasonable explanation to the temperature-induced structural shifts from lamellar through cubic to reverse hexagonal phases, requiring greater curvature than in the lamellar phase. Increasing the thermal motion of both the hydrocarbon chains and the water molecules would increase the CPP values via expanding the volume of the lipophilic moiety but decrease the chain length and the head-group area. This leads to an increase in curvature and therefore induces the formation of cubic and hexagonal mesophases. (Gin et al., 2008; Israelachvili et al., 1976).

Bearing in mind the thermodynamic and structural considerations noted above, systematic research was conducted to decrease the cubic to hexagonal temperature transition and stabilize the GMO-based H_{II} mesophase at room temperature (Amar-Yuli and Garti, 2005). To achieve this goal, experiments relative to incorporation of triglycerides (TAGs) with various chain lengths to the binary GMO–water system were conducted. Amar-Yuli and Garti surmised that immobilization of a TAG between GMO tails would lead to a change in the geometry of monoolein molecules from cylindrical to wedge-shaped and thereby an increase in the CPP value of the system (Amar-Yuli and Garti, 2005). This should encourage transition from lamellar or cubic phases to hexagonal structures. In addition, the immobilization of TAG in the system was expected to reduce the packing frustration, stabilizing the hexagonal LLC at room temperature. These experimental results showed that a critical and optimal chain length of the triglyceride is required to induce the formation of H_{II} at room temperature. Among the examined TAGs, tricaprylin (C_8) was the most successful, flexibly accommodating between the tails of the GMO.

The structural properties of ternary hexagonal mesophases composed of GMO, tricaprylin, and water were extensively and systematically studied as was shown in numerous publications (Achrai et al., 2011; Amar-Yuli et al., 2007a,b, 2008a,b, 2009). Several additives, including dermal penetration enhancers, were solubilized to control the physical properties of these carriers (Amar-Yuli et al., 2007b, 2008a,b), such as viscosity and thermal stability. For instance, the synergistic solubilization of two major hydrophilic (vitamin C, ascorbic acid, AA) and lipophilic (vitamin E, D-α-tocopherol, VE) antioxidants within H_{II} mesophases was reported by Bitan-Cherbakovsky et al. (2009, 2010). This enabled expanding conditions to obtain stable H_{II} mesophases at room temperature. In addition, it was shown that phosphatidylcholine (PC) can be embedded into the ternary GMO–TAG–water system (Ben Ishai et al., 2009; Libster et al., 2007, 2008, 2009a,b). PC was incorporated into the ternary mesophases because it is known that phospholipid-based structures possess relatively high thermal stability and enhance transdermal drug permeation (Kurosaki et al., 1991; Touitou et al., 1994; Spernath and Aserin, 2006) and transmembrane transport across the digestive tract (Brondsted et al., 1995; Liu et al., 1999).

Incorporation of PC to the ternary system caused competition for water binding between the hydroxyl groups of GMO and the phosphate groups of the PC, leading to dehydration of the GMO hydroxyls in favor of the phospholipid hydration (Libster et al., 2009a). On the macroscopic level, this was correlated with the improvement of elastic properties and thermal stability of the H_{II} mesophase (Libster et al., 2007). Structural flexibility and stability is essential for a drug delivery system, especially for tuning its physical properties (such as viscosity) and composition.

Libster et al. (2008) also explored and demonstrated correlations between the microstructural and mesostructural properties of the reverse hexagonal LLC, using the environmental scanning electron microscopy (ESEM) technique. It was shown that the mesoscopic organization of these systems is based on an alignment of polycrystalline domains. The topography of H_{II} mesophases, imaged directly in their hydrated state, as a function of aqueous phase concentration, was found to possess fractal characteristics, indicating a discontinuous and disordered alignment of the corresponding internal water rods on the mesoscale. Fractal analysis indicated that the mesoscale topography of the H_{II} phase was likely to be influenced by the microstructural parameters and the water content of the samples (Libster et al., 2008). Garti and coworkers also made considerable progress elucidating the solubilization of therapeutic peptides into the H_{II} mesophase (Cohen-Avrahami et al., 2010; Libster et al., 2007, 2009a,b; Mishraki et al., 2010a,b).

Lately, for the first time Cohen-Avrahami et al. (2011) used cell-penetrating peptides (CPPs) as enhancers to overcome the stratum corneum barrier.

Sodium diclofenac (Na-DFC) is a common NSAID (nonsteroid anti-inflammatory drug) used to treat mild to moderate pain, particularly when inflammation is also present. This study utilized the advantages of the H_{II} mesophase as a transdermal vehicle, in addition to those of the CPPs as skin penetration enhancers, for the development of improved drug delivery vehicles in which the drug diffusion rate might be carefully controlled.

The selected CPPs, representing prominent members of these families, were chosen for their great efficiency and short length; their amino acid sequences are the following: RALA″, RALARALARALRALAR, is a 16-amino-acid peptide belonging to a synthetic family of CPPs, based on GALA, amphipatic peptides named after their alanine–leucine–alanine repeats that exhibit improved membrane permeability. "PEN," penetratin (RQIKI-WFQNRRMKWKK), is a 16-amino-acid sequence based on the active penetrating peptide produced by the homeodomain of the antennapedia homeoprotein. "NONA," nonaarginine, RRRRRRRRR. Arginine was shown to play a key role in CPP attachment to membranes; thus, much research was done concerning the penetrating activity of polyarginine through membranes. The maximal solubilization load of Na-DFC, RALA, and PEN into the H_{II} structures were found to be 5 wt %, while NONA solubilization capacity was 6 wt %.

Small-angle X-ray scattering (SAXS) data of the Na-DFC-loaded systems revealed that the drug caused shrinkage of the hexagonal channel diameter. The lattice parameter sharply decreases from 57.55 ± 0.5 Å in the blank system to 51.55 ± 0.5 Å in the 5-wt % loaded system (Fig. 8.7), leading at to denser packing of the system. The investigators assumed that such a "kosmotropic effect" of the Na-DFC might originate from the drug's location between the GMO molecules, causing an enhancement of the interactions between them. This assumption was supported by the slight increase in the domain lengths of the hexagonal clusters in the presence of Na-DFC from 392 to 493 ± 50 Å (Fig. 8.7), hinting a stabilizing effect. The main effect appeared in the low concentration regime, whereas the additional Na-DFC quantities had a weaker influence. These effects were noticed until the system transformed into a micellar structure.

In addition, differential scanning calorimety (DSC) results suggested enhanced water binding in the presence of the drug. Shrinkage of the aqueous columnar cylinders (as detected also by SAXS) caused a change in the structure curvature (curvature more concave toward the inner water phase) and an increase in the fraction of the bound water at the expense of the free water that populated the inner channels. It also indicated that Na-DFC interacts with the surfactant tails, reducing the number of molecules that are free to participate in the melting process. Combined DSC measurements and attenuated total reflection (ATR)–Fourier transform infrared (FTIR) analysis led the authors to conclude that Na-DFC solubilization effect was primarily manifested at the interfacial region, when the drug intercalated between the GMO molecules. Na-DFC influenced the intermolecular bonding at this molecular region. It strengthened the interactions between GMO tails, loosened the

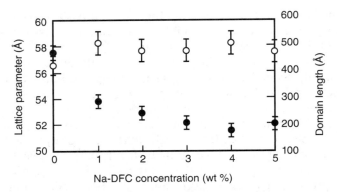

Figure 8.7   Lattice parameter (●) and the domain length (○) obtained from the SAXS measurements versus solubilized Na-DFC concentration at the H_{II} mesophases (Cohen-Avrahami et al., 2011).

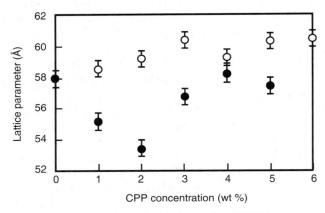

Figure 8.8   Lattice parameter, calculated from the SAXS measurements, versus NONA (○) and PEN (●) concentration in the mesophase (Cohen-Avrahami et al., 2011).

repulsion tension between the carbon atoms at the head groups, and enabled the formation of a tighter structure with higher curvature.

Each CPP was separately solubilized within the H_{II} for identification of its specific location and chemical interactions within the mesophase. The SAXS measurements indicated two trends in the effect of the peptides. Up to 2 wt % of PEN solubilization and initial shrinkage of the aqueous cylinders to a minimum of 53.4 ± 0.5 Å was detected. With increasing solubilization loads of PEN to 5 wt %, the shrinkage changed, and a swelling effect of the cylinders was detected until their diameter reached their original radius of 57.9 ± 0.5 Å (Fig. 8.8).

NONA solubilization had a minor, yet consistent swelling effect on the cylinders. The initial lattice parameter in the blank system was 57.9 ± 0.5 Å, and it slightly increased with NONA loading to a maximal value of 60.5 ± 0.5 Å at 6 wt % NONA. Its effect was found to be focused mostly on the water and the GMO carbonyls. The stretching mode of the water was gradually shifted from 3361 to 3346 cm^{-1} with increasing NONA concentration in the mesophase, indicating stronger hydrogen bonding of the water with the peptide.

It was summarized that structural investigation revealed that the solubilization sites of the different guest molecules depend on their molecular structure and differ significantly. Na-DFC populates the interfacial region, enhances the interactions between GMO tails, and shrinks and stabilizes the H$_{II}$ mesophase (Fig. 8.9c). PEN solubilization is concentration dependent. The initial PEN loads populate the hydrophilic head-group area, whereas the higher PEN loads pack closer to the GMO tails (Figs. 8.9e and 8.9f). The hydrophilic NONA populated the inner channels region and swelled the mesophase (Fig. 8.9d). RALA acted as a chaotropic agent at the H$_{II}$ mesophase, interacting mostly with the water within the channels and enhancing the hydration of the GMO head groups.

Furthermore, the effect of the various peptides on skin permeation efficiency was determined by diffusion experiments of Na-DFC through porcine skin using Franz diffusion cells. All systems were composed of H$_{II}$ mesophases loaded with 1 wt % Na-DFC and 1 wt % of each CPP (the control was a system with Na-DFC and no CPP). These experiments indicated that all three peptides increased significantly the diffusion of Na-DFC through the skin (Fig. 8.10a).

NONA was found to be the most efficient CPP, significantly enhancing the transdermal penetration (a 2.2-fold increase in the total amount of Na-DFC that diffused through the skin). PEN and RALA caused 1.9- and 1.5-fold increases, respectively, compared with the control system. In all tested systems, there was a gradual and linear increase in the cumulative penetration of Na-DFC with time.

A calculation of Na-DFC percentage that permeated through the skin showed that at the blank systems the total amount that diffused through the skin during the 24-h experiment was 0.9% from the initial applied dose. The amount released at the NONA, PEN, and RALA systems was 1.9, 1.8, and 1.3%, respectively. The permeability coefficients (K_p, calculated as cm·h^{-1}), derived from the steady-state flux of Na-DFC, revealed the same tendency (Fig. 8.10b), a 1.5-fold increase in the presence of RALA and 2.3- and 2.2-fold increases with PEN and NONA, respectively. Since the skin permeation studies revealed different profiles for each of the three peptides, a more detailed analysis using the well-established technique of "emptying experiments" was conducted by the authors. These experiments aimed to determine whether the rate-determining step of Na-DFC skin permeation from the mesophase is migration out of the mesophase or permeation through the skin. In addition,

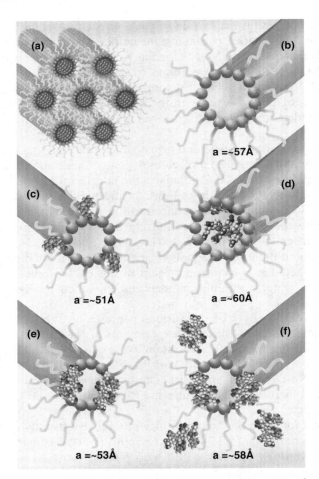

Figure 8.9 (a) Schematic illustration of the H_{II} mesophase general structure. The water populates the inner region of the cylinders; the GMO tails as well as the TAG point outward. (b) Single "blank," empty system of the H_{II} mesophase and (c) Na-DFC-loaded system. The drug populates the interfacial area and causes channel shrinkage. (d) NONA-loaded H_{II} mesophase. The peptide populates the aqueous channels and swells them. (e, f) PEN-loaded systems. At low PEN concentration (e), it is adsorbed on the GMO head groups and causes channel shrinkage. With increasing PEN concentration (f), it populates an additional hosting region at the interface (Cohen-Avrahami et al., 2011).

this analysis provides information on the rate-determining step in the CPP enhancement.

The emptying experiments revealed a surprising result. The control system, without CPP, released the largest quantities of Na-DFC with time. The release from the PEN systems was the second in order, whereas NONA and RALA systems released the lowest quantities of Na-DFC (Fig. 8.11a).

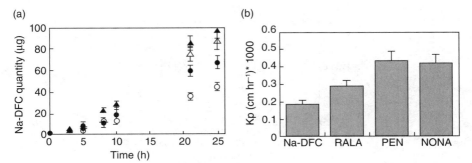

Figure 8.10 (a) Na-DFC cumulative skin permeation quantity vs. time in the different systems: control (Na-DFC only) (○), RALA-loaded (●), PEN-loaded (Δ), and NONA-loaded (▲). (b) The calculated permeability coefficient (K_p) based on the diffusion rate of 1.0 wt % Na-DFC via Franz diffusion cells as affected by the different peptides added to the mesophase (Cohen-Avrahami et al., 2011).

The Higuchi diffusion equation was applied to analyze Na-DFC release from the examined mesophases. The Higuchi model implies that drug release is primarily controlled by diffusion through the matrix and can be described by the following equation:

$$Q = [D_m(2A - C_d)C_d t]^{1/2}$$

where Q is the mass of drug released at time t and is proportional to the apparent diffusion coefficient of the drug in the matrix, D_m; the initial amount of drug in the matrix, A; and the solubility of the drug in the matrix, C_d. The slope of a linear fit of the data from this plot is proportional to the "apparent diffusion coefficient" for the drug in the matrix and permits preliminary assessment of diffusion as the primary means of drug release from the correlation coefficient for the linear fit and also as a means to compare the diffusion of a drug from the different matrices into the release medium. The linearity of the plots was found to be >0.95 for all CPPs, indicating the existence of a diffusion-controlled transport mechanism (Fig. 8.11). The slope was the greatest for the blank system, 2.8 $h^{-1/2}$, and it decreased to 2.0, 1.3, and 1.2 in the PEN, NONA, and RALA systems, respectively.

The "blank" systems released the highest amounts of Na-DFC, and the release from the CPP-loaded systems was much lower. The order of Higuchi slopes was exactly the same; blank > PEN > NONA > RALA, and their values were 3.0, 0.9, 0.7, and 0.6 $h^{-1/2}$, respectively.

Considering the Na-DFC's amphiphilic nature and the H_{II} mesophase structure, in which the aqueous phase is tightly packed within the lipid structure, with the limited accessibility to the surrounding media, one can assume that the drug diffusion out of the mesophase occurs through the lipophilic oily regions. The CPPs incorporated within the mesophase slowed the

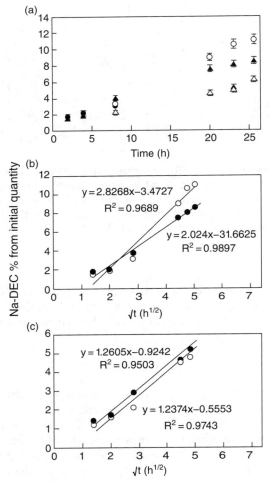

Figure 8.11 (a) Na-DFC release to water from the H_{II} mesophase containing: no CPP (○), RALA (●), NONA (Δ), and PEN (▲). Na-DFC release vs. the square root of time from (b) the "blank" (no CPP), (○) and PEN (●) systems and (c) RALA (●) and NONA (○) systems (Cohen-Avrahami et al., 2011).

drug diffusion rate by decreasing its mobility through the mesophase by any specific molecular interfacial interactions.

Nevertheless, the emptying experiments showed that the main enhancing effect of the CPPs on skin penetration does not take place in the step of drug migration out of the mesophase. Apparently, it is due to the enhanced permeation capability of the skin membrane and via any kinetic diffusion-controlled process. It should be stressed that RALA and NONA, which populate the inner aqueous channels, revealed a similar diffusion profile with the same slope, 1.2 ± 0.1 h$^{-1/2}$. PEN, which populates the outer interfacial region, is less disturbed in its diffusion, causing a release profile with a slope of 2 h$^{-1/2}$. Nevertheless, this interesting mechanism of the drug migration from the H$_{II}$ mesophase in the presence of CPP molecules co-solubilized within the mesophase should be further investigated.

8.4 SOLUBILIZATION AND DELIVERY OF BIOMACROMOLECULES

Solubilization of biomacromolecules into LLCs is of great practical interest. Mezzenga et al. (2005) investigated the effects of confining polymeric forms of glucose in the water domains of liquid crystals composed of monoglyceride and water solutions. Using rheological and SAXS methods, the authors studied the effect of sugar concentration and molecular weight on cubic-to-H$_{II}$ transitions by further solubilization of maltose and dextran I in the system. It was found that higher concentrations of sugars resulted in a decrease of the Pn3m–H$_{II}$ transition temperature, as a consequence of the Hofmeister effect. However, no dependence on molecular weight of the sugar was observed. In the case of a polymeric sugar, as soon as the end-to-end distance of the polymer approached the diameter of water channels, its molecular weight made it more difficult for the guest molecule to fit within narrow water channels. Dextrans of larger molecular weights than dextran I did not fit into the water channels of the cubic phases, inducing a phase separation. Moreover, it was noticed that dextran I, solubilized into the Ia3d phase channels, induced the appearance of a Pn3m phase, which possesses larger water channels. Hence, when the chain end-to-end distance of the polysaccharide approached the diameter of the water channels, it induced phase transitions toward structures with different topologies, thus enhancing the solubilization capacity of large sugars.

Peptides and proteins are increasingly considered for the development of new therapeutic compounds (Dass and Choong, 2006; Kumar et al., 2006). Peptides of various sizes are currently tested for a broad spectrum of diseases, including skin cancer, acne, psoriasis, hypertension, hepatitis, and rheumatism. These biomacromolecules are usually found to be very effective, and in most cases low doses of them are needed for good medical treatment. Nevertheless, when delivered orally, these peptides and proteins are susceptible to cleavage by various enzymes, mainly the human digestive proteases, leading to practically poor bioavailability and poor pharmaceutical efficacy. As a result of these critical limitations, excessively large doses of the peptide-based drugs are usually required to obtain therapeutic effects in vivo, which in turn often cause a wide range of hazardous side effects (Dass and Choong, 2006; Kumar et al., 2006). This is the main reason the medical application of these otherwise high potential therapeutics is currently very limited, and even those few that are used at the present time are far from being efficient. In this respect, LLCs seem to be promising candidates as alternative delivery means for various pharmaceuticals. Such vehicles can provide enhanced drug solubility, relative protection of the solubilized drugs, and controlled release of drugs, while avoiding substantial side effects (Ericsson et al., 1991; Fong et al., 2009; Shah et al., 2001). The efficiency of the LLC delivery system can be further enhanced in transdermal delivery, where peptide degradation and low absorption rates are much less relevant.

Significant progress has been made during the last decade in the characterization of the interactions of peptides and proteins with LLCs, mainly for crystallographic and drug delivery purposes. LLCs have been shown to provide sustained release of drug molecules with a range of physicochemical properties (Drummond and Fong, 2000; Shah and Paradkar, 2005; Shah et al., 2001). The cubic phase was shown to host a range of water-soluble biomacromolecules for use in controlled-release applications. Clogston and Caffrey (2005) systematically examined biomacromolecules ranging from a single amino acid (tryptophan) to complex proteins and nuclear deoxyribonucleic acid (DNA). These investigators found that for a given cubic phase, the rate of diffusion depends on the molecular size of the specific diffusing molecule. Shah and Paradkar (2005) prepared in situ a cubic phase system of GMO. This system provided protection to the metalloenzyme seratiopeptidase (STP) in the gastric environment and gave delayed and controlled release with no initial burst after oral administration.

Although the cubic phase has been proposed and studied as a drug delivery vehicle, there is relatively little information about the interactions of peptides and proteins with the H_{II} mesophases and the therapeutic potential of these structures. Libster and co-workers explored and controlled the physical properties of H_{II} mesophases to use these systems as drug delivery vehicles for biologically active peptides and proteins. The results of this structural research enabled significant expansion of the application spectrum of hexagonal LLCs, utilizing them for the solubilization of peptides and proteins (Libster et al., 2007, 2009a,b, 2011), into this mesophase and its utilization as sustained drug delivery vehicle. Two model cyclic peptides, cyclosporin A (11 amino acids) and desmopressin (9 amino acids), of similar molecular weight but with very different hydrophilic and lipophilic properties, were chosen to demonstrate the feasibility of using the H_{II} mesophase (Libster et al., 2011). In addition, a larger peptide, RALA (Cohen-Avrahami et al., 2010) (16 amino acids), was solubilized into the H_{II} structures as a model skin penetration enhancer. Finally, a larger macromolecule, LSZ protein (129 amino acids), was directly incorporated into a GMO-based H_{II} mesophase (Mishraki et al., 2010a,b).

With the aim of designing a biologically inspired carrier in which the encapsulation and the delivery of DNA can be efficiently controlled, Amar-Yuli et al. (2011a) have designed two lipid-based columnar hexagonal LLCs, which can accomplish two opposite roles while maintaining the same liquid crystalline symmetry. The first system was based on a nonionic lipid, such as monoolein, while the second system was modified by a low additional amount of the oleyl amine cationic surfactant. DNA was enzymatically treated to generate a broad distribution of contour lengths and diffusion characteristic times.

The impact of DNA confinement on these two columnar hexagonal structures was investigated by SAXS. Both the neat neutral and cationic LLC systems had a columnar hexagonal symmetry, the main difference being the reduced lattice of 49.8 Å in the cationic system compared to the 55.5 Å of the neutral formulation. This difference was interpreted to be the result of the kosmotropic effect of the oleyl amine, which, due to its charged nature, dehydrates the surfactant polar heads and reduces the LLC lattice.

A very different effect on the H_{II} lattice parameter was found when the DNA (1.4 wt % from the aqueous phase) was incorporated into the two systems. In the nonionic LLC system, the lattice parameter decreased from 55.5 to 50.8 Å (± 0.5 Å) in the presence of DNA. In contrast, in the cationic columnar phase the lattice parameter increased from 49.8 to 59.2 Å. These effects can be rationalized by considering the negative charge of the DNA: When added to the neutral formulation, DNA leads to a negative effective charge that dehydrates the lipid polar heads and reduces the lattice parameter; on the other hand, when added to the cationic formulation, DNA partially neutralizes the overall positive charge, moderates the dehydration effect caused by the cationic surfactants, and induces, at least partially, swelling of the lattice.

Amar-Yuli et al. (2011a) postulated two different possibilities of arrangements for the DNA within the nonionic and cationic columnar hexagonal phases. Being strongly hydrophilic, the DNA double strands must be either segregated outside the water channels, within the grain boundaries, or confined within the relative narrow aqueous cylinders of the H_{II} structure. The strong impact of DNA on the lattice of the mesophases provides a robust indication that the DNA molecules were indeed confined within the water channels of the reverse hexagonal phases, although the different trends observed for the lattice evolution in the nonionic and cationic case suggest two different types of interactions among the lipids and the DNA.

This issue was directly addressed by performing ATR-FTIR analysis on the two formulations and indirectly by following the release of the DNA from each LLC system by ultraviolet (UV) spectrophotometry. In the nonionic-based H_{II} mesophase the results provided an evidence for a breakage of GMO–GMO and GMO–D_2O hydrogen bonds due to incorporation of DNA molecules. Furthermore, the carbonyl absorption revealed a slight modification in the area of their peaks, when the DNA was present. In the absence of DNA, the number of hydrogen-bonded carbonyls was calculated to be 72%, and it slightly increased to 78% in the presence of DNA. It was concluded that the DNA molecules interfere with the interfacial region of the cylinders, with tangible effects on the water (D_2O) and GMO molecules, including the hydroxyl and carbonyl groups and the ester moiety. The decrease in lattice parameter upon the addition of DNA, as detected by the SAXS measurements, is consistent with a dehydration of the GMO hydroxyl groups, that is, a breakage of hydrogen bonds between GMO polar groups caused by the presence of the highly polar DNA. Additionally, the modifications within the vicinity of the GMO–water interface imply a less ordered state with the increased content of *gauche* conformations, which is again coherent with the shrinkage of the H_{II} cylinders.

According to Amar-Yuli et al. (2011a), these results indicated that the DNA is confined within the aqueous domains of the H_{II} cylinders, stabilized by hydrogen bonds with the water and/or GMO head groups, and leading to an observable rupture of the H bonds in the H-bond donor part of the GMO polar heads (OH) and a decrease in the frequency of the H-bond acceptor bands of the GMO polar heads (CO–O). The role of DNA in this case is thus consistent with moving away bound water from the interfacial region of the water channels.

Examination of the second H_{II} system, which contains the cationic surfactant oleyl amine, shows very moderate changes upon incorporation of DNA in all the bands considered when compared to the nonionic mesophase. These data clearly suggested that in the presence of a cationic surfactant decorating the interface of the water channels, the interactions between the lipid polar heads and DNA change in nature, as the hydrogen bonding of D_2O and the GMO polar heads, is mostly unaffected by the presence of DNA. It was inferred that the DNA is confined in the water channels as well as in the cationic H_{II} mesophases, and electrostatically bound to their surfaces.

The DNA cumulative release profile from the two mesophases is depicted in Figure 8.12, which illustrates the difference in release rate from the nonionic and cationic H_{II} host mesophases. While practically no release of DNA mol-

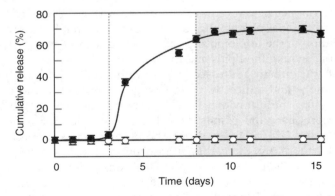

Figure 8.12 Release profile of DNA from the two columnar H_{II} mesophases considered: release from the nonionic system (filled circle, GMO–tricaprylin–water–ascorbic acid) and release from the cationic-based columnar hexagonal phase (open circle, GMO/oleyl amine–tricaprylin–water/ascorbic acid) at room temperature (Amar-Yuli et al., 2011a).

ecules was found in the presence of oleyl amine even after 15 days, three main steps on the release profile can be identified when DNA is confined within the pure nonionic columnar system. During the first 3 days, no release is detected, and this can be attributed to the lag time needed for DNA molecules to diffuse out of the H_{II} cylinders. In the range of 4–9 days the release of DNA from the LLC system into the excess water begins, progressing at a relatively high rate, as indicated by the sharp slope observed in Fig. 8.12. After 9 days the release of DNA slowed down significantly and reached a plateau. At this point a significant amount of DNA, approximately 65% of the initial content, has already been released by the nonionic H_{II} mesophase. These results are consistent with the conclusions based on SAXS and ATR-FTIR analyses stating that hydrogen bonds between DNA, water, and GMO head-groups are responsible for the DNA confinement within the water channels, in the case of the nonionic hexagonal mesophase, which allows transport of DNA into the excess water. In contrast, the presence of cationic oleyl amine in the ionic hexagonal mesophase leads to such a strong electrostatic confinement of the DNA at the water–lipid interface that release into the excess water is prevented: This formulation is then more reliable for permanent encapsulation and protection of the DNA rather than a controlled release.

In order to gain insight on the correlations between the release loads and the molecular conformation of the DNA molecules released, single-molecule atomic force microscopy (AFM) was used to provide the needed information on DNA contour length distributions (Fig. 8.13). Figure 8.13a depicts the DNA contour length distribution prior to the release experiment. As can be seen, several DNA fragments are generated upon the enzymatic treatment and can be divided into three main lengths: relatively small fragments (=100 nm contour length), medium size DNA strands (130–160 nm), and large macromolecules (~190 to 200 nm). The contour length distribution of DNA after its release from the nonionic system was examined on the third, fourth, and the fifteenth days of release, corresponding to the lag time, the sharp slope, and plateau regions as identified in the release profile (Figs. 8.13b–8.13d).

During the lag time, on the third day of the release experiment, only very short-length DNA fragments with contour lengths below 40 nm could be observed by AFM analysis (Fig. 8.13b). These very short strands were too small and in too low concentration to be detected by the spectrophotometer, resulting in the flat absorption signal. The following day (day 4), in correspondence with the sharp increase in the diffusion as measured by the UV–visible absorption, the medium-size DNA strands, mainly 40–100 nm in contour length, started to appear, coexisting with the very short DNA fragments (Fig. 8.13c). Finally, after 15 days, when the release profile has reached a plateau, the longest fragments, 190–200 nm in contour length, also appeared in the AFM analysis of the excess water (Fig. 8.13d). Interestingly, after 15 days, but only after this time, the contour length distribution of the DNA released matches closely the initial distribution, indicating that release is nearly entirely completed (Fig. 8.13d).

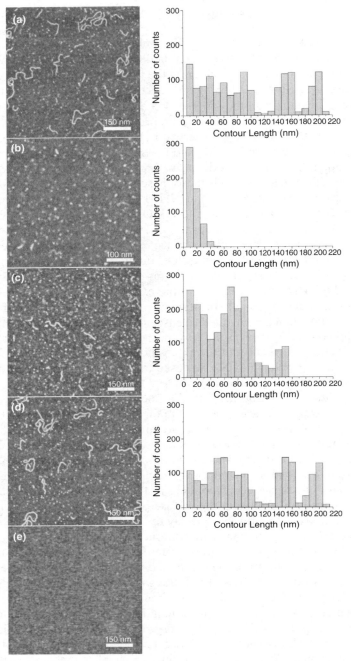

Figure 8.13 AFM image and the corresponding contour length distribution of DNA fragments: (a) before incorporating DNA into the H_{II} nonionic LLC system; (b) after 3 days of release in the excess water environment; (c) after 4 days release, and (d) after 15 days release at room temperature. (e) AFM image after 15 days of DNA release from the cation-based H_{II} LLC system (Amar-Yuli et al., 2011a).

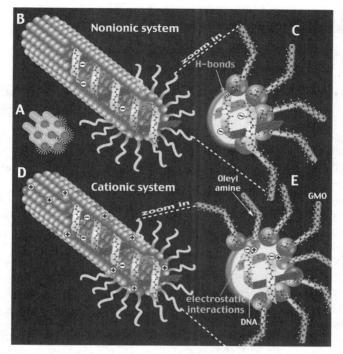

Figure 8.14 Schematic summary illustrating a (A) typical columnar hexagonal phase, the (B, C) DNA entrapment and interactions with the surfactants forming the nonionic, and (D, E) cationic H_{II} systems. The molecular ratio between all the components is not in scale in the image (Amar-Yuli et al., 2011a).

Quite in contrast, AFM examination of the excess water in the case of the cationic hexagonal phase reveals no traces of DNA, including the smallest fragments detectable by high-resolution AFM (Fig. 8.13e), confirming the strong binding of the DNA to the cationic lipids.

Figure 8.14 summarizes the main findings of this work. When confined within nonionic columnar hexagonal phases, DNA interacts with the polar heads of the lipids via hydrogen bonding (as illustrated in the figure), but this enables a controlled release of the DNA in excess water following three main release stages. When positive charges belonging to cationic lipids are decorating the water channels, binding between the lipid and DNA becomes strong, and any release of DNA is completely suppressed at the charge ratio used. It is reasonable to anticipate that further adjustment of the positive:negative charge ratio or use of variable ionic strengths to partially screen electrostatic attraction can be used as additional means to fine tune the DNA release kinetics.

8.5 "ON DEMAND" LYOTROPIC LIQUID CRYSTAL-BASED DRUG DELIVERY SYSTEMS

Based on the principle that the H_{II} phases released model hydrophilic and hydrophobic drugs more slowly than the GMO cubic phase matrix, Fong et al. (2009) designed phytantriol and GMO-based bicontinuous cubic (Q_2) and H_{II} nanostructures, designed to allow change to the nanostructure in response to an external change in temperature, with an intention of eventual control of drug release rates in vivo. Using glucose as a model hydrophilic drug, drug diffusion was shown to be reversible on switching between the H_{II} (very low release) and Q_2 nanostructures (high release), at temperatures above and below the physiological temperature, respectively (Fig. 8.15).

However, the matrix used by Fong et al. (2009) required the inclusion of a modifier (vitamin E acetate) to reduce the transition temperature close to physiological temperature for in vivo application, thereby enabling control over the structure by application of, for example, a heat pack to the skin surface after subcutaneous administration. This is a major limitation as there is no specificity in the heat source; hence, exposure to extremes of temperature may unintentionally induce drug release. Consequently, an alternative means to induce the phase transition was necessary that did not require direct heating and did not require a reduction in the temperature at which the transition occurs, ideally occurring at the inherent transition temperature without additive (approximately 55°C for phytantriol), removing the potential for accidental activation.

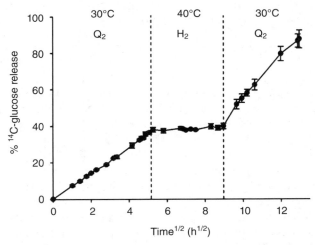

Figure 8.15 Dynamic release profiles for glucose into phosphate-buffered saline from phytantriol +3% vitamin E acetate with changing temperature, plotted against square root of time. Temperature was switched from 30°C → 40°C → 30°C at the times indicated by the dashed lines in order to form the phase transition cubic → hexagonal → cubic (Fong et al., 2010).

In this context, toward the design of an advanced drug delivery system based on light-triggered phase transition of liquid crystalline phases, Fong et al. (2010) reported the design of novel liquid crystalline matrix–gold nanorod hybrid materials. Hydrophobized gold nanorods (GNRs) have been incorporated within the liquid crystalline matrix, composed of phytantriol and water, to provide remote heating and trigger the phase transitions on irradiation at close to their resonant wavelength. The surface of plasmonic metal nanoparticles delivers heat into surrounding material on exposure to an appropriate light source at the plasmon resonance. The application of near-infrared-sensitive nanorods has significance in designing systems for ophthalmic, subcutaneous, or deeper tissue applications. These investigators used plasmonic nanoparticles to achieve reversible phase transitions, hence offering a novel practical solution. The presence of nanorods at concentrations up to 3 nM did not change the lattice parameter of cubic (v_2), hexagonal (H_{II}), and micellar (L_2) systems in the absence of laser irradiation.

Irradiation of the system [at 808 nm near-infrared (NIR) diode laser], on inclusion of a low concentration of GNRs (0.3 nM) in the matrix, did not induce a change in phase structure away from the Pn3m bicontinuous cubic phase. However, a small decrease in lattice dimension occurred, indicating that heating of the matrix has occurred. The structure relaxed back to the original position when the laser was off. On repeated application of the 5-s laser pulse, the system again displayed the heating effect and relaxed back to the starting position when irradiation was complete (Fig. 8.16). This contraction and

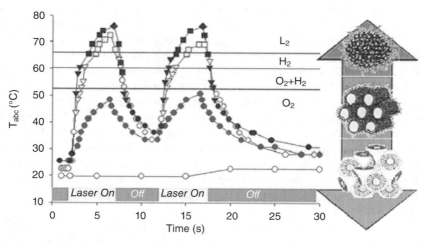

Figure 8.16 Effect of laser irradiation on apparent temperature of the phytantriol + water matrix (T_{app}) with change in GNR concentration. [GNR] = 0 nM are white symbols, 0.3 nM blue symbols, 1.5 nM yellow symbols, and 3 nM black symbols. Circles indicate v_2 phase, triangles indicate $v_2 + H_2$, squares indicate $H_2 + L_2$, and diamonds indicate L_2. The cartoon on the right indicates the type of phase structure present with increasing temperature (Fong et al., 2010).

expansion of the lattice on heating and cooling (the "breathing mode") was accompanied by concurrent expulsion and uptake of water from the matrix to satisfy the changes in lattice dimension.

In the case of higher concentrations of GNR, the 5-s on pulse for both the 1.5- and 3-nM systems did induce a phase change to the H_2 and L_2 phase structures. At 3 nM, complete transformation to the L_2 phase occurred by the end of the 5-s irradiation, while the lower 1.5-nM concentration resulted in a mixed $H_2 + L_2$ phase, indicating a reduced heating effect (Fig. 8.16). Again, when the laser was switched off, the system ultimately returned to the initial v_2 (Pn3m) phase structure. Interestingly, on conversion from the L_2 or $L_2 + H_2$ state back to the v_2 structure, the "gyroid" bicontinuous cubic phase with Ia3d space group was encountered. The gyroid phase coexisted with the H_2 phase initially and then with the v_2 (Pn3m) phase for approximately 5–6 s after the laser was switched off.

In addition, Fig. 8.16 clearly demonstrated that in the case where no GNRs were added to the matrix, there was no significant change in apparent sample temperature (T_{app}) upon laser irradiation. The reversibility of the heating effect in the presence of the GNR is evident from the T_{app} profiles for the three samples containing the nanorods. The sample containing 0.3 nM GNR is heated to approximately 50°C, just below the transition above which the coexisting H_2 phase occurs. Increasing the nanorod concentration to 1.5 nM induces heating to approx 70°C, while 3 nM GNR heated the matrix to an apparent temperature of 75°C. The repeat irradiation provided the same peak temperature within 1–2°C and reproducible kinetics of heating and cooling. The heating effect observed in these experiments is clearly a function of nanorod concentration, although the relationship between concentration and maximum temperature at differing irradiation conditions requires further investigation and is likely complicated by concurrent cooling. The "breaking wave" shape of the profiles indicates nonlinear heating/cooling gradients in the material.

The authors (Fong et al., 2010) concluded that GNRs embedded in liquid crystalline matrices produce localized plasmonic heating of the hybrid matrix, enabling fine control over the nanostructure. The phase transitions resulting from photothermal heating were fully reversible and specific to the GNR/laser wavelength combination. Localized plasmonic heating of the liquid crystal did not compromise the integrity of the lipid molecules in any of the mesophases. Undoubtedly, this research represented a significant advance toward effective, light-activated drug delivery systems with potential to solve unmet medical needs.

The most straightforward stimulus, which can be exploited for in vivo controlled release, is certainly pH because of the large pH changes occurring spontaneously within the mouth–stomach–intestine tract.

In this context Negrini and Mezzenga (2011) presented a pH-responsive lipid-based LLC, which has a number of significant advantages for real oral-administration-controlled delivery. First, the system is based on a simple formulation of monolinolein and linoleic acid, which maintains it entirely food

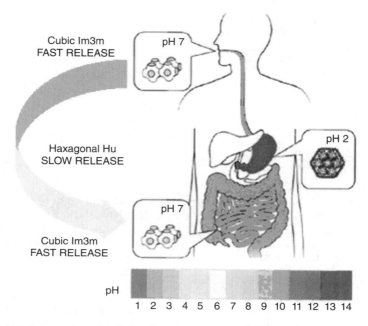

Figure 8.17 Schematics of the proposed pH-responsive drug delivery strategy across the gastrointestinal tract. At pH 7, the drug nanocarrier has a reverse bicontinuous cubic phase (Im3m) symmetry; at pH 2, in the stomach, the symmetry of the mesophase changes to reverse hexagonal phase, slowing down diffusion and preventing the premature release of the drug; and at neutral pH in the intestine, the symmetry reverts to bicontinuous cubic phase (Im3m), triggering the release of the active ingredients loaded in the mesophase (Negrini and Mezzenga, 2011).

grade. Second, it offers a general, tunable release and diffusion strategy that is not specific to the particular drug ingredients. Finally, the release can be controlled in a fully reversible way and makes it suitable for a targeted delivery to specific points (pH) of the gastrointestinal tract.

The lipid (monolinolein), the pH-responsive molecule (linoleic acid), and the model hydrophilic polyphenol drug (phloroglucinol) were used in this study, and the main idea behind the strategy proposed is highlighted in Fig. 8.17. Because of the presence of linoleic acid, which can be in either the deprotonated or protonated state when changing the pH from neutral to acidic conditions, respectively, the LLC undergoes a structural change from reverse bicontinuous cubic phase to reverse columnar hexagonal phase, which is accompanied by a change in the release rate by a factor of 4, thus preventing the release in the stomach and making this system an ideal candidate for the targeted delivery of active ingredients in the basic environment typical of the intestinal tract.

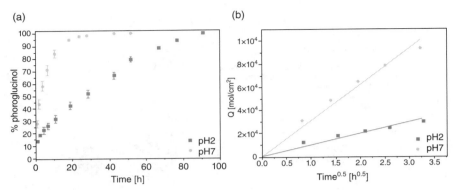

Figure 8.18   pH-induced changes in the release of phloroglucinol from the reverse bicontinuous cubic and hexagonal phases at 37°C: (a) Phloroglucinol released from the bicontinuous cubic phase at pH 7 (●) and reverse hexagonal phase at pH 2 (■) plotted against time. (b) Moles released per unit area plotted against the square root of time, illustrating the different diffusion-controlled behaviors in the two mesophases (Negrini and Mezzenga, 2011).

This order–order transition can be explained by the presence of the ionizable carboxylic group of the linoleic acid (intrinsic $pK_a \sim 5$). The linoleic acid is negatively charged at pH 7; the electrostatic repulsions between the negatively charged head groups stabilize the Im3m bicontinuous cubic phase. When the pH decreases below the pK_a value, the carboxyl group reprotonates to a large extent, the surface charge density on the water channels at the water–lipid interface decreases, and the linoleic acid became highly hydrophobic, acting as an oil and stabilizing the hexagonal phase at 37°C.

This order–order transition is well-rationalized by the concept of the CPP expressed as v/Al, which is the ratio between the volume of the hydrophobic lipid tail, v, and the product of the cross-sectional lipid head area, A, and the lipid chain length, l. When linoleic acid is deprotonated (pH 7), the effective area A is large because of the electrostatic repulsive interactions among different lipid heads. When, however, the linoleic acid is mostly neutral (pH 2), A decreases and the CPP increases, promoting the transition from flat to reverse (water-in-oil) interfaces and inducing a bicontinuous cubic → reverse columnar hexagonal transition.

In vitro release studies were carried out first to establish the influence of the liquid crystalline symmetry on the release behavior. Figure 8.18a illustrates the drug release profiles from the liquid crystalline matrices, plotted as a percentage of the released drug against time. As can be observed, the release from Im3m is much more rapid and is nearly completed after 20 h; at this time, H_{II} has not yet released half of the initially loaded drug.

In Figure 8.18b the profiles of drug released are plotted against the square root of time; the linear behavior confirms the Fickian diffusion release. Using the Higuchi equation, the diffusion coefficients are calculated to be

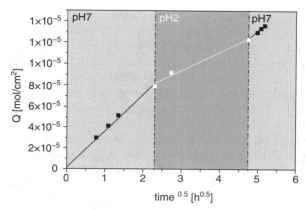

Figure 8.19 Release profiles of phloroglucinol plotted against the square root of time, upon sequential switch in the pH of the excess solution from pH 7 → pH 2 → pH 7 (Negrini and Mezzenga, 2011).

$D_{pH2} = 2.2 \times 10^{-8}$ cm^2/s and $D_{pH7} = 30.6 \times 10^{-8}$ cm^2/s (slopes = 1.34×10^{-8} at pH 2 and 5×10^{-8} at pH 7), respectively. It is therefore possible to conclude that the bicontinuous cubic phase releases almost four times faster than the reverse hexagonal phase, consistent with previous reported work. To demonstrate a pH-triggered on–off release behavior and tunable release profiles in dynamic conditions, the release study was also carried out by switching the pH between the values of 7 and 2 using the same setup, and the resulting release data are given in Figure 8.19. The values of the slopes obtained in these conditions are 5.7×10^{-8} in the initial pH 7 conditions, 2.58×10^{-8} after switching the solution at pH 2, and 5.3×10^{-8} upon reverting the pH to 7 again. The values of the slopes at pH 7 agree well with those obtained by the static release experiment, whereas that measured at pH 2 shows a slightly faster release; this results from the relatively long time needed to induce a complete change of Im3m into H$_{II}$ in bulk mesophases, leading to a long-living coexistence of Im3m and H$_{II}$ upon the switch of the pH from 7 to 2. It can be easily anticipated that, when applying these concepts to cubosomes and hexosomes of a few hundred nanometers in diameter, the order–order transitions will occur much faster, leading to a sharper pH-induced on–off–on release along the gastrointestinal tract.

8.6 MOLECULAR INTERACTIONS OF LYOTROPIC LIQUID CRYSTALS WITH PROTEINS AND NUCLEOTIDES

Murgia et al. (2009) entrapped several nucleotides into cubic and lamellar monoolein-based mesophases in order to protect them and enable their release. Nucleoside and nucleotide analogs represent a class of innovative therapeutics for the treatment of viral diseases and cancer. Simultaneously, nucleotides can constitute a promising choice to promote molecular recognition. However, these substances need to be protected since they can be easily recognized and degraded by different extracellular nucleases, resulting in poor in vivo pharmacokinetic properties. The usually severe cytotoxic side effects of nucleoside and nucleotide analogs, particularly myelotoxicity, can be partially relieved by lipid derivatization of the phosphate group of the active nucleotide. The resulting drug is amphiphilic and can be delivered to the therapeutic target by incorporation in lipid-based nanovectors.

In this regard these investigators used simple models for both hydrophilic and hydrophobic modified/functionalized nucleotide-based drugs, adenosine monophosphate (AMP), guanosine monophosphate (GMP), uridine monophosphate (UMP), and cytidine monophosphate (CMP) (xanthosine monophosphates [XMPs]), along with two hydrophobically functionalized nucleotides (nucleolipids), that is, the 1-palmitoyl-2-oleoyl-*sn*-glycerol-3-phosphoadenosine (POPA) and the hexadecylphosphoadenosine (HPA) (Fig. 8.20). In these nucleolipids a nucleic base is enzymatically exchanged with the choline head group of a lipid precursor. The resulting systems were investigated mainly through SAXS and ^{31}P–nuclear magnetic resonance (NMR) techniques.

The GMO-based LC phases used here can be easily regarded as membrane models as a result of lipid type and bilayer occurrence. These lipid LC systems have been shown to constitute suitable matrices to entrap either hydrophilic nucleotides or amphiphilic nucleolipids and are therefore promising lipid-based nanovectors to protect and deliver nucleotide–analog drugs. However, a drastically different behavior regarding the long-term stability of the nanodevice is displayed. All these molecules contain a phosphate moiety. It has been ascertained that in the case of the amphiphilic additives (POPA and HPA), which locate at the GMO polar–apolar interface, the LC systems were found endowed of high stability, more than 2 years. Differently, the hydrophilic XMP molecules, which are located in the aqueous domain of the GMO LC phases, undergo a hydrolysis process at the phosphoester bond due to a preferential orientation with respect to the MO interface. The XMP degradation, within about 4 months, causes an impressive alteration of the interface arrangement, resulting in structural transition from cubic to hexagonal mesophase. It was demonstrated that the relatively small amount of HPO_4^{2-} anion resulting from the XMP hydrolysis were sufficient to induce a cubic-to-hexagonal phase transition as a result of different, and specific, interactions at the polar–apolar interface that favored a reverse curvature. Murgia et al. (2009) concluded that the presence of unesterified phosphate groups in fully hydrophilic drug molecules (e.g., XMPs) may have dramatic effects on the stability of the lipid bilayer of a biological membrane.

Amar-Yuli et al. (2011b) combined the potential of both liquid crystalline structure as well as glycerol as cosolvent to enhance insulin thermal stability and moderate the aggregation progress. Insulin was incorporated into several modified reverse hexagonal systems based on friendly surfactant and polyols

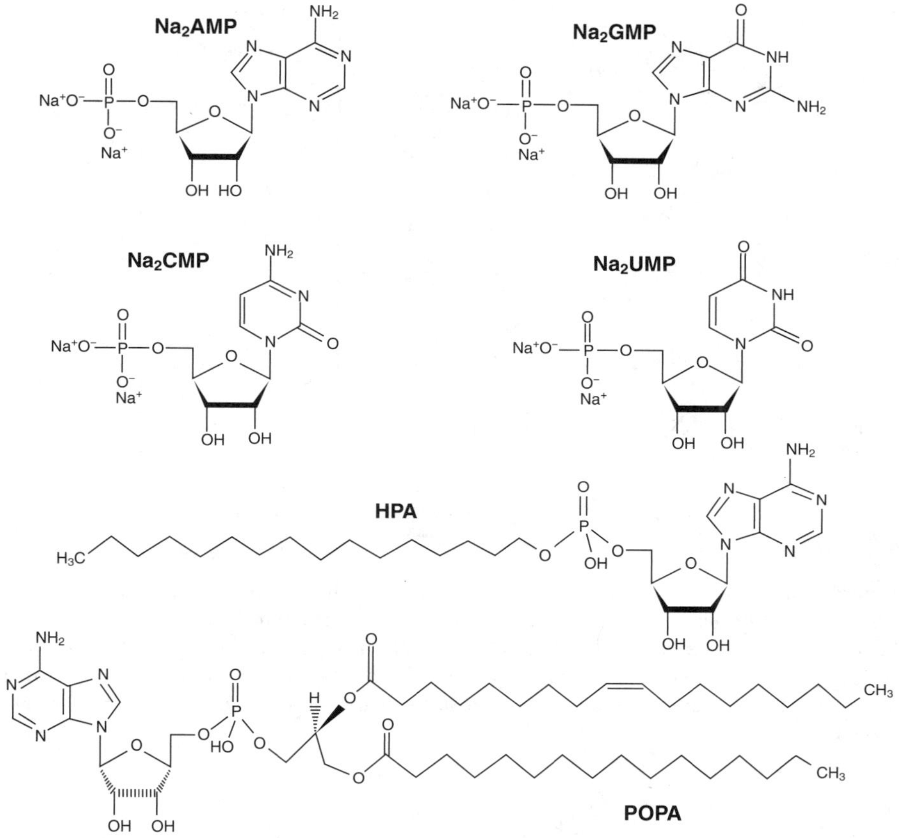

Figure 8.20 Mononucleotides disodium salts and nucleolipids (Murgia et al., 2009).

to explore the impact of the protein confinement on its stability, unfolding behavior, and morphology with severe external conditions, low pH, and higher temperatures (up to 70°C).

The investigators focused on solubilizing insulin in GMO–decane–water, GMO–decane–glycerol–water, and GMO–PC–decane–glycerol–water mixtures and compared them to bulk solution (water).

The incorporation of insulin within the water and especially in the glycerol–water cylinders compels the protein to be assembled in a restricted, confined configuration compared to its rearrangement in solution. The phenomenon of a greater degree of compactness of the tertiary fold of insulin in the presence of glycerol is known and here is overstressed upon its incorporation within the water- and glycerol-filled cylinders, which limits its free volume. The ability to obtain more confined protein structure, suggesting a more compact configuration of solvated dimeric insulin, is expected to be mirrored by an enhancement of the protein stability.

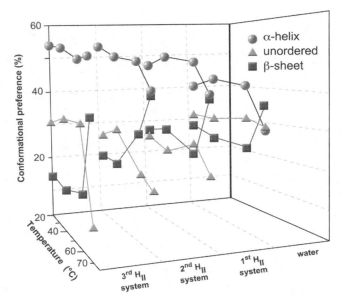

Figure 8.21   Insulin secondary structure (α-helix, β-sheets, and unordered conformations, turn together with random coils) during heating from 25 to 70°C in different hosting media: water, water-filled H$_{II}$ phase (first H$_{II}$ system), water–glycerol-filled H$_{II}$ phase (second system), and water–glycerol-filled H$_{II}$ phase containing PC (third system). Each system contains 4 wt % insulin in the polar phase (Amar-Yuli et al., 2011b).

Comparison of insulin secondary structure during heating from 25 to 70°C in different hosting media: water, water-filled H$_{II}$ phase (first H$_{II}$ system), glycerol/water-filled H$_{II}$ phase (second system), and glycerol/water-filled H$_{II}$ phase containing PC (third system) is shown in Figure 8.21. Each system depicts the corresponding content of α helix, β sheets, and unordered conformations (turns together with random coils) of insulin. The dependence of the insulin secondary structure on prolonged heating (for 60 min) at 60 and 70°C was tested. The pronounced modifications were observed only at 70°C. These temperatures were selected on the basis of previous reports demonstrating that the insulin aggregation process begins at 55°C (but is surmised to be too slow to be detected); however, at 60°C, immediately after the dimer dissociation, the aggregation process was detectable and accelerated with increasing temperature (to 70°C) or upon prolonged heating.

All four matrices did not reveal any significant changes in content of the conformational elements up to 60°C (Fig. 8.21). However, prolonged incubation of insulin at 60°C for 60 min caused an increase in the β-sheet configurations from 25 to 36% only when insulin was confined in the water-filled H$_{II}$ phase (first H$_{II}$ system, data not shown). Once the temperature reached 70°C,

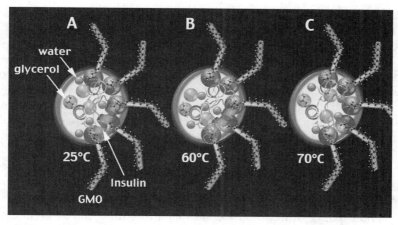

Figure 8.22 Schematic illustration of the interactions between the components, representing the second and third H_{II} systems (water–glycerol-filled H_{II} phases, in the absence and presence of PC, respectively) at three temperatures: (a) 25°C, (b) 60°C, (c) 70°C. For simplification PC and tricaprylin are not shown in the image (Amar-Yuli et al., 2011b).

marked increase in the parallel and antiparallel β-sheet bands (Fig. 8.21) were detected in all systems, suggesting an initiation of insulin aggregation, with noticeable variations between the hosting systems. The total β sheets of insulin in water increased from ~25 to 37% at 70°C, and after 60 min treatment further increased to 56%. Similar quantities of β sheets were obtained at 70°C when insulin was confined within the H_{II} cylinders (39, 40, and 36% in the water-, glycerol/water-H_{II} phase, and glycerol/water-filled H_{II} phase containing PC, respectively). On the other hand, heat treatment of insulin for 60 min revealed the differences in system efficiency in intermolecular β-sheet growth. While an increase of ~20% in β-sheet content was recorded in the bulk solution, only a slight increase of 5% was observed when insulin was incorporated in the water-filled H_{II} channels. Moreover, when glycerol was present in these narrow channels (in the second and third systems) insulin retained its β-sheet content, although severe conditions (70°C for 60 min) were applied. The addition of PC contributed to the stability of the secondary structure of insulin.

Figure 8.22 illustrates the interactions between the components, representing the second and third H_{II} systems. It should be stressed that in these new systems, the dehydration process ended at 60°C, whereas further increase in the temperature to 70°C caused the formation of stronger hydrogen bonds between the GMO hydroxyls and water. Above 60°C, owing to the protein–solvation effect, the water and glycerol molecules were probably excluded from the protein surface and redistributed themselves in the vicinity of the protein molecule with a preferential binding of the solvent components. The increase in the strength of the hydrogen bonds as detected by FTIR is considered to be responsible for the more stable insulin secondary structure, which is probably the reason for hampering the formation of fibrillar aggregates, associated with several neurodegenerative diseases.

The results presented in this study provide valuable insights into the protection abilities and mechanism of the used colloid structures, containing co-solvent as a medium for long-term protein stability. Moreover, these findings yield valuable information regarding the effect of confinement of these colloid structures on insulin fibrillation and/or aggregation.

8.7 SUMMARY

This chapter demonstrated that LC mesophases with complex architectures were rationally designed and prepared for the solubilization and potential administration of bioactive molecules. It was shown that comprehensive understanding of the structural properties of the carriers is imperative for rational and successful tailoring of delivery vehicles. The principal strategy was to first characterize the different levels of organization of these special materials and then to explore the detailed relationships and possible correlations between them, their structures, and the macroscopic properties.

The results lately obtained, although very interesting and promising, definitely do not cover all the aspects and do not answer all the questions addressed by the fascinating field of lyotropic liquid crystals and their applications as drug delivery systems. Clearly, many additional experiments need to be carried out in order to clarify the detailed structure, the exact properties, and specific potential of these systems. We hope that we at least opened a new window and provided new thoughts and interest into this rapidly growing field of research.

REFERENCES

Achrai, B., Libster, D., Aserin, A., and Garti, N. (2011). Solubilization of gabapentin into H_{II} mesophases. *Journal of Physical Chemistry B, 115*, 825–835.

Amar-Yuli, I. (2008). Hexagonal liquid crystals and hexosomes structural modifications and solubilization. Ph.D. Dissertation, The Hebrew University of Jerusalem, Israel.

Amar-Yuli, I., and Garti, N. (2005). Transitions induced by solubilized fat into reverse hexagonal mesophases. *Colloids and Surfaces B, 43*, 72–82.

Amar-Yuli, I., Wachtel, E., Ben Shoshan, E., Danino, D., Aserin, A., and Garti, N. (2007a). Hexosome and hexagonal phases mediated by hydration and polymeric stabilizer. *Langmuir, 23*, 3637–3645.

Amar-Yuli, I., Wachtel, E., Shalev, D., Moshe, H., Aserin, A., and Garti, N. (2007b). Thermally induced fluid reversed hexagonal (H-II) mesophase. *Journal of Physical Chemistry B, 111*, 13544–13553.

Amar-Yuli, I., Wachtel, E., Shalev, D., Aserin, A., and Garti, N. (2008a). Low viscosity reversed hexagonal mesophases induced by hydrophilic additives. *Journal of Physical Chemistry B*, *112*, 3971–3982.

Amar-Yuli, I., Aserin, A. and Garti, N. (2008b). Solubilization of nutraceuticals into reverse hexagonal mesophases. *Journal of Physical Chemistry B*, *112*, 10171–10180.

Amar-Yuli, I., Libster, D., Aserin, A., and Garti, N. (2009). Solubilization of food bioactives within lyotropic liquid crystalline mesophases. *Current Opinion in Colloid & Interface Science*, *14*, 21–32.

Amar-Yuli, I., Adamcik, J., Blau, S., Aserin, A., and Mezzenga, R. (2011a). Controlled embedment and release of DNA from lipidic reverse columnarhexagonal mesophases. *Soft Matter*, *7*, 8162–8168.

Amar-Yuli, I., Azulay,D., Mishraki, T., Aserin, A., and Garti, N. (2011b). The role of glycerol and phosphatidylcholine in solubilizing and enhancing insulin stability in reverse hexagonal mesophases. *Journal of Colloid & Interface Science*, *364*, 379–387.

Ben Ishai, P., Libster, D., Aserin, A., Garti, N., and Feldman, Y. (2009). Molecular interactions in lyotropic reverse hexagonal liquid crystals: A dielectric spectroscopy study. *Journal of Physical Chemistry B*, *113*, 12639–12647.

Bitan-Cherbakovsky, L., Amar-Yuli, I., Aserin, A., and Garti, N. (2009). Structural rearrangements and interaction within H_{II} mesophase induced by cosolubilization of vitamin E and ascorbic acid. *Langmuir*, *25*, 13106–13113.

Bitan-Cherbakovsky, L., Amar-Yuli, I., Aserin, A., and Garti, N. (2010). Solubilization of vitamin E into H_{II} LLC mesophase in the presence and in the absence of vitamin C. *Langmuir*, *26*, 3648–3653.

Boyd, B. J., Whittaker, D. V., Khoo, S.-M., and Davey, G. (2006). Lyotropic liquid crystalline phases formed from glycerate surfactants as sustained release drug delivery systems. *International Journal of Pharmaceutics*, *309*, 218–226.

Boyd, B. J., Khoo, S.-M., Whittaker, D. V., Davey, G., and Porter, C. J. H. (2007). A lipid-based liquid crystalline matrix that provides sustained release and enhanced oral bioavailability for a model poorly water soluble drug in rats. *International Journal of Pharmaceutics*, *340*, 52–60.

Brondsted, H., Nielsen, H. M., and Hovgaard, L. (1995). Drug-delivery stuties in caco-2 monolayers. 3. Intestinal transport of various vasopressin analogs in the presence of lysophosphatidylcholine. *International Journal of Pharmaceutics*, *114*, 151–157.

Cherezov, V., Clogston, J., Papiz, M. Z., and Caffrey, M. (2006). Room to move: Crystallizing membrane proteins in swollen lipidic mesophases. *Journal of Molecular Biology*, *357*, 1605–1618.

Clogston, J., and Caffrey, M. (2005). Controlling release from the lipidic cubic phase. Amino acids, peptides, proteins and nucleic acids. *Journal of Controlled Release*, *107*, 97–111.

Cohen-Avrahami, M., Aserin, A., and Garti, N. (2010). H_{II} mesophase and peptide cell-penetrating enhancers for improved transdermal delivery of sodium diclofenac. *Colloids and Surfaces B*, *77*, 131–138.

Cohen-Avrahami, M., Libster, D., Aserin, A., and Garti, N. (2011). Sodium diclofenac and cell-penetrating peptides embedded in H_{II} mesophases: Physical characterization and delivery. *Journal of Physical Chemistry B*, *115*, 10189–10197.

Dass, C. R., and Choong, P. F. M. (2006). Biophysical delivery of peptides: Applicability for cancer therapy. *Peptides*, *27*, 3479–3488.

Drummond, C. J., and Fong, C. (2000). Surfactant self-assembly objects as novel drug delivery vehicles. *Current Opinions in Colloid & Interface Science*, *4*, 449–456.

Ericsson, B., Eriksson, P. O., Löfroth, J. E., and Engström, S. (1991). Cubic phases as delivery systems for peptide drugs, polymeric drugs and drug delivery system. *American Chemical Society*, Symposium Series, *469* (Polym. Drugs Drug Delivery Systems), 251–265.

Fong, W. K., Hanley, T., and Boyd, B. J. (2009). Stimuli responsive liquid crystals provide "on-demand" drug delivery in vitro and in vivo. *Journal of Controlled Release*, *135*, 218–226.

Fong, W. K., Hanley, T., Thierry, B., Kirby, N., and Boyd, B. J. (2010). Plasmonic nanorods provide reversible control over nanostructure of self-assembled drug delivery materials. *Langmuir*, *26*, 6136–6139.

Gin, D. L., Pecinovsky, C. S., Bara, J. E., and Kerr, L. (2008). Functional lyotropic liquid crystal materials. *Structure & Bonding*, *128*, 181–222.

Gradzielski, M. (2004). Investigations of the dynamics of morphological transitions in amphiphilic systems. *Current Opinions in Colloid & Interface Science*, *9*, 256–263.

Gruner, S. M. (1989). Stability of lyotropic phases with curved interfaces. *Journal of Physical Chemistry*, *93*, 7562–7570.

Gruner, S. M., Tate, M. W., Kirk, G. L., So, P. T. C., Turner, D. C., Keane, D. T., Tilcock, C. P. S., and Cullis, P. R. (1988). X-ray-diffraction study of the polymorphic behavior of *N*-methylated dioleoylphosphatidylethanolamine. *Biochemistry*, *27*, 2853–2866.

Israelachvili, J. N., Mitchell, D. J., and Ninhan, B. W. (1976). Theory of self-assembly of hydrocarbon amphiphiles into micelles and bilayers. *Journal of the Chemical Society, Faraday Transactions 2*, *72*, 1525–1568.

Kumar, T. R. S., Soppimath, K., and Nachaegari, S. K. (2006). Novel delivery technologies for protein and peptide therapeutics. *Current Pharmaceutical Biotechnology*, *7*, 261–276.

Kurosaki, Y., Nagahara, N., Tanizawa, T., Nishimura, H., Nakayama, T., and Kimura, T. (1991). Use of lipid disperse systems in transdermal drug delivery: Comparative study of flufenamic acid permeation among rat abdominal skin, silicon rubber membrane and stratum corneum sheet isolated from hamster cheek pouch. *International Journal of Pharmaceutics*, *67*, 1–9.

Larsson, K. (1989). Cubic lipid-water phases: Structures and biomembrane aspects. *Journal of Physical Chemistry*, *93*, 7304–7314.

Libster, D., Aserin, A., Wachtel, E., Shoham, G., and Garti, N. (2007). An H_{II} liquid crystal-based delivery system for cyclosporin A: Physical characterization. *Journal of Colloid & Interface Science*, *308*, 514–524.

Libster, D., Ben Ishai, P., Aserin, A., Shoham, G., and Garti, N. (2008). From the microscopic to the mesoscopic properties of lyotropic reverse hexagonal liquid crystals. *Langmuir*, *24*, 2118–2127.

Libster, D., Ben Ishai, P., Aserin, A., Shoham, G., and Garti, N. (2009a). Molecular interactions in reverse hexagonal mesophase in the presence of cyclosporin A. *International Journal of Pharmaceutics*, *367*, 115–126.

Libster, D., Aserin, A., Yariv, D., Shoham, G., and Garti, N. (2009b). Concentration- and temperature-induced effects of incorporated desmopressin on the properties of reverse hexagonal mesophase. *Journal of Physical Chemistry B, 113*, 6336–6346.

Libster, D., Aserin, A., and Garti, N. (2011). Interactions of biomacromolecules with reverse hexagonal liquid crystals: Drug delivery and crystallization applications. *Journal of Colloid & Interface Science, 356*, 375–386.

Liu, D.-Z., LeCluyse, E. L., and Thakker, D. R. (1999). Dodecylphosphocholine-mediated enhancement of paracellular permeability and cytotoxicity in Caco-2 cell monolayers. *Journal of Pharmaceutical Science, 88*, 1161–1168.

Lopes, L. B., Speretta, F. F. F., Vitoria, M., and Bentley, L. B. (2007). Enhancement of skin penetration of vitamin K using monoolein-based liquid crystalline systems. *European Journal of Pharmaceutical Sciences, 32*, 209–215.

Mezzenga, R., Grigorov, M., Zhang, Z., Servais, C., Sagalowicz, L., and Romoscanu, A. I. (2005). Polysaccharide-induced order-to-order transitions in lyotropic liquid crystals. *Langmuir, 21*, 6165–6169.

Mishraki, T., Libster, D., Aserin, A., and Garti, N. (2010a). Lysozyme entrapped within reverse hexagonal mesophases: Physical properties and structural behavior. *Colloids and Surfaces B, 75*, 47–56.

Mishraki, T., Libster, D., Aserin, A., and Garti, N. (2010b). Temperature-dependent behavior of lysozyme within the reverse hexagonal mesophases (H_{II}). *Colloids and Surfaces B, 75*, 391–397.

Murgia, S., Lampis, S., Angius, R., Berti, D., and Monduzzi, M. (2009). Orientation and specific interactions of nucleotides and nucleolipids inside monoolein-based liquid crystals. *Journal of Physical Chemistry B, 113*, 9205–9215.

Negrini, R., and Mezzenga, R. (2011). pH-responsive lyotropic liquid crystals for controlled drug delivery. *Langmuir, 27*, 5296–5303.

Rizwan, S. B., Hanley, T., Boyd, B. J., Rades, T., and Hook, S. (2009). Liquid crystalline systems of phytantriol and glyceryl monooleate containing a hydrophilic protein: Characterisation, swelling and release kinetics. *Journal of Pharmaceutical Science, 98*, 4191–4204.

Sagalowicz, L., Mezzenga, R., and Leser, M. E. (2006a). Investigating reversed liquid crystalline mesophases. *Current Opinions in Colloid & Interface Science, 11*, 224–229.

Sagalowicz, L., Leser, M. E., Watzke, H. J., and Michel, M. (2006b). Monoglyceride self-assembly structures as delivery vehicles. *Trends in Food Science & Technology, 17*, 204–214.

Seddon, J. M. (1990). Structure of the inverted hexagonal (H_{II}) phase, and non-lamellar phase-transitions of lipids. *Biochimica et Biophysica Acta, 1031*, 1–69.

Seddon, J. M., Zeb, N., Templer, R. H., McElhaney, R. N., and Mannock, D. A. (1996). An Fd3m lyotropic cubic phase in a binary glycolipid/water system. *Langmuir, 12*, 5250–5253.

Seddon, J. M., Robins, J., Gulik-Krzywicki, T., and Delacroix, H. (2000). Inverse micellar phases of phospholipids and glycolipids. *Physical Chemistry Chemical Physics, 2*, 4485–4493.

Shah, M. H., and Paradkar, A. (2005). Cubic liquid crystalline glyceryl monooleate matrices for oral delivery of enzyme. *International Journal of Pharmaceutics, 294*, 161–171.

Shah, J. C., Sadhale, Y., and Dakshina, M. C. (2001). Cubic phase gels as drug delivery systems. *Advanced. Drug Delivery Review*, *47*, 229–250.

Shearman, G. C., Ces, O., Templer, R. H., and Seddon, J. M. (2006). Inverse lyotropic phases of lipids and membrane curvature. *Journal of Physics: Condensed Matter*, *18*, S1105–S1124.

Spernath, A., and Aserin, A. (2006). Microemulsions as carriers for drugs and nutraceuticals. *Advances in Colloid & Interface Science*, *128–130*, 47–64.

Touitou, E., Junginger, H.E., Weiner, N.D., Nagai, T., and Mezei, M. (1994). Liposomes as carriers for topical and transdermal delivery, *Journal of Pharmaceutical Sciences*, *83*, 1189–1203.

Yang, L., Ding, L., and Huang, H. W. (2003). New phases of phospholipids and implications to the membrane fusion problem. *Biochemistry*, *42*, 6631–6635.

■■■■■■ **CHAPTER 9**

Stimuli-Responsive Lipid-Based Self-Assembled Systems

BEN J. BOYD and WYE-KHAY FONG

Drug Delivery, Disposition, and Dynamics, Monash Institute of Pharmaceutical Sciences, Monash Universtiy, Parkville, Victoria, Australia

Abstract

Lipid based self assembled matrices have been proposed as novel drug delivery systems, as the phases formed by these materials have been shown to reliably encapsulate and control the release of actives. The unique nanostructures formed by many lipids are dictated by the specific local packing of the amphiphiles in the matrix. Lipid packing is determined by molecular geometry of the lipids, which can be reversibly or irreversibly modified by changes in environmental factors such as temperature and pH. In recent times, much research has tapped into the rich vein of liquid crystalline nanostructures and their manipulations for "on-demand" drug release. This chapter details the use of self assembled systems in drug delivery with a particular focus on how they have been rendered stimuli responsive and their potential uses in pharmaceutical therapies.

Self-Assembled Supramolecular Architectures: Lyotropic Liquid Crystals, First Edition.
Edited by Nissim Garti, Ponisseril Somasundaran, and Raffaele Mezzenga.
© 2012 John Wiley & Sons, Inc. Published 2012 by John Wiley & Sons, Inc.

9.1   LIPID SELF-ASSEMBLY AND TRANSITIONS BETWEEN SELF-ASSEMBLED STRUCTURES

9.1.1   Lipid Self-Assembly

Amphiphilic lipids and surfactants self-assemble in aqueous environments to form a variety of possible structures, depending on the solubility and effective geometry of the assembling amphiphiles. These structures form spontaneously and in the absence of physical or chemical changes to the system are thermodynamically stable. The spectrum of frequently encountered structures is represented in Figure 9.1, which shows the usual progression of these structures as water content is decreased from left to right. Not all amphiphile systems show all of the structures represented.

Micelles are the simplest geometry, and in their "normal" configuration they are denoted as an L_1 phase and typically occur at relatively high hydration. The addition of water to relatively soluble surfactants and lipids usually results in the observation of a range of the structures in Figure 9.1 from right to left, with an eventual transition to the L_1 micellar phase. However, oftentimes the addition of water to surfactant does not induce a transition through to the normal L_1 micellar phase. If the matrix has a finite capacity to incorporate additional water, the extra water coexists as a separate phase. Such systems form the basis of bilayer structures that coexist in excess water, as in the case of cellular membranes and liposomes. Such systems are typically formed by lipids such as phospholipids, which have an inherent geometric shape suitable for the formation of bilayer structures.

The addition of excess water to poorly water-soluble amphiphilic lipids can also result in the formation of finite swelling inverse nonlamellar geometries,

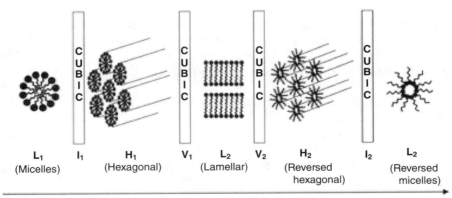

Increasing amphipnile concentration

Figure 9.1   Spectrum of commonly encountered amphiphile self-assembly structures in water. [Reproduced from Kaasgaard and Drummond (2006) with permission of the PCCP Owner Societies.]

such as the inverse hexagonal (H_2) and inverse bicontinuous cubic (v_2, also often denoted as Q_2) phases analogous to the lamellar phase described above. The lipids that typically form these nonlamellar phase structures have been recently reviewed (Kaasgaard and Drummond, 2006). Thus lipid and surfactant self-assembly not only provides access to a range of thermodynamically stable structures but also provides systems with very different classes of behavior that can be translated into different applications.

The unique nanostructures formed by these lipids dictate their use, whether it be to control the rate of release of actives from within (Lee et al., 2009a), matrix-guided synthesis (Hegmann et al., 2007), or more classical surfactant applications such as wetting and detergency applications. In addition to their thermodynamically stable long-range order, they also possess excellent biocompatibility due to their composition—lipid and water.

The self-assembled structures formed by these lipids are determined by the specific local packing of the amphiphiles in the matrix. In addition to molecular geometry, their packing is dictated by the presence of additives, solution conditions, for example, pH, and environmental conditions such as temperature and pressure. The addition of known amounts of specific additives can result in quantifiable modifications in phase behavior—for example, cholesterol in liposomes affects bilayer fluidity (Bisby et al., 1999b), oleic acid (OA) in glycerol monooleate (GMO) (Aota-Nakano et al., 1999; Borne et al., 2001) and vitamin E acetate in phytantriol (Dong et al., 2006) cause the reduction of the v_2 to H_2 to L_2 phase transition temperatures, allowing for the ability to fine tune the desired properties of lipid-based systems. These parameters, and hence lipid packing, can be manipulated providing specific stimuli. For example, in the case of liquid crystalline structures, an increase in temperature will tend to induce a higher degree of chain splay, therefore increasing the spontaneous mean curvature of the monolayer and forcing the structure to switch to a more energetically favorable phase with potential to thereby manipulate drug release.

9.1.2 Transitions between Self-Assembled Structures

Transitions between different self-assembled geometries for lipids in aqueous environments are ubiquitous in nature and have received a huge amount of attention from biophysical chemistry researchers (Brink-van der Laan et al., 2004; Conn et al., 2006). The formation and structure of cellular membranes, membrane fusion, and some cellular uptake mechanisms all involve manipulation of lipid self-assembly at the molecular and mesolength scales. The rich diversity of structures that lipids may adopt lends itself as the basis of complex biological functions. There is potential to utilize the reversible phase transitions in lipid systems to impart responsiveness into materials. With the advent of advanced synchrotron and neutron scattering facilities, we are only now coming to grips with the mechanisms by which transformations between lipid self-assembled structures take place (Kriechbaum and Laggner, 1996; Rappolt et al., 2003). Harnessing this information is key to utilizing the transformation between lipid systems.

9.2 APPLICATION OF SELF-ASSEMBLED SYSTEMS IN DRUG DELIVERY AND THE CASE FOR STIMULI-RESPONSIVE SYSTEMS

9.2.1 Benefits of Self-Assembled Systems in Drug Delivery

The use of self-assembled systems based on lipids and surfactants in drug delivery applications is predicated primarily on three main features.

1. *Enhance Solubilization*   The self-assembly provides an internal hydrophobic environment for enhanced solubilization of drug molecules that are otherwise poorly water soluble and limits the dose able to be administered. Emulsions, micelles, and liposomes have all been utilized for this purpose in a range of pharmaceutical products already on the market. While micelles have a disadvantage of potential loss of solubilization on dilution beyond the critical micelle concentration of the surfactant, emulsions and liposomes do not.

2. *Manipulate Pharmacokinetic Behavior*   Side effects due to drug administration are most often associated with high blood concentrations immediately after administration. Incorporation of drug into a carrier system can slow down the availability of drug to induce side effects without reducing the total dose of drug administered. One example of this is the use of liposomes to encapsulate amphotericin B, an antifungal drug.

3. *Target Drug to Particular Tissues*   The functionalization of micelles and liposomes has been utilized to direct the carrier to the desired tissue site, either by passive or active targeting. Passive targeting involves surface coverage with polyethylene glycol (PEG) chains to prevent nonspecific removal from the circulatory system, leading to preferential deposition in tumor tissues and sites of inflammation. Active targeting involves functionalizing the surface of the particle with a tissue-specific moiety such as an antibody or antigen, to encourage specific interaction with diseased tissues that differentially express the complementary entity to the functional group compared to healthy tissues.

While the three drivers for using a self-assembled system in drug delivery are important in providing improved therapeutic outcomes using self-assembled systems, there are still a number of unmet therapeutic needs. In particular, external control over the release of incorporated agents is currently not achievable. There are a number of therapeutic scenarios in which external control over drug release may be beneficial including:

- Frequently injected compounds with short half-lives: Reducing the frequency of administration of drugs that require multiple injections per day to, for example, once daily with external activation when required, such as in the case of peptide hormones such as insulin, would greatly improve patient quality of life and compliance.

- Imaging guided therapy: There is much research being undertaken into novel imaging contrast agents, and there is potential to use the same carrier to carry active compounds so that imaging reveals the ideal location for remote activation of the release mechanism.
- Chronically administered compounds with high risk of complications from administration: For example, the necessity for frequent intravitreal injection of steroids or anti-vascular endothelial growth factor (VEGF) compounds for macular degeneration is associated with a high risk of infection and retinal detachment.
- Symptom-responsive drug release: The ability to rapidly administer a dose of drug on the development of symptoms without going through an administration procedure, via external activation of a prior injected dose form, for example, where a patient loses consciousness as a result of a disease condition, making regular drug administration difficult.

It is proposed that self-assembled systems have an unrealized potential in providing external control over drug delivery, through the manipulation of self-assembled structure and hence optimization of drug release, disposition, and bioavailability. The thermodynamic stability of self-assembled systems is a distinct advantage over, for example, polymer-based systems, because in theory the self-assembly mode may be reversibly changed from one structure to another to modify drug release—in effect acting as an inherent on–off "switch." This chapter will provide a view of the landscape of stimuli-responsive self-assembled systems with a focus on surfactants and lipids, particularly for drug delivery applications.

9.3 STIMULI-RESPONSIVE SYSTEMS

For the purpose of this chapter, we propose that there are two main categories of stimuli-responsive systems for drug delivery applications, namely endogenous and external stimuli. The specific stimuli-responsive mechanisms in each category are summarized in Table 9.1. This chapter, therefore, discusses the

TABLE 9.1 Categories of Stimuli-Responsive Systems in Self-Assembled Lipid-Based Drug Delivery Vehicles

Endogenous Stimuli	External Stimuli
• Ionic, osmotic, and pH changes • Increased expression and activity of proteins	• Temperature • Light • Electromagnetic field • Ultrasound

general aspects of each of these stimuli-responsive areas. Advances in stimuli-responsive drug delivery from self-assembled systems have been recently reviewed (Bayer and Peppas, 2008; Couvreur and Vauthier, 2006; Torchilin, 2009), with a focus on cancer therapy and on liposomes and other lipid systems (Andresen et al., 2005; Kaasgaard and Andresen, 2010).

It is somewhat obvious that for application in drug delivery, the aim is to control the availability of the drug from the carrier to the general circulation if systemic availability is desired or the intended site of action for localized delivery. It is, therefore, useful to briefly discuss drug release mechanisms so that subsequent discussion of stimuli responsiveness can be understood in terms of the particular mechanism at play.

Release of a drug from self-assembled systems is recognized as being via three means, namely, (i) passive diffusion, (ii) nonspecific degradation of the carrier through normal physiological processes such as phagocytic uptake and/or hydrolytic destruction, or (iii) through covalent release. Passive diffusion relies on a chemical potential gradient to drive the drug from the self-assembled system into the bulk liquid into which it is introduced, such as the bloodstream. Drug is maintained inside the self-assembled system prior to administration purely by partitioning or equivalence of concentration in the internal and external aqueous compartments—dilution is the trigger by which drug is released by reducing the concentration of a drug in the external surrounding fluid. Degradation is the usual mechanism controlling release from solid matrices made from, for example, biodegradable polymer or solid lipid nanoparticles, in which hydrolysis induces chemical degradation of the polymer structure and erosion of the particle. Covalent drug release is usually achieved through incorporation of a functional linker within the self-assembled system that is cleaved in the particular environment of the target tissue. These latter systems often are achieved through pH-sensitive functional groups that decouple drug from the carrier in hypoxic environments or enzymatically activated systems. Both of these types of systems are reviewed later in the chapter.

9.3.1 Endogenous Stimuli

The "holy grail" of therapy is for there to be absolutely no intervention in the management and treatment of disease—the development of biochemical markers or symptoms that are not noticeable to the patient would trigger drug release automatically in response to some measured indicator. Such systems are already in use for the treatment of diabetes—implantable mechanical pumps that release insulin in response to falling glucose levels in the blood are on the market. Hence, drug delivery systems that are responsive to variation in physiological conditions are sought after. The ability to induce variation in lipid-based systems (rather than mechanical-based systems) using physiological stimuli has not been used to such great advantage, particularly in consideration of the vast capability of cellular structures to utilize changes in lipid assembly to modulate function.

The use of physiological triggers to date in drug delivery, particularly in polymers, has largely exploited the difference in physiological conditions between pathological and normal tissues in conditions such as cancer and inflammatory diseases.

9.3.1.1 *pH Variation*

Differences in pH in tissue and cellular components in pathological conditions provide points of potential exploitation. Normal pH gradients exist naturally throughout the body, for instance, inside some parts of the gastrointestinal tract, and cellular components differ from the extracellular pH of 7.4. This is commonly utilized to enhance oral drug bioavailability through the formulation of drug molecules as salts of base forms. In drug delivery it is utilized in selective drug release through preparation of enteric coatings for tablets, which protect a drug from the harsh acidic environment of the stomach, allowing a drug to be released at a more acceptable pH in the intestines.

Abnormal pH gradients exist between normal tissues and those affected by conditions such as cancer, inflammation, and infection. Sites of infection, site of inflammation, metastasized tumors and primary tumors have a significantly lower pH to normal tissues (Gerweck and Seetharaman, 1996). For instance, 60 min after an inflammatory infection, the pH at the site of inflammation drops to 6.5. These disease states have been the target of many lipid-based self-assembly systems.

pH-Sensitive systems have been extensively investigated since the early 1980s as a means of increasing drug delivery to tumors. Liposomal systems have been designed to be systemically dosed and accumulate in tumors where they are designed to release their contents under mildly acidic conditions (Yatvin et al., 1980). These are made responsive by the inclusion of a modified lipid as a sensitizer, and the mechanisms of release have been suggested to be one or a mixture of the following:

1. The pH change triggers a morphology change from the lamellar lipid bilayer (L_α) to, for example, micellar or hexagonal phase (Gerasimov et al., 1999). The liposome contains negatively charged lipids, for example, lipids with a carboxyl group that neutralize in acidic pH and so causes the collapse of L_α into a hexagonal phase. This concept is exploited in the enteric coating of pharmaceuticals to protect them from degradation in the stomach at low pH where the carboxylate groups of the coating polymer are protonated, imparting low solubility, but ionize in the small intestine to allow polymer dissolution. In self-assembled systems this concept has been rarely employed but has recently been demonstrated in lyotropic liquid crystalline systems prepared using monoglycerides doped with fatty acid to induce a morphology change from the hexagonal phase to the bicontinuous cubic phase (Negrini and Mezzenga, 2011). The hexagonal phase present under simulated gastric conditions at low pH released the model drug probucol slowly, compared to the cubic phase structure present at intestinal pH.

2. The pH change causes the ionization of bilayer surfactants. Cholesterol hemisuccinate (Kirpotin et al., 1996; Slepushkin et al., 1997; Zhang et al., 2004) and oleic acid (Hong et al., 2002; Torchilin et al., 1992) are two mildly acidic amphiphiles that become protonated when in acidic conditions. When exposed to the acidic environment, a decrease in hydration of the lipid head groups results, and, as a consequence, there is the formation of a nonbilayer structure that can fuse with the endosome membrane and so release contents into the cytoplasm (Chernomordik, 1996).

3. The pH change results in acid-catalyzed hydrolysis of specifically engineered lipids resulting in the formation of detergents or fusion. Acid-labile PEG-conjugated vinyl ether lipids have been used by Thompson et al. to stabilize liposomes. At pH <5, hydrolysis of the ether bond results in the removal of the PEG moiety and consequently allowing the liposome to become fusogenic (Boomer et al., 2003; Shin et al., 2003). Other groups have employed acid-labile and ester (Guo and Szoka, 2001; Heller et al., 1985) linkages in order to render liposomes pH sensitive.

4. The pH change leads to the destabilization of the bilayer through lysis, phase separation, pore formation, or fusion. This can be achieved in vesicles with incorporated pH-sensitive polymers such as poly(2-ethylacrylic acid) (Mills et al., 1999) and poly(glycidol)s (Yuba et al., 2010).

In the presence of serum, however, some liposomes lose their pH sensitivity or become unstable (Collins et al., 1990; Roux, Lafleur et al., 2003). This has been somewhat overcome through the use of a sterically stabilizing PEG coating, which is removable (i.e., dePEGylation) in order to maximize release response on encountering a low pH.

However, two limiting issues exist for the application of pH-sensitive liposomes in tumor treatment. First, the narrow pH window of between pH 7.4 in normal tissues and tumor pH, which is typically at a minimum at 6.5, provides only a narrow range over which the stimuli must be tuned to occur. Second, the region of highest acidity in the tumor is in the tumor interstitium, which is not close to the vasculature, thereby limiting its effectiveness. In light of the issues of application to tumor treatment, the focus on potential clinical applications of pH-sensitive liposomes has shifted somewhat to the control of drug release in cellular components such as endosomes and lysosomes, where the pH may be lower than 5. The pH-sensitive liposomes have been reviewed recently for these applications (Drummond et al., 2000; Guo and Szoka, 2003).

9.3.1.2 Redox, Ionic, and Osmotic Variation

Drug delivery systems have been designed to take advantage of cellular reduction potential at different locations within and around cells (Saito et al., 2003; West and Otto, 2005). Disulfide linkages (–S–S–) exist most commonly in cellular material as oxidized sulfhydryl (SH) groups of cysteine moieties in peptides and proteins. They are unique in that they form reversible covalent bonds, a property that has been manipulated in drug delivery to the cytosolic space, cellular components such as endosomes, lysosomes, endoplasmic reticulum, and mitochondria, and tumors. The reduction of disulfide linkages can also be a consequence of exposure to a reductase enzyme. Entry of molecules to these compartments is limited to small molecules; however, techniques have been developed to overcome this, including the use of liposomes. In liposomal drug delivery strategies, disulfides act as linkers for targeting conjugates (Ishida et al., 2001; Kirpotin et al., 1996; Zhang et al., 2004) or as disulfide-bridged lipids (Bhavane et al., 2003; Karathanasis et al., 2005), which are critical to liposome stability. Kirpotin et al. synthesized a disulfide-linked PEG conjugate that had two functions in two different approaches with these materials; first, thiolytic cleavage led to destabilization of the vesicle and release of its content, and second it provided pH sensitivity in the system (Kirpotin et al., 1996). Karathanasis and co-workers (2005) have made cross-linked liposomes toward creating a drug delivery system suitable for nebulization for inhalation delivery. These modified liposomes are cleavable on exposure to cysteine, thereby altering the size distribution and stimulating drug release.

Lipid self-assembled systems have not been commonly reported in which responsiveness to ionic or osmotic variation are used to stimulate self-assembly changes leading to control over drug release. These two stimuli present an interesting option for controlling self-assembled systems, as pressure and ionic strengths are known to affect the internal packing of the amphiphiles (Czeslik et al., 1995; Greaves and Drummond, 2008; Winter et al., 1999; Yaghmur et al., 2009). A possible application of pressure dependence of a drug delivery system could arise in the treatment of eye conditions such as glaucoma, where the "intelligent" drug delivery system could act as a sensor to changes in intraocular pressure and so automatically adjust the release of an encapsulated drug. The pressure is elevated inside tumor tissues due to compromised lymphatic drainage that also presents an albeit limited opportunity for using pressure as a trigger for drug release from such systems (Baxter and Jain, 1990).

The addition of specific ions or change in counterions has been reported as a means by which to manipulate lipid self-assembly and hence could be used as a drug release trigger. Shen reported that the self-assembly of an amphiphilic ionic liquid that forms micelles in water was disrupted on exchange of the bromide counterion with hexafluorophosphate, with subsequent dye release (Shen et al., 2008). Yaghmur et al. have demonstrated a direct vesicle to inverted hexagonal phase transition for mixtures of glyceryl monooleate and dioleyl phosphatidyl glycerol when exposed to increasing concentration of calcium ions (Yaghmur et al., 2008b). The divalent counterion reduced the apparent area of the head group, leading to a change in molecular packing and subsequent phase transition. Although not resulting in a change in self-assembly behavior, the release of a cationic ruthenium complex, "rubipy," from cubic phase on addition of NaCl to the aqueous compartment was demonstrated when the glyceryl monooleate cubic phase contained a small amount of oleic acid (Clogston and Caffrey, 2005). The shielding of the electrostatic

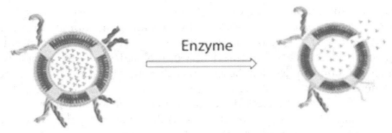

Figure 9.2 Schematic for enzymatically triggered cleavage of phospholipid structure resulting in release of internal payload. [Adapted from Lan-rong et al. (1986) with permission from Elsevier.]

interaction between the cationic compound and the negatively charged acid provided the trigger for release, highlighting the potential for specific salt effects to provide triggers for changes in self-assembly and or release in lipid-based systems.

9.3.1.3 Increased Expression and Activity of Proteins The up-regulation of proteins in tumors and inflammation and the specificity of protein location throughout tissues in the body have afforded a tissue-specific target for drug release. This has already been exploited in polymer systems for selective delivery in tumor tissues. Due to their quick and unusual growth, tumors often overexpress macromolecules, in particular enzymes, which can recognize and cleave specific linker groups between drug and carrier or within the carrier structure itself. Inflammatory conditions such as cystic fibrosis, rheumatoid arthritis, and emphysema are accompanied by an increase in the release of elastase from phagocytic cells into extracellular compartments.

Enzymatically activated drug delivery from liposomes has been investigated and reviewed (Meers, 2001). Liposomes have been designed to be activated by many macromolecules resulting in the collapse of the liposome or to induce membrane fusion under the action of the enzyme. A common strategy is to incorporate an enzymatically sensitive lipid to facilitate conversion of the lipid to a fusogenic derivative, such as dioleylphosphatidylethanolamine (DOPE), or to otherwise disrupt the bilayer structure of the liposome (see Fig. 9.2). Phospholipids are natural substrates for phospholipase enzymes—with phospholipase A2 (Andresen et al., 2005; Davidsen et al., 2001, 2003) and C (Montes et al., 2004; Nieva et al., 1989; Ruiz-Arguello et al., 1998) having been investigated for liposome destabilization. A more common and specific strategy is to build a custom phospholipid or lipopeptide with a molecular feature that is a substrate for specific enzymes overexpressed in diseased tissues. Lipids sensitive to elastase were prepared by incorporation of the N-Ac-Ala-Ala sequence into the head group (Pak et al., 1998, 1999). Similarly, matrix metalloproteinase (MMP-9), often overexpressed in tumors, was found to cleave lipopeptides containing the specific motif sensitive to this gelatinase enzyme, resulting in release of the liposome cargo (Sarkar et al., 2005; Terada et al., 2006). With a view to activation by quinone reductase, Ong et al. (2008) have reported the reductive cleavage of quinone head groups from phospholipid derivatives to reveal DOPE, leading to liposome release of incorporated model drug molecules. An alternative approach has been to sensitize liposomes through the use of cholesterol esters sensitive to alkaline phosphatase (Davis and Szoka, 1998). Incorporation of transmembrane proteins into DOPE particles has been known for decades to induce bilayer formation, providing liposomes stabilized by incorporated proteins. The degradation of the protein by trypsin has been long proposed as a means of inducing liposome destabilization (Lan-rong et al., 1986).

9.3.2 External Stimuli

The clinical advantage of physiological triggers inducing phase changes in materials and thus stimulate drug release centers on the simplicity and specificity of the process. The problem with this approach, however, is the potential for nonspecific activation and release of material in tissues other than the target tissues.

Accidental release may occur, for example, in inflamed tissues when targeting tumors or tissues where enzymes are expressed albeit at much lower levels than the target tissues, which may lead to side effects. Having specific temporal and spatial control over exactly where the delivery system is activated overcomes these concerns, although it introduces increased complexity and likely requires health professional intervention. Hence systems that are responsive to external stimuli that can be controlled in a noninvasive manner have attracted significant attention in the literature.

9.3.2.1 Temperature Change Temperature is a popular trigger mechanism to control the structure of self-assembled systems as the assembly of molecules is inherently affected by changes in temperature, and temperature is a very accessible trigger for many therapeutic applications. Hyperthermic conditions have been used to induce cytotoxicity (Dewhirst et al., 1997) and can augment some treatments (Hettinga et al., 1997). The primary means of inducing the hyperthermic environment are summarized in Table 9.2. Laser heating using a near-infrared (NIR) laser is commonly reported. An electric field is also reportedly able to induce localized heating and release of contents from "magnetoliposomes." The simple application of heat pack or cool pack to manipulate self-assembled structure and modulate drug release has also been reported. Hence there are a variety of potential approaches available to induce drug release using hyperthemia as the common trigger.

Thermoliposomes As is the case for enzymatically activated systems, liposomal systems sensitive to temperature have received the most attention in

TABLE 9.2   Means of Inducing Hyperthermic Condition in Tissues for Activation of Thermally Sensitive Drug Delivery Systems

Heat Source	Lipid System and Route of Administration	Activation Site
Laser: NIR lasers can penetrate to depths of 50 mm though the skin.	Liposome, systemic	Subcutaneous (Mackanos et al., 2009)
	Liposome, systemic	Tumor (Gaber, 2002; Gaber, et al., 1996; Ponce et al., 2006; Weinstein et al., 1979)
	Liposome, intravenous (IV)	Ocular (Desmettre et al., 1999)
	Liposome, systemic	Liver (Mordon et al., 1996)
Electromagnetic field (EMF)	Liposome incorporating dextran magnetite or colloid; iron oxide; systemic	Site accessed by EMF (Babincová et al., 2002; Viroonchatapan et al., 1997; Zhu et al., 2009)
Ultrasound		
Microwave	Liposome incorporating dextran magnetite or colloid; iron oxide; systemic	Site accessed by EMF (Babincová, 1993)
Heat/cool pack	Bulk liquid crystal; subcutaneous	Subcutaneous (Fong et al., 2009)

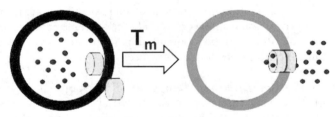

Figure 9.3   Schematic showing the formation of a pore by pore-forming amphiphiles above the melting transition temperature of the lipids comprising the liposome, resulting in drug release. [Adapted from Petrov et al. (2009) with permission from the American Chemical Society © 2009.]

the literature compared to other self-assembled structures. The application of hyperthermia takes advantage of the acyl-chain melting phase transition in phospholipids, which results in an increase in permeability across the lipid bilayer, and consequent drug release (as illustrated in Fig. 9.3). The transition temperature of the lipids and the heat applied is tuned to be just above physiological temperature but low enough so as to not perturb cell function in surrounding tissues. The first systems were reported by Yatvin and co-workers (Weinstein et al., 1979; Yatvin et al., 1978), who designed liposomes to release drug in this narrow temperature range. More recently, liposome formulations have been modified with polymers, most commonly poly(N-isopropylacrylamide) or a derivative thereof, which become water insoluble above a critical solution temperature (CST) and so destabilize the liposome bilayer (Kono, 2001; Kono et al., 1999, 2002, 2005; Yoshino et al., 2004). Interestingly, Regan and co-workers have created pore-forming amphiphiles comprised of cholic acid, lysine, and spermine, which act as thermal gates when heat melts the liposome bilayer, thereby changing the release rate of glucose in vitro (Chen and Regen, 2005; Petrov et al., 2009).

Other Thermoresponsive Structures Reversibility of the system, that is, the ability of the system to return to its original state on removal of the stimulus, is a potentially important attribute for responsive systems for the treatment of chronic conditions. Reversible activation may reduce the required frequency of administration by providing repeat doses of drug on activation, and switch drug release off when the stimulus is removed.

Nonlamellar liquid crystalline systems are gaining increasing attention as potential reversible responsive materials, as the specific geometries adopted by the lipids are thermodynamically stable, and transitions between structures are generally reversible (de Campo et al., 2004). This is in contrast to most liposomal or polymer systems where drug release behavior is often irreversible. The complexity of potential structures that may be adopted, some of which are illustrated in Figure 9.1, also offers control over drug release rates through the selection of specific amphiphile/phase structure combinations (Lee et al., 2009a). Fong et al. (2009) have taken advantage of the reversibility of the transition between the bicontinuous cubic (Q_2) and inverse hexagonal phase (H_2) structures exhibited by the phytantriol + vitamin E acetate liquid crystal system to switch drug release between fast and slow release rates, respectively. Figure 9.4 shows the in vitro release of glucose and use of

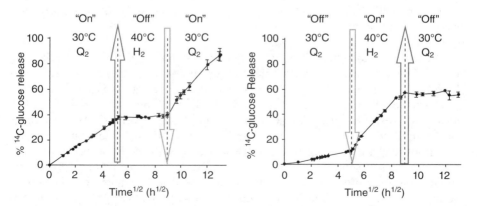

Figure 9.4 Release of glucose from liquid crystalline matrix prepared from phytantriol containing 3% vitamin E acetate. [Modified from Fong et al. (2009).]

temperature change to reversibly modify drug release rates from the matrix through control of the nanostructure.

The responsive phytantriol + vitamin E acetate system was taken into a proof-of-concept in vivo study in rats. The subcutaneous injection of the liquid crystal matrix and subsequent application of a heat or cool pack to the site of injection provided control over the matrix nanostructure, which consequently modified drug absorption from the implanted delivery system in vivo. Along similar lines, the phase transformation of vesicles to nonlamellar phase structures has also been proposed as a means to stimulate drug release from self-assembled lipid systems. One such system is monoelaidin as it is known to undergo a lamellar-to-cubic phase transition at physiological temperatures (Yaghmur et al., 2008a).

However, practicalities of using direct heat as the stimulus will likely limit the ultimate development of such a system as the basis of an actual product. The use of temperature as a stimulus would be limited to applications where the site of required treatment is readily accessible and amenable to direct heating or cooling, essentially limiting applicability to subcutaneous depot application. Further problems arise when one considers that the body encounters local temperature changes in everyday situations, such as a hot shower, leaning against cold surfaces, and the like, with potential to lead to accidental activation. Hence more selective means of inducing hyperthermic conditions are desirable, and one possible solution is considered when discussing photothermal activation later.

9.3.2.2 Light   Light-responsive systems have the advantage of being non-invasive and provide a broad range of adjustable parameters such as wavelength, duration, and intensity that can be optimized in a nondisruptive manner. Light-sensitive self-assembling systems have been reviewed recently (Eastoe and Vesperinas, 2005) as have light-activated drug delivery systems (Alvarez-Lorenzo et al., 2009; Shum et al., 2001). The photoactivation mechanism in self-assembled systems can be distributed into three categories: photochromics, irreversible photoactivation, and photothermal, each of which are discussed further below.

Photochromic Systems   Photochromic molecules reversibly isomerize on exposure to a light source, in most cases in the UV–visible range. When incorporated into a self-assembled system, the activation and deactivation of photochromics cause a steric disturbance in the packing in the self-assembled structures resulting in a change in, for example, drug release. The response of the lipid system can be manipulated by the composition of the lipid host and photochromic concentration (Bisby et al., 1999a,b). Research into photochromic self-assembled systems has focused on azobenzene and spiropyran moieties among others (Eastoe and Vesperinas, 2005). In some cases, photoisomerization does not compromise interface integrity, and a pulsatile delivery may be achieved (Khoukh et al., 2005; Tong et al., 2005; Zhang et al., 2009).

(a)   (b)

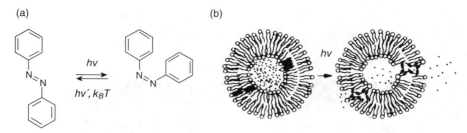

Figure 9.5 Azobenzene photoisomerization: (a) Reversible trans to cis isomerization on exposure to an appropriate wavelength of UV light. (b) Diagrammatic representation of the disruption of a liposome through the activation of the azo moiety with UV light and subsequent drug release. [Adapted from Morgan et al. (1995) with permission from Elsevier.]

Azobenzene   Aromatic azo compounds are a group of molecules whose photochemistry is determined by an azobenzene moiety; an azo group –N=N– in conjugation with two phenyl substituents. Upon UV irradiation, the azobenzene moiety undergoes a considerable change in length with the distance between the 4 and 4′ carbons decreasing from 9.0 Å in the trans form to 5.5 Å in the cis form (Natansohn and Rochon, 2002). Azobenzene photoisomers also differ in polarity; the cis isomer is more hydrophilic than the trans. The N=N isomerization between trans and cis azobenzene isomers is shown in panel (a) in Fig. 9.5. The extent of trans to cis isomerization is dependent on strength and time of irradiation, solvent effects, and is slightly affected by substituent type (Knoll, 2004).

The main uses of azobenzene are as yellow/orange synthetic dyes; however, molecules and surfactants containing an azobenzene moiety have been used to impart control over membrane permeability in aggregates such as emulsions (Bufe and Wolff, 2009; Cicciarelli et al., 2007; Han and Hara, 2005; Khoukh et al., 2005), Langmuir–Blodgett films (Haruta et al., 2008; Kumar et al., 2009b; Oki and Nagasaka, 2009; Sorensen et al., 2008), liposomes and vesicular membranes (Eastoe et al., 2008; Kuiper and Engberts, 2004; Kuiper et al., 2008; Lei and Hurst, 1999; Liu et al., 2005; Morgan et al., 1995; Song et al., 1995; Zou et al., 2008), polymeric vesicles (Liu et al., 2008; Tong et al., 2005), hydrogels (Zhao and Stoddart, 2009), silica materials (Tanaka et al., 2007), and liquid crystals (Corvazier and Zhao, 1999; Yamamoto et al., 2006, 2009; Yi et al., 2009). The isomerization of the azo-moiety is sufficient to destabilize the surrounding area and, as a consequence, induce a phase transformation or disruption of the bilayer as schematically demonstrated in panel (b) of Figure 9.5.

Spiropyrans   Upon exposure to UV light, spiropyrans undergo a heterocyclic ring cleavage that results in the relatively hydrophilic, open cationic merocyanine form. The spiropyran (the relatively hydrophobic closed form) is

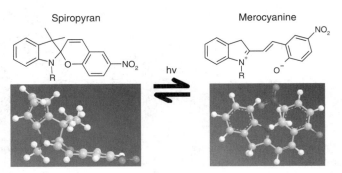

Figure 9.6 Isomerization of benzospiropyran to merocyanine on exposure to UV light, demonstrating its change in state of ionization and molecular shape.

colorless, whereas the open merocyanine structure has a strong absorption in the visible region and so is a purple color. The photoisomerization of the spiropyran moiety is shown in Figure 9.6. Spiropyrans have been used to alter the packing of amphiphiles in bilayer membranes (Khairutdinov and Hurst, 2001; Wohl and Kuciauskas, 2005) and micelles (Takumi et al., 2007) and so modifying the permeation of solutes through the interface.

Cinnamic Acid (CA)   On exposure to UV light, cinnamic acid and its derivatives undergo a trans/cis isomerization. These photoisomers are also ionizable with a change in pH. There have been no light-switchable drug delivery systems reported; however, Baglioni et al. (2009) have reported that the trans-to-cis isomerization of *o*-methoxyCA modifies the molecular packing of a surfactant and, as a consequence, induces a reduction in fluid viscosity.

Irreversible Photoactivated Systems   Light-induced polymerization, fragmentation, and oxidation are irreversible processes that disrupt the integrity of the lipid structure. As the reactions of the photosensitive components discussed in this section are irreversible, the use of these in drug delivery systems is effective for single use but not pulsatile release.

Photo-induced oxidation utilizes the reaction of a photosensitizer to a specific wavelength of light with oxygen, which generates free radicals and singlet oxygen ($^1O_2$) to mediate their effects. Photodynamic therapy (PDT) uses these free radicals to cause cell death. These radical oxygen species cause localized oxidative damage to the surrounding tissue (Dolmans et al., 2003; Zeimer and Goldberg, 2001). The consequent physiological effect depends on the type of photosensitizer and where it is applied and activated. Current PDTs are focused on treating tumors by either targeting the tumor microvasculature, tumor cells, or the inflammatory and immune host system. However, PTD is limited by the need to keep doses of the photosensitizer and light low enough to avoid collateral damage. The formation of free radicals has also been employed to induce drug release from vesicles such as liposomes. The free radicals cause membrane destabilization through lipid hydrolysis. Thompson and co-workers have manipulated the effects of plasmalogen photooxidation on membrane permeability in order to release the drug from liposomes. The photooxidation of light-sensitive components promotes bilayer fusion and contents release (Anderson and Thompson, 1992; Collier et al., 2001; Gerasimov et al., 1999).

Systems have also been synthesized that undergo photofragmentation or polymerization of the stabilizing element on exposure to light. Photocleavable lipid derivatives in conjunction with a lipid, for example, DOPE, have been used to form liposomes. On exposure to UV irradiation, the fragmentation of the modified lipid destabilizes the bilayer and contents release ensues. Zhang and Smith (1999) achieved 50% calcein release through the fragmentation of NVOC-DOPE, a photocleavable nitroveratryloxycarbonyl derivative of dioleoylphosphatidylethanolamine. Activation of photopolymerizing components in self-assembled systems has also been used to cause drug release. Polymerization of these components condenses some of the interface, resulting in the formation of pores. For instance, Spratt et al. (2003) achieved a 28,000-fold increase in liposome bilayer permeability on activation of their synthesized photoreactive lipid bis-SorbPC$_{17,17}$.

Photothermal Systems   Light can be used to impart heat into systems in two ways. First, NIR lasers can penetrate tissues into the posterior segment of the eye and as deep as 10 cm under the skin (Weissleder, 2001). A major focus of the application of photothermal systems concentrates on drug delivery to the posterior section of the eye, especially the retina and choroid, in order to treat degenerative conditions such as choroidal neovascularisation (CNV), a cause of age-related macular degeneration (AMD). Current methods of drug delivery are based on topical, periorbital, intravitreal, and systemic administration. These are problematic as low penetration is gained through topical and periorbital administration, periorbital and intravitreal administration is invasive, and systemic administration of the drugs presents toxicity issues as the whole body is exposed to large concentrations of drug in order to get a therapeutic amount into the target tissue. An interesting method developed by Zeimer et al., called light-targeted drug delivery (LTD), uses a noninvasive laser source to gently heat liposomes and so trigger drug release (Zeimer and Goldberg, 2001; Zeimer et al., 1988), the intended target of this system being the posterior section of the eye. The method involves the intravenous administration of drug encapsulated in a heat-sensitive liposome, and release of its contents at the target tissue by gently warming up the target tissue to 41°C with a directed laser light pulse. LTD, like PTD, is limited by the need to keep doses of the photosensitizer and light low enough to avoid collateral damage.

Second, light can be used to activate metallic nanoparticles (NPs), which are able to act as "nanoheaters" on a highly localized scale. The evolved heat

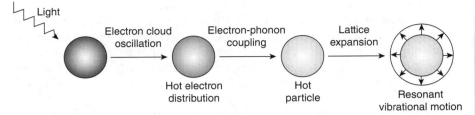

Figure 9.7 Interaction of light with metallic nanoparticles. Light absorption heats the electrons of the particles, which then equilibrate through electron–phonon coupling. This initiates coherent vibrational motion in the particles called surface plasmon resonance. The heat created can then be transferred to its surroundings through transduction. [Adapted from Hartland (2007) with permission from Macmillan Publishers Ltd.]

can induce bond cleavage, phase transitions, or mechanical damage to biological systems. In recent times, there has been a flurry of research into the optical properties and potential applications of metallic nanoparticles. Metallic nanoparticles (mainly gold but also silver and copper) are of interest as on exposure to radiation of a specific wavelength and intensity, can heat NPs to their vaporization point within femtoseconds due to the activation of surface plasmon resonances, a phenomenon demonstrated in Figure 9.7 (Govorov and Richardson, 2007; Murray and Barnes, 2007). Wavelengths at which the NPs absorb is tuneable according to their size, shape, composition, and state of aggregation (Jain et al., 2006; Volodkin et al., 2009). The use of spheres, rods, and shells has been reported. The wavelengths at which the NP absorb are usually in the NIR region, wavelengths not significantly absorbed by cellular material. In addition to their specific dimensions, metallic nanoparticles can also be functionalized and made more biocompatible through the facile binding of various types of biomolecules, for example, phosphatidylcholine (Takahashi et al., 2006).

Therapeutic applications of current research has focused on the use of metallic nanoparticles to cause tumor cell death by localized heating of the tumor (Dickerson et al., 2008; Lee et al., 2009b; Maeda et al., 2003; Tong et al., 2007) or to elicit remote release of entrapped materials via photothermal conversion (Skirtach et al., 2005) from within liposomes (Paasonen et al., 2007; Park et al., 2006; Troutman et al., 2009; Volodkin et al., 2009; Wu et al., 2008), polyelectrolyte capsules (Munoz et al., 2008), hydrogels (Das et al., 2007, 2008; Kim and Lee, 2006), polymer vesicles (Murphy and Orendorff, 2005), or from a NP–drug conjugate (Agasti et al., 2009; Barhoumi et al., 2009). Plasmon resonant heating has also been used to trigger phase transitions in self-assembled lipid systems with a view to elucidate the mechanisms behind photothermal activated responses (Orendorff et al., 2009; Urban et al., 2009; Yaghmur et al., 2010). Recently, we have shown that incorporation of hydrophobized gold nanorods can be utilized to switch liquid crystalline structure

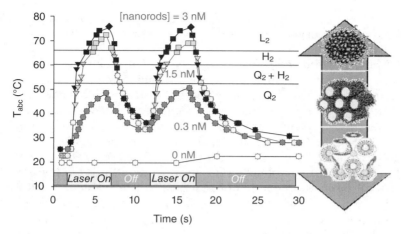

Figure 9.8   Incorporation of gold nanorods at increasing concentration into phytantriol-based cubic phase facilitates reversible optical control over the apparent temperature of the matrix (T_{app}), and hence over phase transitions. Transitions from bicontinuous cubic to inverse hexagonal to inverse micellar phase were in agreement with equilibrium heating studies. The phase structures formed on irradiation are illustrated schematically on the right. [Adapted from Fong et al. (2010).]

between non-lamellar liquid crystalline phases on irradiation, with the matrix returning to its original state when irradiation ceases (Fig. 9.8) (Fong et al., 2010), opening up the potential for reversible, multidose drug delivery systems from such materials.

9.3.2.3 Electromagnetic Field-Responsive Systems   Electromagnetic fields (EMF) can be used in drug delivery for two purposes: to focus delivery to a target (Arruebo et al., 2007; Blasi et al., 2007; Jain et al., 2008), for example, tumor site, where the system can cause cell death through intracellular hyperthermia (Yanase et al., 1997) or release drug or imaging dyes (Kumar et al., 2009a). Carrier particles are magnetically sensitive by virtue of being magnetic nanoparticles encapsulated within the membrane or in the aqueous interior of polymers or liposomes, which are known as magnetoliposomes. After local or systemic dosing of magnetoliposomes, the tumor-affected area can either be placed between two poles of a magnet (Nobuto et al., 2004; Viroonchatapan et al., 1996; Zhu et al., 2009) or the magnet is imbedded in the tumor (Kubo et al., 2000), where the application of an EMF for some time results in the increased accumulation of drug concentration in the tumor where the drug is released from the liposome by diffusion or by heating of the bilayer by the EMF. The use of low-frequency magnetic fields to disrupt the liposome loaded with magnetic nanoparticles in the aqueous chamber has also been investigated (Nappini et al., 2010) where magnetic nanoparticle motion causes disruptions in the bilayer and consequently drug release is achieved. EMF-sensitive

polymeric hydrogels (Satarkar and Hilt, 2008) and vesicles (Krack et al., 2008) have also been reported.

9.3.2.4 Ultrasound-Responsive Systems   Ultrasound is an attractive local physical stimulus as it can be easily focused in the body and is able to penetrate into deep tissue (Wu and Nyborg, 2008). Low-frequency (LFU) and high-frequency (HFU) ultrasound have been employed to release drugs from acoustically active systems such as polymeric micelles and matrices (Husseini et al., 2000; Kost and Langer, 2001), and more recently liposomes (Chen and Wu, 2010; Lin and Thomas, 2003; Schroeder et al., 2007, 2009). It is proposed that the presence of microbubbles in the structure affords the liposome sensitivity to ultrasound. Insonation of acoustically active systems results in the cavitation of nanoparticles and consequently the leakage of the vesicle contents into its environment. Heating of the tissues through ultrasound may also play a role in triggering drug release.

Research into creating lipid-based systems has focused on forming LFU-responsive liposomes as contrast agents or for drug delivery as they have been found to be more effective in contents release from liposomes than HFU (Huang and MacDonald, 2004; Liu et al., 2006). However, the potential clinical applications of ultrasound-activated liposomes may be more compatible with HFU as they may integrate with readily available instruments. In lipid-based self-assembled systems, it is well established that insonation and sonication of sufficient intensity can be employed to destabilize self-assembled systems and so make multilamellar liposomes into unilamellar ones (Huang, 1969; Zasadzinski, 1986) and viscous cubic phase gels into cubosomes (Landh and Larsson, 1996; Ljusberg-Wahern et al., 1996). As ultrasound has the ability to disrupt self-assembled bilayers, ultrasound can also penetrate through cell membranes, which can result in cell death (Clarke and Hill, 1970; Liu et al., 1998). This must be accounted for in ultrasound-responsive systems for use in drug delivery applications. Lin and Thomas, achieved this through the incorporation of PEG lipids and oligo(ethylene glycol) surfactants (Lin and Thomas, 2003), thereby tuning the liposome to be more sensitive to insonation than cell membranes. Huang and MacDonald (2004) have formulated liposomes with different lipids, which increase their sensitivity to LFU.

9.4 FUTURE DIRECTIONS OF STIMULI-RESPONSIVE SYSTEMS

As discussed, there are many types of stimuli that can be chosen to induce phase transitions in lipid-based self-assembly systems. However, most work has revolved around imparting responsiveness into liposomes. Other self-assembled nanostructures such as nonlamellar liquid crystals have received much less attention despite the potential for reversibility of structural changes that could permit repeated on-demand administration of a drug with a single injection. In particular, cubic and hexagonal phase dispersions may provide future directions in which this research will head.

It is notable that no drug delivery products have yet made it to market based on stimuli-responsive systems. A heat-sensitive liposome system containing doxorubicin for heat-selective tumor delivery (Thermodox owned by Celsion) has entered clinical trials. However, the potential for accidental activation of the delivery system needs to be carefully controlled to avoid inadvertent activation in nontarget tissues, consequently it might be expected that alternative means of activation such as optical, will ultimately be viewed as a safer approach to on-demand delivery systems. The major hurdle to progress in light-activated delivery systems is the need to include light-responsive elements that may themselves be prohibitively toxic and require their own highly costly preclinical and clinical toxicological evaluation. Hence the search continues for the ideal selective responsive drug delivery system that meets the demands of functionality, safety, and toxicity.

REFERENCES

Agasti, S. S., Chompoosor, A., You, C.-C., Ghosh, P., Kim, C. K., and Rotello, V. M. (2009). Photoregulated release of caged anticancer drugs from gold nanoparticles. *Journal of the American Chemical Society*, *131*(16), 5728–5729.

Alvarez-Lorenzo, C., Bromberg, L., and Concheiro, A. (2009). Light-sensitive intelligent drug delivery systems. *Photochemistry and Photobiology*, *85*(4), 848–860.

Anderson, V. C., and Thompson, D. H. (1992). Triggered release of hydrophilic agents from plasmologen liposomes using visible light or acid. *Biochimica et Biophysica Acta (BBA)—Biomembranes*, *1109*(1), 33–42.

Andresen, T. L., Jensen, S. S., and Jørgensen, K. (2005). Advanced strategies in liposomal cancer therapy: Problems and prospects of active and tumor specific drug release. *Progress in Lipid Research*, *44*(1), 68–97.

Aota-Nakano, Y., Li, S. J., and Yamazaki, M. (1999). Effects of electrostatic interaction on the phase stability and structures of cubic phases of monoolein/oleic acid mixture membranes. *Biochimica et Biophysica Acta (BBA)—Biomembranes*, *1461*(1), 96–102.

Arruebo, M., Fernández-Pacheco, R., Ibarra, M. R., and Santamaría, J. (2007). Magnetic nanoparticles for drug delivery. *Nano Today*, *2*(3), 22–32.

Babincová, M. (1993). Microwave induced linkage of magnetoliposomes. Possible clinical implications. *Bioelectrochemistry and Bioenergetics*, *32*(2), 187–189.

Babincová, M., Cicmanec, P., Altanerová, V., Altaner, C., and Babinec, P. (2002). AC-magnetic field controlled drug release from magnetoliposomes: Design of a method for site-specific chemotherapy. *Bioelectrochemistry*, *55*(1–2), 17–19.

Baglioni, P., Braccalenti, E., Carretti, E., Germani, R., Goracci, L., Savelli, G., and Tiecco, M. (2009). Surfactant-based photorheological fluids: Effect of the surfactant structure. *Langmuir*, *25*(10), 5467–5475.

Barhoumi, A., Huschka, R., Bardhan, R., Knight, M. W., and Halasa, N. J. (2009). Light-induced release of DNA from plasmon-resonant nanoparticles: Towards light-controlled gene therapy. *Chemical Physics Letters*, *482*, 171–179.

Baxter, L. T., and Jain, R. K. (1990). Transport of fluid and macromolecules in tumors. 2. Role of heterogeneous perfusion and lymphatics. *Microvascular Research*, *40*(2), 246–263.

Bayer, C. L., and Peppas, N. A. (2008). Advances in recognitive, conductive and responsive delivery systems. *Journal of Controlled Release*, *132*(3), 216–221.

Bhavane, R., Karathanasis, E., and Annapragada, A. V. (2003). Agglomerated vesicle technology: A new class of particles for controlled and modulated pulmonary drug delivery. *Journal of Controlled Release*, *93*(1), 15–28.

Bisby, R. H., Mead, C., Mitchell, A. C., and Morgan, C. G. (1999a). Fast laser-induced solute release from liposomes sensitized with photochromic lipid: Effects of temperature, lipid host, and sensitizer concentration. *Biochemical and Biophysical Research Communications*, *262*(2), 406–410.

Bisby, R. H., Mead, C., and Morgan, C. G. (1999b). Photosensitive liposomes as "cages" for laser-triggered solute delivery: The effect of bilayer cholesterol on kinetics of solute release. *FEBS Letters*, *463*(1–2), 165–168.

Blasi, P., Giovagnoli, S., Schoubben, A., Ricci, M., and Rossi, C. (2007). Solid lipid nanoparticles for targeted brain drug delivery. *Advanced Drug Delivery Reviews*, *59*(6), 454–477.

Boomer, J. A., Inerowicz, H. D., Zhang, Z.-Y., Bergstrand, N., Edwards, K., Kim, J.-M., and Thompson, D. H. (2003). Acid-triggered release from sterically stabilized fusogenic liposomes via a hydrolytic dePEGylation strategy. *Langmuir*, *19*(16), 6408–6415.

Borne, J., Nylander, T., and Khan, A. (2001). Phase behavior and aggregate formation for the aqueous monoolein system mixed with sodium oleate and oleic acid. *Langmuir*, *17*(25), 7742–7751.

Brink-van der Laan, E. V., Killian, J. A., and de Kruijff, B. (2004). Nonbilayer lipids affect peripheral and integral membrane proteins via changes in the lateral pressure profile. *Biochimica et Biophysica Acta—Biomembranes*, *1666*(1–2), 275–288.

Bufe, M., and Wolff, T. (2009). Reversible switching of electrical conductivity in an AOT-isooctane-water microemulsion via photoisomerization of azobenzene. *Langmuir*, *25*(14), 7927–7931.

Chen, W.-H., and Regen, S. L. (2005). Thermally gated liposomes. *Journal of the American Chemical Society*, *127*(18), 6538–6539.

Chen, D., and Wu, J. (2010). An in vitro feasibility study of controlled drug release from encapsulated nanometer liposomes using high intensity focused ultrasound. *Ultrasonics*, *50*, 744–749.

Chernomordik, L. (1996). Non-bilayer lipids and biological fusion intermediates. *Chemistry and Physics of Lipids*, *81*(2), 203–213.

Cicciarelli, B. A., Elia, J. A., Hatton, T. A., and Smith, K. A. (2007). Temperature dependence of aggregation and dynamic surface tension in a photoresponsive surfactant system. *Langmuir*, *23*(16), 8323–8330.

Clarke, P. R., and Hill, C. R. (1970). Physical and chemical aspects of ultrasonic disruption of cells. *Journal of the Acoustical Society of America*, *47*(2B), 649–653.

Clogston, J., and Caffrey, M. (2005). Controlling release from the lipidic cubic phase. Amino acids, peptides, proteins and nucleic acids. *Journal of Controlled Release*, *107*(1), 97–111.

Collier, J. H., Hu, B. H., Ruberti, J. W., Zhang, J., Shum, P., Thompson, D. H., and Messersmith, P. B. (2001). Thermally and photochemically triggered self-assembly of peptide hydrogels. *Journal of the American Chemical Society, 123*(38), 9463–9464.

Collins, D., Litzinger, D. C., and Huang, L. (1990). Structural and functional comparisons of pH-sensitive liposomes composed of phosphatidylethanolamine and 3 different diacylsuccinylglycerols. *Biochimica et Biophysica Acta, 1025*(2), 234–242.

Conn, C. E., Ces, O., Mulet, X., Finet, S., Winter, R., Seddon, J. M., and Templer, R. H. (2006). Dynamics of structural transformations between lamellar and inverse bicontinuous cubic lyotropic phases. *Physical Review Letters, 96*(10), 108102.

Corvazier, L., and Zhao, Y. (1999). Induction of liquid crystal orientation through azobenzene-containing polymer networks. *Macromolecules, 32*(10), 3195–3200.

Couvreur, P., and Vauthier, C. (2006). Nanotechnology: Intelligent design to treat complex disease. *Pharmaceutical Research, 23*(7), 1417–1450.

Czeslik, C., Winter, R., Rapp, G., and Bartels, K. (1995). Temperature- and pressure-dependent phase behavior of monoacylglycerides monoolein and monoelaidin. *Biophysical Journal, 68*(4), 1423–1429.

Das, M., Sanson, N., Fava, D., and Kumacheva, E. (2007). Microgels loaded with gold nanorods: Photothermally triggered volume transitions under physiological conditions. *Langmuir, 23*(1), 196–201.

Das, M., Mordoukhovski, L., and Kumacheva, E. (2008). Sequestering gold nanorods by polymer microgels. *Advanced Materials, 20*(12), 2371–2375.

Davidsen, J., Vermehren, C., Frokjaer, S., Mouritsen, O. G., and Jørgensen, K. (2001). Drug delivery by phospholipase A2 degradable liposomes. *International Journal of Pharmaceutics, 214*(1–2), 67–69.

Davidsen, J., Jørgensen, K., Andresen, T. L., and Mouritsen, O. G. (2003). Secreted phospholipase A2 as a new enzymatic trigger mechanism for localised liposomal drug release and absorption in diseased tissue. *Biochimica et Biophysica Acta (BBA)—Biomembranes, 1609*(1), 95–101.

Davis, S. C., and Szoka, F. C. (1998). Cholesterol phosphate derivatives: Synthesis and incorporation into a phosphatase and calcium-sensitive triggered release liposome. *Bioconjugate Chemistry, 9*(6), 783–792.

de Campo, L., Yaghmur, A., Sagalowicz, L., Leser, M. E., Watzke, H., and Glatter, O. (2004). Reversible phase transitions in emulsified nanostructured lipid systems. *Langmuir, 20*(13), 5254–5261.

Desmettre, T. J., Soulie-Begu, S., Devoisselle, J. M., and Mordon, S. R. (1999). Diode laser-induced thermal damage evaluation on the retina with a liposome dye system. *Lasers in Surgery and Medicine, 24*(1), 61–68.

Dewhirst, M. W., Prosnitz, L., Thrall, D., Prescott, D., Clegg, S., Charles, C., MacFall, J., Rosner, G., Samulski, T., Gillette, E., and LaRue, S. (1997). Hyperthermic treatment of malignant diseases: Current status and a view toward the future. *Seminars in Oncology, 24*(6), 616–625.

Dickerson, E. B., Dreaden, E. C., Huang, X., El-Sayed, I. H., Chu, H., Pushpanketh, S., McDonald, J. F., and El-Sayed, M. A. (2008). Gold nanorod assisted near-infrared plasmonic photothermal therapy (PPTT) of squamous cell carcinoma in mice. *Cancer Letters, 269*(1), 57–66.

Dolmans, D. E. J. G. J., Fukumura, D., and Jain, R. K. (2003). Photodynamic therapy for cancer. *Nature Reviews Cancer*, *3*(5), 380–387.

Dong, Y.-D., Larson, I., Hanley, T., and Boyd, B. J. (2006). Bulk and dispersed aqueous phase behavior of phytantriol: Effect of vitamin E acetate and F127 polymer on liquid crystal nanostructure. *Langmuir*, *22*(23), 9512–9518.

Drummond, D. C., Zignani, M., and Leroux, J.-C. (2000). Current status of pH-sensitive liposomes in drug delivery. *Progress in Lipid Research*, *39*(5), 409–460.

Eastoe, J., and Vesperinas, A. (2005). Self-assembly of light-sensitive surfactants. *Soft Matter*, *1*(5), 338–347.

Eastoe, J., Zou, A., Espidel, Y., Glatter, O., and Grillo, I. (2008). Photo-labile lamellar phases. *Soft Matter*, *4*(6), 1215–1218.

Fong, W.-K., Hanley, T., and Boyd, B. J. (2009). Stimuli responsive liquid crystals provide "on-demand" drug delivery *in vitro* and *in vivo*. *Journal of Controlled Release*, *135*(3), 218–226.

Fong, W.-K., Hanley, T. L., Thierry, B., Kirby, N., and Boyd, B. J. (2010). Plasmonic nano-rods provide reversible control over nanostructure of self-assembled drug delivery materials. *Langmuir*, *26*(9), 6136–6139.

Gaber, M. H. (2002). Modulation of doxorubicin resistance in multidrug-resistance cells by targeted liposomes combined with hyperthermia. *Journal of Biochemistry, Molecular Biology, and Biophysics*, *6*(5), 309–314.

Gaber, M. H., Wu, N. Z., Hong, K. L., Huang, S. K., Dewhirst, M. W., and Papahadjo-poulos, D. (1996). Thermosensitive liposomes: Extravasation and release of contents in tumor microvascular networks. *International Journal of Radiation Oncology Biology Physics*, *36*(5), 1177–1187.

Gerasimov, O. V., Boomer, J. A., Qualls, M. M., and Thompson, D. H. (1999). Cytosolic drug delivery using pH- and light-sensitive liposomes. *Advanced Drug Delivery Reviews*, *38*(3), 317–338.

Gerweck, L. E., and Seetharaman, K. (1996). Cellular pH gradient in tumor versus normal tissue: Potential exploitation for the treatment of cancer. *Cancer Research*, *56*(6), 1194–1198.

Govorov, A. O., and Richardson, H. H. (2007). Generating heat with metal nanopar-ticles. *Nano Today*, *2*(1), 30–38.

Greaves, T. L., and Drummond, C. J. (2008). Ionic liquids as amphiphile self-assembly media. *Chemical Society Reviews*, *37*(8), 1709–1726.

Guo, X., and Szoka, F. C. (2001). Steric stabilization of fusogenic liposomes by a low-pH sensitive PEG-diortho ester-lipid conjugate. *Bioconjugate Chemistry*, *12*(2), 291–300.

Guo, X., and Szoka, F. C. (2003). Chemical approaches to triggerable lipid vesicles for drug and gene delivery. *Accounts of Chemical Research 36*(5), 335–341.

Han, M., and Hara, M. (2005). Intense fluorescence from light-driven self-assembled aggregates of nonionic azobenzene derivative. *Journal of the American Chemical Society*, *127*(31), 10951–10955.

Hartland, G. V. (2007). Is perfect better? *Nature Materials*, *6*(10), 716–718.

Haruta, O., Matsuo, Y., and Ijiro, K. (2008). Photo-induced fluorescence emission enhancement of azobenzene thin films. *Colloids and Surfaces A: Physicochemical and Engineering Aspects*, *313–314*, 595–599.

Hegmann, T., Qi, H., and Marx, V. (2007). Nanoparticles in liquid crystals: Synthesis, self-assembly, defect formation and potential applications. *Journal of Inorganic and Organometallic Polymers and Materials*, *17*(3), 483–508.

Heller, J., Himmelstein, K. J., Kenneth, J. W., and Ralph, G. (1985). Poly(ortho ester) biodegradable polymer systems. In *Methods in Enzymology*, Academic, New York, Vol. 112, pp. 422–436.

Hettinga, J. V. E., Konings, A. W. T., and Kampinga, H. H. (1997). Reduction of cellular cisplatin resistance by hyperthermia—a review. *International Journal of Hyperthermia*, *13*(5), 439–457.

Hong, M. S., Lim, S. J., Oh, Y. K., and Kim, C. K. (2002). pH-sensitive, serum-stable and long-circulating liposomes as a new drug delivery system. *Journal of Pharmacy and Pharmacology*, *54*, 51–58.

Huang, C.-H. (1969). Phosphatidylcholine vesicles. Formation and physical characteristics. *Biochemistry*, *8*(1), 344–352.

Huang, S.-L., and MacDonald, R. C. (2004). Acoustically active liposomes for drug encapsulation and ultrasound-triggered release. *Biochimica et Biophysica Acta (BBA)—Biomembranes*, *1665*(1–2), 134–141.

Husseini, G. A., Myrup, G. D., Pitt, W. G., Christensen, D. A., and Rapoport, N. Y. (2000). Factors affecting acoustically triggered release of drugs from polymeric micelles. *Journal of Controlled Release*, *69*(1), 43–52.

Ishida, T., Kirchmeier, M. J., Moase, E. H., Zalipsky, S., and Allen, T. M. (2001). Targeted delivery and triggered release of liposomal doxorubicin enhances cytotoxicity against human B lymphoma cells. *Biochimica et Biophysica Acta (BBA)—Biomembranes*, *1515*(2), 144–158.

Jain, P. K., Lee, K. S., El-Sayed, I. H., and El-Sayed, M. A. (2006). Calculated absorption and scattering properties of gold nanoparticles of different size, shape, and composition: Applications in biological imaging and biomedicine. *Journal of Physical Chemistry B*, *110*(14), 7238–7248.

Jain, T. K., Richey, J., Strand, M., Leslie-Pelecky, D. L., Flask, C. A., and Labhasetwar, V. (2008). Magnetic nanoparticles with dual functional properties: Drug delivery and magnetic resonance imaging. *Biomaterials*, *29*(29), 4012–4021.

Kaasgaard, T., and Andresen, T. L. (2010). Liposomal cancer therapy: Exploiting tumor characteristics. *Expert Opinion on Drug Delivery 7*(2), 225–243.

Kaasgaard, T., and Drummond, C. J. (2006). Ordered 2-D and 3-D nanostructured amphiphile self-assembly materials stable in excess solvent. *Physical Chemistry Chemical Physics*, *8*(43), 4957–4975.

Karathanasis, E., Ayyagari, A. L., Bhavane, R., Bellamkonda, R. V., and Annapragada, A. V. (2005). Preparation of in vivo cleavable agglomerated liposomes suitable for modulated pulmonary drug delivery. *Journal of Controlled Release*, *103*(1), 159–175.

Khairutdinov, R. F., and Hurst, J. K. (2001). Photocontrol of ion permeation through bilayer membranes using an amphiphilic spiropyran. *Langmuir*, *17*(22), 6881–6886.

Khoukh, S., Perrin, P., Berc, F. B. D., and Tribet, C. (2005). Reversible light-triggered control of emulsion type and stability. *ChemPhysChem*, *6*(10), 2009–2012.

Kim, J.-H., and Lee, T. R. (2006). Discrete thermally responsive hydrogel-coated gold nanoparticles for use as drug-delivery vehicles. *Drug Development Research*, *67*(1), 61–69.

Kirpotin, D., Hong, K., Mullah, N., Papahadjopoulos, D., and Zalipsky, S. (1996). Liposomes with detachable polymer coating: Destabilization and fusion of dioleoylphosphatidylethanolamine vesicles triggered by cleavage of surface-grafted poly(ethylene glycol). *FEBS Letters*, *388*(2–3), 115–118.

Knoll, H. (2004). Photoisomerism of azobenzenes. In W. M. Horspool and F. Lenci (Eds.), *CRC Handbook of Organic Photochemistry and Photobiology*, 2nd ed., CRC Press, Boca Raton, FL.

Kono, K. (2001). Thermosensitive polymer-modified liposomes. *Advanced Drug Delivery Reviews*, *53*(3), 307–319.

Kono, K., Nakai, R., Morimoto, K., and Takagishi, T. (1999). Thermosensitive polymer-modified liposomes that release contents around physiological temperature. *Biochimica et Biophysica Acta (BBA)—Biomembranes*, *1416*(1–2), 239–250.

Kono, K., Yoshino, K., and Takagishi, T. (2002). Effect of poly(ethylene glycol) grafts on temperature-sensitivity of thermosensitive polymer-modified liposomes. *Journal of Controlled Release*, *80*(1–3), 321–332.

Kono, K., Murakami, T., Yoshida, T., Haba, Y., Kanaoka, S., Takagishi, T., and Aoshima, S. (2005). Temperature sensitization of liposomes by use of thermosensitive block copolymers synthesized by living cationic polymerization: Effect of copolymer chain length. *Bioconjugate Chemistry*, *16*(6), 1367–1374.

Kost, J., and Langer, R. (2001). Responsive polymeric delivery systems. *Advanced Drug Delivery Reviews*, *46*(1–3), 125–148.

Krack, M., Hohenberg, H., Kornowski, A., Lindner, P., Weller, H., and Förster, S. (2008). Nanoparticle-loaded magnetophoretic vesicles. *Journal of the American Chemical Society*, *130*(23), 7315–7320.

Kriechbaum, M., and Laggner, P. (1996). States of phase transitions in biological structures. *Progress in Surface Science*, *51*(3), 233–261.

Kubo, T., Sugita, T., Shimose, S., Nitta, Y., Ikuta, Y., and Murakami, T. (2000). Targeted delivery of anticancer drugs with intravenously administered magnetic liposomes in osteosarcoma-bearing hamsters. *International Journal of Oncology*, *17*(2), 309–315.

Kuiper, J. M., and Engberts, J. B. F. N. (2004). H-Aggregation of azobenzene-substituted amphiphiles in vesicular membranes. *Langmuir*, *20*(4), 1152–1160.

Kuiper, J. M., Stuart, M. C. A., and Engberts, J. B. F. N. (2008). Photochemically induced disturbance of the alkyl chain packing in vesicular membranes. *Langmuir*, *24*(2), 426–432.

Kumar, A., Jena, P. K., Behera, S., Lockey, R. F., Mohapatra, S., and Mohapatra, S. (2009a). Multifunctional magnetic nanoparticles for targeted delivery. *Nanomedicine: Nanotechnology, Biology and Medicine* 6(1), 64–69.

Kumar, B., Prajapati, A. K., Varia, M. C., and Suresh, K. A. (2009b). Novel mesogenic azobenzene dimer at air-water and air-solid interfaces. *Langmuir*, *25*(2), 839–844.

Landh, T., and Larsson, K. (1996). Particles, method of preparing said particles and uses thereof. USPTO. Sweden, GS Biochem AB. 5531925.

Lan-rong, H., Ho, R. J. Y., and Huang, L. (1986). Trypsin induced destabilization of liposomes composed of dioleoylphosphatidylethanolamine and glycophorin. *Biochemical and Biophysical Research Communications*, *141*(3), 973–978.

Lee, K. W. Y., Nguyen, T.-H., Hanley, T., and Boyd, B. J. (2009a). Nanostructure of liquid crystalline matrix determines in vitro sustained release and in vivo oral absorption kinetics for hydrophilic model drugs. *International Journal of Pharmaceutics*, *365*(1–2), 190–199.

Lee, S. E., Liu, G. L., Kim, F., and Lee, L. P. (2009b). Remote optical switch for localized and selective control of gene interference. *Nano Letters*, *9*(2), 562–570.

Lei, Y., and Hurst, J. K. (1999). Photoregulated potassium ion permeation through dihexadecyl phosphate bilayers containing azobenzene and stilbene surfactants. *Langmuir*, *15*(10), 3424–3429.

Lin, H.-Y., and Thomas, J. L. (2003). PEG-lipids and oligo(ethylene glycol) surfactants enhance the ultrasonic permeabilizability of liposomes. *Langmuir*, *19*(4), 1098–1105.

Liu, J., Lewis, T. N., and Prausnitz, M. R. (1998). Non-invasive assessment and control of ultrasound-mediated membrane permeabilization. *Pharmaceutical Research*, *15*(6), 918–924.

Liu, X.-M., Yang, B., Wang, Y.-L., and Wang, J.-Y. (2005). Photoisomerisable cholesterol derivatives as photo-trigger of liposomes: Effect of lipid polarity, temperature, incorporation ratio, and cholesterol. *Biochimica et Biophysica Acta (BBA)—Biomembranes*, *1720*(1–2), 28–34.

Liu, Y., Miyoshi, H., and Nakamura, M. (2006). Encapsulated ultrasound microbubbles: Therapeutic application in drug/gene delivery. *Journal of Controlled Release*, *114*(1), 89–99.

Liu, J., He, Y., and Wang, X. (2008). Azo polymer colloidal spheres containing different amounts of functional groups and their photoinduced deformation behavior. *Langmuir*, *24*(3), 678–682.

Ljusberg-Wahern, H., Nyberg, L., and Larsson, K. (1996). Dispersion of the cubic liquid crystalline phase—Structure preparation and functionality aspects. *Chimica Oggi—Chemistry Today*, *14*(6), 40–43.

Mackanos, M. A., Larabi, M., Shinde, R., Simanovskii, D. M., Guccione, S., and Contag, C. H. (2009). Laser-induced disruption of systemically administered liposomes for targeted drug delivery. *Journal of Biomedical Optics*, *14*(4), 8.

Maeda, H., Fang, J., Inutsuka, T., and Kitamoto, Y. (2003). Vascular permeability enhancement in solid tumor: Various factors, mechanisms involved and its implications. *International Immunopharmacology*, *3*(3), 319–328.

Meers, P. (2001). Enzyme-activated targeting of liposomes. *Advanced Drug Delivery Reviews*, *53*(3), 265–272.

Mills, J. K., Eichenbaum, G., and Needham, D. (1999). Effect of bilayer cholesterol and surface grafted poly(ethylene glycol) on pH-induced release of contents from liposomes by poly(2-ethylacrylic acid). *Journal of Liposome Research*, *9*(2), 275–290.

Montes, L. R., Goni, F. M., Johnston, N. C., Goldfine, H., and Alonso, A. (2004). Membrane fusion induced by the catalytic activity of a phospholipase C/sphingomyelinase from Listeria monocytogenes. *Biochemistry*, *43*(12), 3688–3695.

Mordon, S., Desmettre, T., Devoisselle, J. M., and Soulie, S. (1996). Fluorescence measurement of 805 nm laser-induced release of 5,6-CF from DSPC liposomes for real-time monitoring of temperature: An in vivo study in rat liver using indocyanine green potentiation. *Lasers in Surgery and Medicine*, *18*(3), 265–270.

Morgan, C. G., Bisby, R. H., Johnson, S. A., and Mitchell, A. C. (1995). Fast solute release from photosensitive liposomes: An alternative to "caged" reagents for use in biological systems. *FEBS Letters*, *375*(1–2), 113–116.

Munoz, J. A., del Pino, P., Bedard, M. F., Ho, D., Skirtach, A. G., Sukhorukov, G. B., Plank, C., and Parak, W. J. (2008). Photoactivated release of cargo from the cavity of polyelectrolyte capsules to the cytosol of cells. *Langmuir*, *24*(21), 12517–12520.

Murphy, C. J., and Orendorff, C. J. (2005). Alignment of gold nanorods in polymer composites and on polymer surfaces. *Advanced Materials*, *17*(18), 2173–2177.

Murray, W. A., and Barnes, W. L. (2007). Plasmonic materials. *Advanced Materials*, *19*(22), 3771–3782.

Nappini, S., Bombelli, F. B., Bonini, M., Norden, B., and Baglioni, P. (2010). Magnetoliposomes for controlled drug release in the presence of low-frequency magnetic field. *Soft Matter*, *6*(1), 154–162.

Natansohn, A., and Rochon, P. (2002). Photoinduced motions in azo-containing polymers. *Chemical Reviews*, *102*(11), 4139–4175.

Negrini, R., and Mezzenga, R. (2011). pH-Responsive lyotropic liquid crystals for controlled drug delivery. *Langmuir*, *27*(9), 5296–5303.

Nieva, J. L., Goni, F. M., and Alonso, A. (1989). Liposome fusion catalytically induced by phospholipase C. *Biochemistry*, *28*(18), 7364–7367.

Nobuto, H., Sugita, T., Kubo, T., Shimose, S., Yasunaga, Y., Murakami, T., and Ochi, M. (2004). Evaluation of systemic chemotherapy with magnetic liposomal doxorubicin and a dipole external electromagnet. *International Journal of Cancer*, *109*(4), 627–635.

Oki, K., and Nagasaka, Y. (2009). Measurements of anisotropic surface properties of liquid films of azobenzene derivatives. *Colloids and Surfaces A: Physicochemical and Engineering Aspects*, *333*(1–3), 182–186.

Ong, W., Yang, Y., Cruciano, A. C., and McCarley, R. L. (2008). Redox-triggered contents release from liposomes. *Journal of the American Chemical Society*, *130*(44), 14739–14744.

Orendorff, C. J., Alam, T. M., Sasaki, D. Y., Bunker, B. C., and Voigt, J. A. (2009). Phospholipid-gold nanorod composites. *ACS Nano*, *3*(4), 971–983.

Paasonen, L., Laaksonen, T., Johans, C., Yliperttula, M., Kontturi, K., and Urtti, A. (2007). Gold nanoparticles enable selective light-induced contents release from liposomes. *Journal of Controlled Release*, *122*(1), 86–93.

Pak, C. C., Ali, S., Janoff, A. S., and Meers, P. (1998). Triggerable liposomal fusion by enzyme cleavage of a novel peptide-lipid conjugate. *Biochimica et Biophysica Acta (BBA)—Biomembranes*, *1372*(1), 13–27.

Pak, C. C., Erukulla, R. K., Ahl, P. L., Janoff, A. S., and Meers, P. (1999). Elastase activated liposomal delivery to nucleated cells. *Biochimica et Biophysica Acta (BBA)—Biomembranes*, *1419*(2), 111–126.

Park, S.-H., Oh, S.-G., Mun, J.-Y., and Han, S.-S. (2006). Loading of gold nanoparticles inside the DPPC bilayers of liposome and their effects on membrane fluidities. *Colloids and Surfaces B: Biointerfaces*, *48*(2), 112–118.

Petrov, R. R., Chen, W. H., and Regen, S. L. (2009). Thermally gated liposomes: A closer look. *Bioconjugate Chemistry*, *20*(5), 1037–1043.

Ponce, A. M., Vujaskovic, Z., Yuan, F., Needham, D., and Dewhirst, M. W. (2006). Hyperthermia mediated liposomal drug delivery. *International Journal of Hyperthermia*, *22*(3), 205–213.

Rappolt, M., Hickel, A., Bringezu, F., and Lohner, K. (2003). Mechanism of the lamellar/inverse hexagonal phase transition examined by high resolution x-ray diffraction. In *Advances in Planar Lipid Bilayers and Liposomes*. Elsevier, Amsterdam, pp. 3111–3122.

Roux, E., Lafleur, M., Lataste, E., Moreau, P., and Leroux, J. C. (2003). On the characterization of pH-sensitive liposome/polymer complexes. *Biomacromolecules*, *4*(2), 240–248.

Ruiz-Arguello, M. B., Goni, F. M., and Alonso, A. (1998). Phospholipase C hydrolysis of phospholipids in bilayers of mixed lipid compositions. *Biochemistry*, *37*(33), 11621–11628.

Saito, G., Swanson, J. A., and Lee, K.-D. (2003). Drug delivery strategy utilizing conjugation via reversible disulfide linkages: Role and site of cellular reducing activities. *Advanced Drug Delivery Reviews*, *55*(2), 199–215.

Sarkar, N. R., Rosendahl, T., Krueger, A. B., Banerjee, A. L., Benton, K., Mallik, S., and Srivastava, D. K. (2005). "Uncorking" of liposomes by matrix metalloproteinase-9. *Chemical Communications*, (8), 999–1001.

Satarkar, N. S., and Hilt, J. Z. (2008). Magnetic hydrogel nanocomposites for remote controlled pulsatile drug release. *Journal of Controlled Release*, *130*(3), 246–251.

Schroeder, A., Avnir, Y., Weisman, S., Najajreh, Y., Gabizon, A., Talmon, Y., Kost, J., and Barenholz, Y. (2007). Controlling liposomal drug release with low frequency ultrasound: Mechanism and feasibility. *Langmuir*, *23*(7), 4019–4025.

Schroeder, A., Kost, J., and Barenholz, Y. (2009). Ultrasound, liposomes, and drug delivery: Principles for using ultrasound to control the release of drugs from liposomes. *Chemistry and Physics of Lipids*, *162*(1–2), 1–16.

Shen, Y., Zhang, Y., Kuehner, D., Yang, G., Yuan, F., and Niu, L. (2008). Ion-responsive behavior of ionic-liquid surfactant aggregates with applications in controlled release and emulsification. *ChemPhysChem*, *9*(15), 2198–2202.

Shin, J., Shum, P., and Thompson, D. H. (2003). Acid-triggered release via dePEGylation of DOPE liposomes containing acid-labile vinyl ether PEG-lipids. *Journal of Controlled Release*, *91*(1–2), 187–200.

Shum, P., Kim, J.-M., and Thompson, D. H. (2001). Phototriggering of liposomal drug delivery systems. *Advanced Drug Delivery Reviews*, *53*(3), 273–284.

Skirtach, A. G., Dejugnat, C., Braun, D., Susha, A. S., Rogach, A. L., Parak, W. J., Mohwald, H., and Sukhorukov, G. B. (2005). The role of metal nanoparticles in remote release of encapsulated materials. *Nano Letters*, *5*(7), 1371–1377.

Slepushkin, V. A., Simões, S., Dazin, P., Newman, M. S., Guo, L. S., De Lima, M. C. P., and Düzgüneş, N. (1997). Sterically stabilized pH-sensitive liposomes. Intracellular delivery of aqueous contents and prolonged circulation in vivo. *Journal of Biological Chemistry*, *272*(4), 2382–2388.

Song, X., Perlstein, J., and Whitten, D. G. (1995). Photoreactive supramolecular assemblies: Aggregation and photoisomerization of azobenzene phospholipids in aqueous bilayers. *Journal of the American Chemical Society*, *117*(29), 7816–7817.

Sorensen, T. J., Kjaer, K., Breiby, D. W., and Laursen, B. W. (2008). Synthesis of novel amphiphilic azobenzenes and X-ray scattering studies of their langmuir monolayers. *Langmuir*, *24*(7), 3223–3227.

Spratt, T., Bondurant, B., and O'Brien, D. F. (2003). Rapid release of liposomal contents upon photoinitiated destabilization with UV exposure. *Biochimica et Biophysica Acta (BBA)—Biomembranes*, *1611*(1–2), 35–43.

Takahashi, H., Niidome, Y., Niidome, T., Kaneko, K., Kawasaki, H., and Yamada, S. (2006). Modification of gold nanorods using phosphatidylcholine to reduce cytotoxicity. *Langmuir*, *22*(1), 2–5.

Takumi, K., Sakamoto, H., Uda, R. M., Sakurai, Y., Kume, H., and Kimura, K. (2007). Photocontrol of anionic micelles containing lipophilic crowned spirobenzopyran. *Colloids and Surfaces A: Physicochemical and Engineering Aspects*, *301*(1–3), 100–105.

Tanaka, T., Ogino, H., and Iwamoto, M. (2007). Photochange in pore diameters of azobenzene-planted mesoporous silica materials. *Langmuir*, *23*(23), 11417–11420.

Terada, T., Iwai, M., Kawakami, S., Yamashita, F., and Hashida, M. (2006). Novel PEG-matrix metalloproteinase-2 cleavable peptide-lipid containing galactosylated liposomes for hepatocellular carcinoma-selective targeting. *Journal of Controlled Release*, *111*(3), 333–342.

Tong, X., Wang, G., Soldera, A., and Zhao, Y. (2005). How can azobenzene block copolymer vesicles be dissociated and reformed by light? *Journal of Physical Chemistry B*, *109*(43), 20281–20287.

Tong, L., Zhao, Y., Huff, T. B., Hansen, M. N., Wei, A., and Cheng, J.-X. (2007). Gold nanorods mediate tumor cell death by compromising membrane integrity. *Advanced Materials*, *19*(20), 3136–3141.

Torchilin, V. (2009). Multifunctional and stimuli-sensitive pharmaceutical nanocarriers. *European Journal of Pharmaceutics and Biopharmaceutics*, *71*(3), 431–444.

Torchilin, V. P., Lukyanov, A. N., Klibanov, A. L., and Omelyanenko, V. G. (1992). Interaction between oleic acid-containing pH-sensitive and plain liposomes fluorescent spectroscopy studies. *FEBS Letters*, *305*(3), 185–188.

Troutman, T. S., Leung, S. J., and Romanowski, M. (2009). Light-induced content release from plasmon-resonant liposomes. *Advanced Materials*, *21*(22), 2334–2338.

Urban, A. S., Fedoruk, M., Horton, M. R., Radler, J. O., Stefani, F. D., and Feldmann, J. (2009). Controlled nanometric phase transitions of phospholipid membranes by plasmonic heating of single gold nanoparticles. *Nano Letters*, *9*(8), 2903–2908.

Viroonchatapan, E., Sato, H., Ueno, M., Adachi, I., Tazawa, K., and Horikoshi, I. (1996). Magnetic targeting of thermosensittve magnetoliposomes to mouse livers in an in situ on-line perfusion system. *Life Sciences*, *58*(24), 2251–2261.

Viroonchatapan, E., Sato, H., Ueno, M., Adachi, I., Tazawa, K., and Horikoshi, I. (1997). Release of 5-fluorouracil from thermosensitive magnetoliposomes induced by an electromagnetic field. *Journal of Controlled Release*, *46*(3), 263–271.

Volodkin, D. V., Skirtach, A. G., and Möhwald, H. (2009). Near-IR remote release from assemblies of liposomes and nanoparticles. *Angewandte Chemie International Edition*, *48*(10), 1807–1809.

Weinstein, J. N., Magin, R. L., Yatvin, M. B., and Zaharko, D. S. (1979). Liposomes and local hyperthermia—selective delivery of methotrexate to heated tumors. *Science*, *204*(4389), 188–191.

Weissleder, R. (2001). A clearer vision for in vivo imaging. *Nature Biotechnology*, *19*(4), 316–317.

West, K. R., and Otto, S. (2005). Reversible covalent chemistry in drug delivery. *Current Drug Discovery Technologies*, *2*, 123–160.

Winter, R., Erbes, J., Templer, R. H., Seddon, J. M., Syrykh, A., Warrender, N. A., and Rapp, G. (1999). Inverse bicontinuous cubic phases in fatty acid/phosphatidylcholine mixtures: The effects of pressure and lipid composition. *Physical Chemistry Chemical Physics*, *1*(5), 887–893.

Wohl, C. J., and Kuciauskas, D. (2005). Isomerization dynamics of photochromic spiropyran molecular switches in phospholipid bilayers. *Journal of Physical Chemistry B*, *109*(46), 21893–21899.

Wu, J., and Nyborg, W. L. (2008). Ultrasound, cavitation bubbles and their interaction with cells. *Advanced Drug Delivery Reviews*, *60*(10), 1103–1116.

Wu, G., Mikhailovsky, A., Khant, H. A., Fu, C., Chiu, W., and Zasadzinski, J. A. (2008). Remotely triggered liposome release by near-infrared light absorption via hollow gold nanoshells. *Journal of the American Chemical Society*, *130*(26), 8175–8177.

Yaghmur, A., Laggner, P., Almgren, M., and Rappolt, M. (2008a). Self-assembly in monoelaidin aqueous dispersions: Direct vesicles to cubosomes transition. *PLoS ONE*, *3*(11), e3747.

Yaghmur, A., Laggner, P., Sartori, B., and Rappolt, M. (2008b). Calcium triggered L-alpha-H-2 phase transition monitored by combined rapid mixing and time-resolved synchrotron SAXS. *PLoS ONE*, *3*(4), 11.

Yaghmur, A., Kriechbaum, M., Amenitsch, H., Steinhart, M., Laggner, P., and Rappolt, M. (2009). Effects of pressure and temperature on the self-assembled fully hydrated nanostructures of monoolein-oil systems. *Langmuir*, *26*(2), 1177–1185.

Yaghmur, A., Paasonen, L., Yliperttula, M., Urtti, A., and Rappolt, M. (2010). Structural elucidation of light activated vesicles. *Journal of Physical Chemistry Letters*, *1*, 962–966.

Yamamoto, T., Tabe, Y., and Yokoyama, H. (2006). Photonic manipulation of topological defects in liquid-crystal emulsions doped with azobenzene derivatives. *Thin Solid Films*, *509*(1–2), 81–84.

Yamamoto, T., Tabe, Y., and Yokoyama, H. (2009). Photochemical transformation of topological defects formed around colloidal droplets dispersed in azobenzene-containing liquid crystals. *Colloids and Surfaces A: Physicochemical and Engineering Aspects*, *334*(1–3), 155–159.

Yanase, M., Shinkai, M., Honda, H., Wakabayashi, T., Yoshida, J., and Kobayashi, T. (1997). Intracellular hyperthermia for cancer using magnetite cationic liposomes: Ex vivo study. *Cancer Science*, *88*(7), 630–632.

Yatvin, M. B., Weinstein, J. N., Dennis, W. H., and Blumenthal, R. (1978). Design of liposomes for enhanced local release of drugs by hyperthermia. *Science*, *202*(4374), 1290–1293.

Yatvin, M., Kreutz, W., Horwitz, B., and Shinitzky, M. (1980). pH-sensitive liposomes: Possible clinical implications. *Science*, *210*(4475), 1253–1255.

Yi, Y., Farrow, M. J., Korblova, E., Walba, D. M., and Furtak, T. E. (2009). High-sensitivity aminoazobenzene chemisorbed monolayers for photoalignment of liquid crystals. *Langmuir*, 25(2), 997–1003.

Yoshino, K., Kadowaki, A., Takagishi, T., and Kono, K. (2004). Temperature sensitization of liposomes by use of N-isopropylacrylamide copolymers with varying transition endotherms. *Bioconjugate Chemistry*, 15(5), 1102–1109.

Yuba, E., Harada, A., Sakanishi, Y., and Kono, K. (2010). Carboxylated hyperbranched poly(glycidol)s for preparation of pH-sensitive liposomes. *Journal of Controlled Release*, 149(1), 72–80.

Zasadzinski, J. A. (1986). Transmission electron microscopy observations of sonication-induced changes in liposome structure. *Biophysical Journal*, 49(6), 1119–1130.

Zeimer, R., and Goldberg, M. F. (2001). Novel ophthalmic therapeutic modalities based on noninvasive light-targeted drug delivery to the posterior pole of the eye. *Advanced Drug Delivery Reviews*, 52(1), 49–61.

Zeimer, R., Khoobehi, B., Niesman, M., and Magin, R. (1988). A potential method for local drug and dye delivery in the ocular vasculature. *Investigative Ophthalmology & Visual Science*, 29(7), 1179–1183.

Zhang, Z. Y., and Smith, B. D. (1999). Synthesis and characterization of NVOC-DOPE, a caged photoactivatable derivative of dioleoylphosphatidylethanolamine. *Bioconjugate Chemistry*, 10(6), 1150–1152.

Zhang, J. X., Zalipsky, S., Mullah, N., Pechar, M., and Allen, T. M. (2004). Pharmaco attributes of dioleoylphosphatidylethanolamine/cholesterylhemisuccinate liposomes containing different types of cleavable lipopolymers. *Pharmacological Research*, 49(2), 185–198.

Zhang, J., Wang, S.-C., and Lee, C. T. (2009). Photoreversible conformational changes in membrane proteins using light-responsive surfactants. *Journal of Physical Chemistry B*, 113(25), 8569–8580.

Zhao, Y.-L., and Stoddart, J. F. (2009). Azobenzene-based light-responsive hydrogel system. *Langmuir*, 25(15), 8442–8446.

Zhu, L., Huo, Z. L., Wang, L. L., Tong, X., Xiao, Y., and Ni, K. Y. (2009). Targeted delivery of methotrexate to skeletal muscular tissue by thermosensitive magnetoliposomes. *International Journal of Pharmaceutics*, 370(1–2), 136–143.

Zou, A., Eastoe, J., Mutch, K., Wyatt, P., Scherf, G., Glatter, O., and Grillo, I. (2008). Light-sensitive lamellar phases. *Journal of Colloid and Interface Science*, 322(2), 611–616.

CHAPTER 10

Nonlamellar Lipid Liquid Crystalline Structures at Interfaces

DEBBY P. CHANG and TOMMY NYLANDER

Physical Chemistry, Lund University, Lund, Sweden

Abstract

Nonlamellar lipid-based liquid crystalline structures, such as cubic, hexagonal, and sponge phases, have potential as delivery systems in pharmaceutical, food, and cosmetic applications. This is due to the space-dividing nature of these phases, which features mono- or bicontinuous networks of both hydrophilic and hydrophobic domains. To utilize these nonlamellar liquid crystalline structures as delivery vehicles, it is crucial to understand how they interact with and respond to different types of interfaces. The progress in the area of liquid crystalline lipid-based nanoparticles opens up new possibilities for preparation of well-defined surface films with well-defined nanostructure. Apart from the relevance to drug delivery, such studies create opportunities for new applications for functionalized and tunable surface coatings as well. This review will focus on recent progress in the formation of nonlamellar dispersion and its interfacial properties at the solid–liquid and biologically relevant interfaces. Various experimental techniques on the study of interfacial interactions of these crystalline structures will be discussed.

Self-Assembled Supramolecular Architectures: Lyotropic Liquid Crystals, First Edition.
Edited by Nissim Garti, Ponisseril Somasundaran, and Raffaele Mezzenga.
© 2012 John Wiley & Sons, Inc. Published 2012 by John Wiley & Sons, Inc.

10.1   INTRODUCTION

The amphiphilic nature of lipids allows it to self-assemble into many different structures depending on the lipid shape, mixture composition, and external conditions such as temperature. This self-association, driven by minimizing the contact between the hydrophobic regions from the aqueous phase and balance by head-group (repulsive) interaction and packing constrains of acyl chains, gives rise to a rich variety of phases (Evans and Wennerström, 1999; Mouritsen, 2005; Seddon, 1990). In particular, the nonlamellar phase-based crystalline structures have been the focus of great interest and study in the last two decades. Unlike the micellar and lamellar structures, nonlamellar crystalline structures, such as cubic, hexagonal, and sponge phases, have extremely high surface area and may solubilize hydrophobic, hydrophilic, and also amphiphilic molecules (Angelova et al., 2011; Larsson, 2000; Larsson et al., 2006; Larsson and Tiberg, 2005; Lawrence, 1994; Malmsten, 2007a; Sagalowicz and Leser, 2010; Sagalowicz et al., 2006b). The crystalline phases can be dispersed into particles with excess water. This was first demonstrated for lamellar phases by Bangham and colleagues in the 1960s (Bangham and Horne, 1964), and today vesicles and liposomes are often used for a range of applications in formulations for drug delivery, food, and consumer products as well as models for biomembranes. (The term *vesicle* is often used for unilamellar aggregate, whereas *liposome* refers to multilamellar structures.) Nearly 20 years ago, Larsson and co-workers demonstrated that it was possible to prepare dispersions of nonlamellar liquid crystalline lipid phases, first as lipid-aqueous cubic phase particle, Cubosome (Gustafsson et al., 1996, 1997; Landh, 1994; Larsson, 1989, 2000; Larsson et al., 2006), which with appropriate dispersion stabilizer could be made monodisperse. The dispersed particles are in equilibrium with water and retain their internal structure. These lipid liquid crystalline nanoparticles (LCNP) have great potential as delivery vehicles and can also be used to functionalize surfaces with nanostructures.

The lipid LCNP recently has gained an increasing interest in the medical field as a drug delivery vehicle because of its (1) space-dividing nature to allow incorporation of different substances and (2) sustained drug release from the crystalline matrix (Drummond and Fong, 1999; Lawrence, 1994). The nonlamellar phases can be used for delivery through topical application to the skin, subcutaneous/intravenous injection, and mucoadhesion. Excellent reviews on the use of lipid liquid crystalline phases as drug delivery vehicles can be found (Drummond and Fong, 1999; Larsson et al., 2006; Lawrence, 1994; Malmsten, 2007b; Shah et al., 2001; Spicer, 2005; Yang et al., 2004). Other than the ability to encapsulate drugs, the nonlamellar phase can also protect peptides from enzymatic digestion (Drummond and Fong, 1999; Ericsson et al., 1983; Larsson, 2009; Larsson et al., 2006; Razumas et al., 1996a,b; Shah et al., 2001), which prompts the idea of using LCNP for surface coating. To enable LCNP for surface coating, it is important to understand the nature of the interaction of the lipid crystalline structure with the interface. For example, the formation of nonlamellar crystalline surfaces may be possible through direct deposition or adsorption with lipid LCNP. In this review, we will focus on recent progress in the formation of nonlamellar dispersions and its interfacial properties at the solid–liquid and biologically relevant interfaces such as biological membrane and mucosa. The effect of surface chemistry, phase structure, and solvent conditions on the LCNP at the solid–liquid interface will be discussed. Key articles on the interfacial behavior of liquid crystalline particles discussed in this review are summarized in Table 10.1.

10.2   BACKGROUND

10.2.1   Lipid Phases

The self-assembly property of lipids is driven by its amphiphilic structure. Lipids aggregate in the aqueous solution exposing their hydrophilic domains to water to maximize the polar interactions and cluster the hydrophobic moieties to minimize interaction with the aqueous phase (Evans and Wennerström, 1994). The phase behavior of lipids depends on many different factors, ranging from the molecular structure, intermolecular forces, to lipid concentration and temperature (Jonsson, 1998; Kaasgaard and Drummond, 2006). The so-called geometric packing properties of the amphiphilic molecule can be used to understand and sometimes even predict the formation of a particular phase (Israelachvili et al., 1976; Mitchell and Ninham, 1981). This concept is based on the cross-section area of the polar head group in relation to that of the acyl chain. This property can be expressed by the so-called packing parameter (v/al), which is defined as the ratio between the volume of the hydrophobic chain (v) and the product of the head-group area (a) and the

TABLE 10.1 Key Articles on the Interfacial Properties of Lipid Liquid Crystalline Particles on Surfaces[a]

Lipid Phase	Particle Formulation	Interfaces	Measurement Technique	References
Reversed cubic	GMO/F-127	Hydrophobic and hydrophilic Si	Ellipsometry	Vandoolaeghe et al. (2006)
Reversed cubic	GMO/F-127	Hydrophilic Si	NR, QCM-D, ellipsometry, fluorescence microscopy	Vandoolaeghe et al. (2009b)
Reversed cubic	GMO/F-127	Model membrane (DOPC bilayers)	NR, QCM-D, ellipsometry	Vandoolaeghe et al. (2009c); Vandoolaeghe et al. (2008)
Reversed cubic	GMO/F-127	Mucin-coated surface	Ellipsometry	Svensson et al. (2008b)
Reversed cubic	GMO/F-127 w/chitosan	Mucin-coated surface	Ellipsometry	Svensson et al. (2008a)
Reversed cubic	GMO/F-127	Leaf, model leaf (tristearin-coated surface) and uncoated zinc Selenide surfaces	ATR-FTIR	Dong et al. (2011)
Reversed hexagonal	PHYT/F-127 PHYT/F-127/VitEA			
Reversed cubic	PHYT/F-127 PHYT/F-127 w/DPPS	Model membrane (POPC bilayers), cell	QCM-D, CLSM	Shen et al. (2010)
Reversed cubic	GMO/P407	Mica	AFM	Neto et al. (1999)
Reversed hexagonal	GMO/GTO/P407			
Reversed cubic	GMO/F-127 w/NBD (fluorescence) CTAB (positively charged surfactant)	Silica	QCM-D, CLSM	Driever et al. (2011)

[a] Abbreviations: CTAB, cetyl trimethylammonium bromide; F-127, Pluronic F-127; GMO, glycerol monooleate; GTO, glycerol trioleate; NBD, 1,2 dipalmitoyl-*sn*-glycero-3-phospho-ethanolamine-*N*-(7-nitro-2-1,3-benzoxadiazol-4-yl); P407, Poloxamer 407; PHYT, phytantriol; VitEA, Vitamin E.

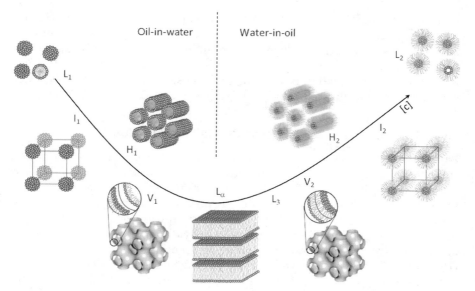

Figure 10.1 Schematic illustration of different liquid crystalline phases. [Reprinted with permission from Barauskas and Nylander (2008). Copyright 2008 by Woodhead Publishing.]

chain length (l). The packing parameter for a particular environment will be reflected in the curvature of the lipid aqueous interface and thus the particular phase. It is important to bear in mind that the packing parameter is not a constant parameter but is dependent on the environment (e.g., water content, ionic strength, pH, temperature, and pressure). In a generalized picture shown in Figure 10.1, the liquid crystal structures mirror at the lamellar phase (L_α), where bilayers of lipid molecules alternate with water layers, and it corresponds to a packing parameter close to unity. At lower lipid concentration, the lipids form the oil-in-water structures: bicontinuous cubic (V_1), hexagonal (H_1), cubic (I_1), and micelles (L_1). At higher lipid concentration, the lipid phases reverse its curvature and form water-in-oil phases: the reversed bicontinuous cubic (V_2), reversed hexagonal (H_2), reversed cubic (I_2), and reversed micelles (L_2). The reversed bicontinuous cubic (V_2) phase formed by curved lipid bilayers can be further broken down into three types according to the infinite periodic minimal surface (IPMS) characterization of the midplane of the lipid bilayers (Hyde, 1996; Hyde et al., 1984; Kaasgaard and Drummond, 2006): gyroid (G, Ia3d), diamond (D, Pn3m), and primitive (P, Im3m). A less common phase, called sponge phase (L_3), can also occur adjacent to the lamellar phase. The sponge phase forms a bicontinuous disordered structure by dividing into two unconnected volumes like the bicontinuous cubic phase.

These interesting lipid phases have the potential to be used as templates and reservoirs. The easiest way to handle and use these structures is to make them into particulate forms, but not all lipid phases can be dispersed into particulate forms in excess water. A lipid that forms, for example, normal micellar phases has sufficient high water solubility to exist as surfactant mono- mers upon dilution. The reversed "water-in-oil" self-assembled lipid aggre- gates, on the other hand, can form stable dispersions in large excess of water with the help of a suitable stabilizer (Kaasgaard and Drummond, 2006).

10.2.2 Formation of Nonlamellar Crystalline Particles

Analogous to liposomes, which are aqueous dispersion of lamellar phase, cubosome (Barauskas et al., 2005b), hexosome (Barauskas et al., 2006a), and sponge (Barauskas et al., 2006b) nonlamellar crystalline particles are disper- sions of reversed bicontinuous cubic, reversed hexagonal, and L$_3$ phases, respectively. Figure 10.2 shows cryo-TEM (transmission electron microscopy) images of these dispersed LCNPs. Detailed reviews on the preparation tech- niques and materials for nonlamellar liquid crystalline dispersion can be found (Boyd et al., 2009; Kaasgaard and Drummond, 2006; Larsson, 1999; Sagalowicz et al., 2006a,b; Yang et al., 2004). In brief, preparation of LCNP ranges from

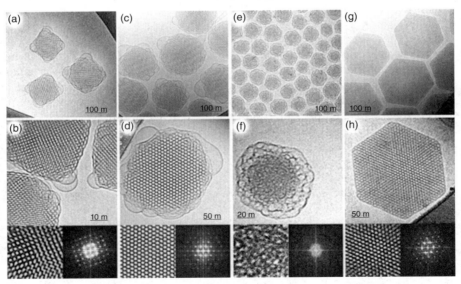

Figure 10.2 Cryo-TEM images of different nonlamellar lipid LCNPs: reversed bicon- tinuous cubic phase [(a)–(d)], sponge or L$_3$ phase [(e)–(f)], and reversed hexagonal phase [(g)–(h)] particles. Fourier transforms of magnified areas in panels (b), (d), (f), and (h) show the structural periodicity of the different nanoparticles consistent with the mesophase structures indicated above. [Reprinted with permission from Barauskas et al. (2005a). Copyright 2005 by the American Chemical Society.]

high-energy dispersion, such as, homogenization and ultrasonication, to low-energy dilution from precursor formulation with additives such as ethanol. The choice of dispersion technique is based on the requirements for the particular application. Less destructive methods such as dilution may be required for the incorporation of sensitive molecules, while high-energy techniques may provide more uniform crystalline structure.

The materials used for the formation of LCNP are categorized into two parts: the lipids that assemble into the nanostructure and the stabilizer that stabilizes the dispersion. It is crucial to choose a lipid system that can form stable reversed phases (cubic, hexagonal, or sponge) in excess of solvent. For example, one of the most widely studied lipid system is the glycerol mono-oleate (GMO)–water system. GMO forms reversed bicontinuous cubic phase in equilibrium with excess water and is commercially available as a food emulsifier (Boyd et al., 2009; Larsson 1989, 1991; Sagalowicz et al., 2006a). To disperse the crystalline phases into particulate form, it is essential to locate a suitable dispersion stabilizer for the particular lipid system. A dispersion stabilizer is needed to reduce the exposure of the hydrophobic domains to aqueous medium, that is, reduce the interfacial tension, and to prevent aggregation of the dispersed particles, which apart from forming large aggregates can also lead to the disruption of the inner crystalline structure. The stabilizer should not interfere with the crystalline phases and at the same time provide stability to particle formation. To satisfy these requirements, a stabilizer is typically amphiphilic with a hydrophobic region that is (partly) miscible with the particular lipid, and hydrophilic unit(s) at the surface of the particle provide steric repulsion against aggregation. One of the most effective and commonly used stabilizers, which works well with GMO, is a block copolymer called Pluronic F-127 (Barauskas et al., 2005b; Boyd et al., 2009; Landh, 1994). Pluronic has a hydrophobic block, polypropylene oxide (PPO), which is sandwiched between two hydrophilic polyethylene glycol (PEO) chains. The PEO chains cover the surface of the nanoparticle and dominate the interfacial interaction. In order to control the interfacial properties of LCNP, it is crucial to understand the properties and surface interaction of the stabilizer because this will control the interfacial behavior of LCNP.

10.3 CHARACTERIZATION TECHNIQUES

A range of techniques are available to study the properties of LCNP. For this review, we focus on those that probe the LCNP behavior at the solid–water interface, such as neutron reflectometry, null ellipsometry, quartz crystal microbalance, florescence microscopy, and Fourier transform infrared spectroscopy. These methods are particularly advantageous to use as they are nonperturbative techniques that are very sensitive to the interfacial properties, thus allowing for the structural and/or compositional information at the surface layer to be extracted. For techniques that probe internal structure and morphology of LCNP in solution or bulk, such as small-angle X-ray scattering (SAXS), small-angle neutron scattering (SANS), cryogenic transmission electron microscopy (cryo-TEM), shear rheology, and nuclear magnetic resonance (NMR), we refer the reader to existing reviews (Binks, 1999; Boyd et al., 2009; Sagalowicz et al., 2006b,c).

10.3.1 Neutron Reflectometry

Neutron reflectometry (NR) is a relatively young, but fast growing, technique for the study of soft matter at an interface, as it can provide quantitative structural and compositional information of the interfacial layer at molecular length scales. The neutron is very sensitive to the nuclear composition as it is scattered by the nucleus of a molecule. Neutrons can distinguish the difference between hydrogen isotopes, ^1H and ^2H (deuterium, D), which makes it an ideal tool to elucidate the composition of multicomponent systems via selective deuteration (Wacklin, 2010). Furthermore, neutrons can penetrate through many materials, allowing it to be used to study buried layers. In NR, a well-collimated beam of neutrons is directed at the sample and reflected from the interface. The intensity of the reflected beam relative the incidence beam (reflectivity) allows information to be gained on the structure normal to the interface as the scattering vector is perpendicular to the surface (Binks, 1999; Thomas, 2004). For specular neutron reflection, where angle of incidence is the same as angle of reflection, the ratio of the intensity of the reflected beam to the incident beam ($R = I/I_0$) is measured as a function of the momentum transfer vector q:

$$q = \frac{4\pi \sin \theta}{\lambda}$$

with θ the angle of incidence and λ the wavelength. Standard optics and scattering theory can be used to obtain layer thickness, interfacial roughness, and scattering length density of the layer normal to the surface from the NR profile. The scattering length density, ρ, is the sum of the coherent nuclear scattering length, b_i, of the atoms in volume V:

$$\rho = \frac{\Sigma_i \, b_i}{V}$$

From the scattering length density, it will then be possible to interpret the physical and compositional structure of the interface. More detailed reviews on the use of neutron reflection to investigate lipid membranes can be found in the literature (Nylander et al., 2008; Thomas, 2004; Wacklin, 2010).

10.3.2 Null Ellipsometry

In soft-matter chemistry, null ellipsometry is commonly used to observe the adsorption of lipids, polymers, or surfactants from aqueous solution. Ellipsometry is an optical technique to evaluate the thickness and the dielectric properties of materials on surfaces by measuring the change of the elliptically polarized light upon reflection (Azzam and Bashara, 1977; Landgren and Jonsson, 1993). By monitoring the change of amplitude (ψ) and phase (Δ), the optical thickness (d) and refractive index (n) of the surface layer can be modeled. The absorbed amount, Γ, then can be calculated using, for instance, the de Feijter approximation (Defeijter et al., 1978):

$$\Gamma = \frac{(n - n_0)d}{dn/dc}$$

where n_0 is the refractive index of the solvent and dn/dc is the concentration dependence of the refractive index of the adsorbed layer. A more detailed description can be found in the literature (Azzam and Bashara, 1977; Landgren and Jonsson, 1993; Tompkins and Irene, 2005).

10.3.3 Quartz Crystal Microbalance with Dissipation

Quartz crystal microbalance with dissipation (QCM-D) is an acoustic technique that has been used to evaluate the adsorption of the lipid nanoparticle on surfaces and monitor the structural and conformational changes of the adsorbed layers (Hook et al., 2001; Rodahl et al., 1997; Vandoolaeghe et al., 2009b). In QCM-D, an alternating current (AC) voltage is applied across a quartz crystal so that the quartz oscillates at its resonance frequencies. When the voltage is turned off, the oscillation decays exponentially. The mass of an adsorbed layer is sensed by the decrease in resonance frequency, and the structural and viscoelastic properties of the layer are deduced by the change in dissipation obtained in the decay of the oscillations. For a rigid film, the adsorbed mass is directly proportional to the change in resonance frequency following the Sauerbrey equation (Sauerbrey, 1959). For a soft film, viscoelastic models, such as the Voigt and Maxwell models, must be used to correct for the energy loss from dissipation. Ellipsometry measures the adsorbed mass from the changes in refractive index and thickness. The measured mass in QCM-D is the adsorbed mass plus the mass of the solvent coupled to the layer. This and the ability to measure the viscoelastic properties of the layer makes it a useful tool to monitor the structural change due to, for instance, vesicle fusion with a surface as well as LCNP adsorption.

10.3.4 Attenuated Total Reflectance and Fourier Transform Infrared

Fourier transform infrared (FTIR) spectroscopy in attenuated total reflectance (ATR) mode has been used to study the interfacial interactions of LCNP at the solid–liquid interface (Dong et al., 2011). In this mode, adsorption at the solid–liquid interface can be determined in situ. In ATR-FTIR, a beam of infrared light is passed through a crystal of high refractive index at an angle that results in total internal reflection. The total internal reflection creates an evanescent wave that extends several microns into the sample, allowing detection at the surface layer. In regions of the infrared spectrum where the sample absorbs energy, the evanescent wave will be attenuated or altered and show up on the reflected IR spectrum (PerkinElmer, 2005). ATR-FTIR measures the changes that occur in the reflected infrared beam to detect and quantify adsorption at the surface layer (Hind et al., 2001). The adsorbed amount can be quantified by comparing the absorbance of C–H stretching against a calibration curve with known deposited mass (Dong et al., 2011).

10.3.5   Fluorescence and Confocal Microscopy

In fluorescence microscopy, the sample is illuminated by light of a certain wavelength and can then be detected at the emission wavelength of the fluorescence molecules. The fluorescence microscope uses a dichroic mirror to filter out the excitation wavelength, allowing detection of the weaker emitted light at the higher wavelength. For the interfacial study of LCNP, fluorescent dye or fluorescent lipid (lipids conjugated with fluorescent molecules) can be used to label the crystalline particles.

For high-resolution imaging deep within a sample, a confocal laser scanning microscope (CLSM) can be used (Prasad et al., 2007). Confocal microscopy uses point illumination with a spatial pinhole to eliminate out-of-focus illumination, such that only fluorescence in the focal plane can be detected. Confocal microscopy reduces blurring of an image from the scattered light to create sharp optical sections. The cross-sectional images can then be combined to create detailed three-dimentional images.

10.4   NONLAMELLAR LIQUID CRYSTALLINE SURFACE LAYERS

10.4.1   Solid Interface

Nonlamellar liquid crystalline surface layers can be formed by adsorption of LCNPs. Recent studies have investigated the adsorption properties of GMO-based bicontinuous cubic phase nanoparticle (CPNP) stabilized by Pluronics F-127, on hydrophobic and hydrophilic solid surfaces by the means of null ellipsometry, QCM-D, atomic force microscopy (AFM), florescence microscopy, and neutron reflectometry (Vandoolaeghe et al., 2006, 2009b). Depending on the surface properties, electrolyte concentrations, and pH, different adsorption scenarios were discerned (Figs. 10.3 and 10.4). The time evolution of the adsorbed amount and thickness shows the structure formation and sheds light on the mechanisms of adsorption. Figure 10.5 schematically

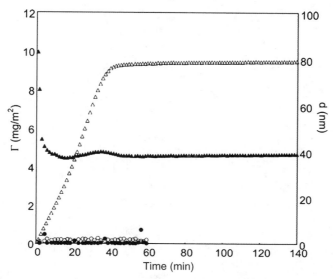

Figure 10.3 Adsorbed amount (open symbols) and layer thickness (filled symbols) as a function of time measured with ellipsometry after addition of 0.05 mg/ml GMO-based LCNP in Milli-Q water (circles) and in the pH 2 water (triangles). [Reprinted with permission from Vandoolaeghe et al. (2006). Copyright 2006 by the American Chemical Society.]

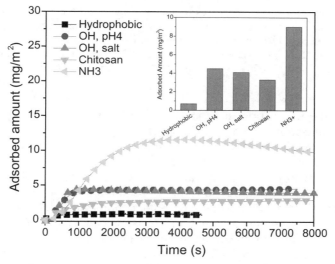

Figure 10.4 Adsorption of LCNP (GDO/SPC/P80) on silica surfaces with different surface chemistry and solvent conditions measured with ellipsometry. Inset shows the final adsorbed amount at equilibrium at different conditions (Chang et al., 2012).

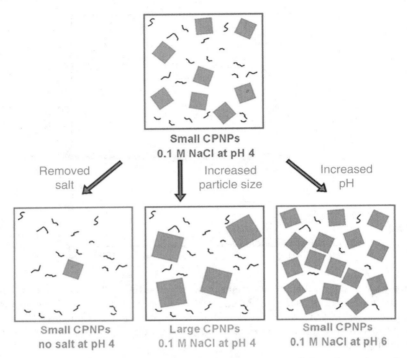

Figure 10.5   Schematic illustration of factors that can control the adsorption of cubosome LCNP (CPNP) stabilized with the Pluronic F-127 molecule relative to adsorption of small CPNP at steady state on silica surface at (a) pH 4 from 0.1 M NaCl aqueous solution. (b) Removing salt increases repulsion between CPNP and surface and favor adsorption of stabilizer; (c) larger particles give less extensive adsorption, while (d) increasing the pH to 6 with smaller particles, leads to a decrease adsorption of free F-127 and therefore increased adsorption of CPNP. [Reprinted with permission from Vandoolaeghe et al. (2009b). Copyright 2009 by the American Chemical Society.]

illustrates the interaction mechanisms of CPNP with hydrophilic silica with changing ionic strength, particle size, and pH.

The adsorption of LCNP on hydrophobic silica is dramatically different from that on bare silica (Vandoolaeghe et al., 2006). On the hydrophobic surface, the hydrophobic attraction results in structural transition from the original particle morphology to a thin 3-nm monolayer coverage of lipid on surface. In contrast, thick layers of lipid nanoparticles can be formed on hydrophilic silica in the presence of electrolyte or at low pH. Figure 10.3 shows that the adsorption on hydrophilic silica is strongly dependent on the pH. At high pH, no adsorption is observed. At low pH close to the isoelectric point of silica, adsorption increases proportionally with time and reaches a saturation value at approximately 10 mg/m². The layer thickness decays initially but stabilizes quickly and plateaus at around 40 nm. The adsorption

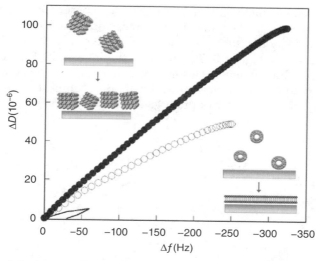

Figure 10.6   QCM-D correlation plot of dissipation difference, ΔD, as a function of frequency shift, Δf, for the adsorption of small ($\circ$) and large ($\bullet$) cubosomes on hydrophilic silica. The solid line represents the adsorption and fusion of vesicles on silica surface. Schematic drawings illustrate the intact crystalline nanoparticles after adsorption (left) and lipid bilayer formed from vesicle fusion (right). [Reprinted with permission from Vandoolaeghe et al. (2009b). Copyright 2009 by the American Chemical Society.]

curve suggests that the interfacial layer is built up by slow attachment of nanoparticles rather than their molecular components. This conclusion is reasonable as random lateral organization of bound nanoparticles is visible from fluorescence microscopy, and QCM-D demonstrates the large change in dissipation and the high stability of the viscoelastic particle at the interface (Vandoolaeghe et al., 2009b).

Different adsorption processes for the LCNP compared to the lamellar vesicles on hydrophilic Si interfaces were revealed by QCM-D (Vandoolaeghe et al., 2009b). The lamellar vesicles adsorb intact on hydrophilic surfaces at low coverage and transform into a bilayer arrangement after a critical coverage (Reimhult et al., 2002, 2003). The collapse of the surface-attached vesicles is shown as a kink in the correlation plot of dissipation change, ΔD, as a function of frequency shift, Δf, as seen with the solid line in Figure 10.6. The drop in dissipation signifies the disruption and spreading of the vesicle and release of solvent mass. For the adsorption of LCNP, a monotonic increase of ΔD versus Δf with no kink shows that the bound particles remain largely intact with significant amount of acoustically coupled water. The schematic drawings in Figure 10.6 illustrate the different adsorption behavior of nonlamellar crystalline particles and lamellar vesicles on the hydrophilic Si surface.

The internal crystalline structure of LCNP is conserved at the interfacial layer upon adsorption as a distinctive Bragg diffraction peak is present in neutron reflectivity measurement (Vandoolaeghe et al., 2009b). The neutron reflectivity curve in Figure 10.7 shows that the intact CPNP binds to hydrophilic surfaces with minimal collapsing and spreading for over 40 h. The curves fit very nicely to the cubic phase model with a repeating structure of 5.2 nm, representing the cubic phase of the organized lipid molecules (Vandoolaeghe

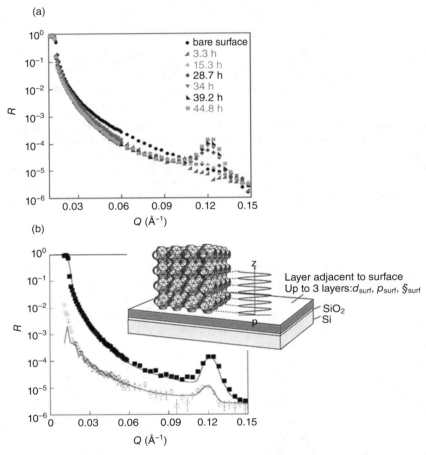

Figure 10.7 Neutron reflectivity, R, as a function of momentum transfer, Q, recorded for adsorption of GMO-based LCNPs on hydrophilic silica: (a) in 0.1 M NaCl at pH 4.4 in D_2O at different incubation times and (b) in D_2O (filled squares) and Si contrast matched water (open circles) at 45 h time. The solid lines correspond to the modeled fits to the data using the cubic phase model, shown in the schematic drawing (inset), where scattering density profile is modeled as a sinusoid. [Reprinted with permission from Vandoolaeghe et al. (2009b). Copyright 2009 by the American Chemical Society.]

et al., 2009b). The NR data agrees well with the AFM result that a slight height deformation occurs upon adsorption (Neto et al., 1999).

10.4.2   Model Biological Interface—Implications for Drug Delivery

The LCNP has a promising future for biological applications, in particular as an agent for intravenous, topical, oral, and nasal delivery. Here we should not only consider the uptake of LCNP through the skin, mucosa or the cell membrane or strive to limit the hemolytic activity of the LCNP as drug delivery vehicles. An often very important factor for the development of drug delivery vehicles is their stability. This means that the vehicles should be in a stable dispersion, and the loss of material in terms of adsorption to vials, catheters, tubes, and other delivery devices should be minimal. Unwanted consequences can occur if the LCNP is disintegrated by contact with an interface, for instance, what happens with the GMO-based CPNP with the hydrophobic surface as discussed above. There are also instances where a maximum possible adsorption is desirable, such as when the LCNP is used as surface coatings for enhancing drug delivery. So it is indeed important to control the LCNP interaction with the surface to achieve selective deposition only on surfaces with certain properties. To date, only a few studies have tried to elucidate the interfacial interaction between LCNP with the type of surfaces they will encounter. Here we will discuss the interaction between LCNP and three model biological interfaces: cell membrane (Vandoolaeghe et al., 2008, 2009c), mucous membrane (Svensson et al., 2008a,b), and leaf surface (Dong et al., 2011).

10.4.2.1 Cell Membrane   Lipid bilayers are the main building block for the cell membrane for which the models have become more and more complex. We recommend the excellent review on the development of the models by Tien and Ottova (2001). Supported phospholipid bilayers have been used as the simplest models for cell membranes and can be prepared in several different ways (Hughes et al., 2008; Richter et al., 2006; Sackmann, 1996; Tiberg et al., 2000; Wacklin, 2010). Spreading of vesicles on surfaces is one method to form a bilayer on a hydrophilic surface and rests on the fact that the phospholipids form a flat bilayer rather than a curved one on the vesicle. It can be regarded as a deposition of spherical vesicle on a surface, which depending on the surface properties and vesicle stability, leads to bursting and spreading of the vesicles (Reimhult et al., 2002, 2003). Another interesting technique used to form lipid bilayers on surfaces is to solubilize the lipid with a nonionic surfactant, for example, DOPC with n-dodecyl-β-D-maltopyranoside (DDM) (Tiberg et al., 2000). The lipid bilayer is prepared by co-adsorbing lipid with a surfactant that could be easily rinsed off the surface after adsorption. A series of subsequent additions of mixed lipid–surfactant solutions of decreasing total concentration, followed by an aqueous rinse after each step, removes the excess of the soluble surfactant. In each dilution step, the system gradually approaches the two-phase region of the

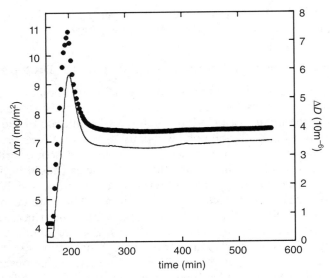

Figure 10.8 Wet mass (Δm, filled circles) and dissipation difference (ΔD, line) measured as a function of time using QCM-D after addition of 0.05 mg/ml GMO-based LCNP to a model cell membrane of DOPC bilayer at pH 4. The wet mass is calculated from the Voigt viscoelastic model and the dissipation corresponds to the third harmonic of the resonance frequency. [Reprinted with permission from Vandoolaeghe et al. (2008). Copyright 2008 by the Royal Society of Chemistry.]

lipid–surfactant aqueous system, and thereby causes the sequentially increased deposition of the phospholipids.

Supported lipid membrane models have been used to model the interaction between LCNP and the cell membrane. Three surface-sensitive techniques, ellipsometry, QCM-D, and NR, provide insight into the interaction mechanism, revealing a kinetically controlled process (Vandoolaeghe et al., 2008, 2009c). The ellipsometry measurements reveal a rapid increase in the adsorbed amount after the introduction of CPNP to a supported DOPC bilayer on silica, indicating a strong attraction between the particle and the bilayer. The adsorbed amount decreases after about an hour, signifying a net release of materials from the surface. The QCM-D measurement (Fig. 10.8) shows that after the addition of CPNP, the surface becomes viscoelastic with a large change in dissipation. The wet mass measured through viscoelastic modeling of the surface layer indicates an initial adsorption of intact nanoparticles followed by relaxation and release. Subsequent NR measurement was used to monitor the composition of the deuterated phospholipid bilayers versus time after LCNP addition. The study shows significant lipid exchange between the nanoparticles comprised of hydrogenated GMO and the deuterated DOPC bilayer (Vandoolaeghe et al., 2008). The exchange takes place regardless of the initial bilayer coverage (Vandoolaeghe et al., 2009c). The rearrangement within the surface layer results in interfacial instability that releases particles after a critical amount of lipids have been exchanged. An SAXS study of the interaction between unilamellar DOPC vesicles and the CPNP particles demonstrated that the exchange leads to local phase change as an additional lamellar phase appears with time (Vandoolaeghe et al., 2009a). In addition, the cubic phase unit cell dimension decreases with time and the peaks corresponding to the cubic phase eventually disappear. The exchange between the CPNP and the supported lipid bilayer is slower, if the bilayer was made up of lipids in the gel state, for example, dipalmitoylphosphatidylcholine (DPPC) (Vandoolaeghe et al., 2009a). These studies demonstrated the usefulness of combining several techniques to reveal the mechanism of interaction. In this case, null ellipsometry revealed the adsorption kinetics, including the appearance of an adsorption maximum; QCM-D revealed the change in interfacial structure and proved that indeed intact CPNP particles adsorbed to the bilayer; and NR revealed the time-dependent change in composition. The SAXS data revealed the change in composition is associated with a phase transition. The interesting and controllable interaction of the nanoparticles with model membranes, where the particles released are triggered by a local phase transition due to the lipid exchange, suggests an interesting concept for the phase change triggered release that might have potential in drug delivery systems.

10.4.2.2 *Mucous Membrane*

To use LCNP as an oral and topical delivery system, it is important to understand the interaction between the mucous gel layer and the crystalline phases. The mucous membrane is the outermost lining of many biological systems, for example, the mouth, stomach walls, intestines, eyes, and genital areas. The mucoadhesion property of GMO has been tested using a flushing system with intestinal surface where the GMO forms liquid crystals once in contact with the wet mucosa (Nielsen et al., 1998). The study shows that the cubic phase is mucoadhesive when formed on wet mucosa.

To facilitate the delivery of the highly viscous cubic phase, particulate dispersion of the cubic phase has been made as a promising vehicle for mucosal drug delivery. The interaction between CPNP and mucin, the main constituent of the mucous layer, was tested using null ellipsometry and particle electrophoresis (Svensson et al., 2008b). This work shows a weak interaction between the particles and mucin that is similar to the interaction between mucin and PEO chains. This finding suggests that the CPNP are covered with the stabilizing PEO chains from the Pluronic stabilizer where the PEO chains dictate the adsorption behavior.

To enhance the interaction between CPNP and the mucin surface, CPNPs were modified with positively charged chitosan (Svensson et al., 2008a). The chitosan-modified CPNPs show substantially larger adsorption on the mucin surface compared to the unmodified particles. The result suggests that the electrostatic attraction between the positively charged chitosan and negatively charged mucin increased the adsorption. This study demonstrates the chitosan-modified CPNP may be of interest for mucosal drug delivery applications. It further confirms that surface modification of LCNP can be tailored to enhance interfacial interaction.

10.4.2.3   Leaf Surface

Inspired by the enhanced delivery of active ingredient using LCNP, a recent study (Dong et al., 2011) examines the possibility of using LCNP to deliver active molecules to hydrophobic plant leaf surfaces. ATR-FTIR was used to monitor the adsorption of CPNP and hexagonal phase nanoparticles (HPNP) on model and real leaf surfaces (Fig. 10.9). The study shows that the adsorption behavior is dependent on the internal nanostructure, with higher adsorption and delivery efficacy for the cubic phase particles than the hexagonal counterpart. The difference in adsorption is explained by the difference in the gain in free energy due to structural changes (relaxation)

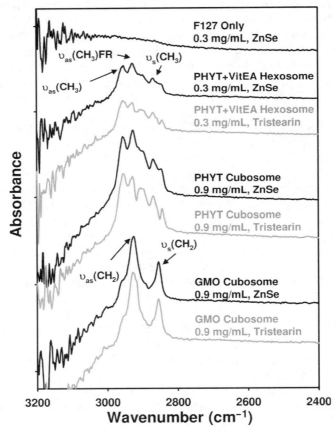

Figure 10.9   ATR-FTIR spectra of cubosome- and hexosome-coated surfaces after 100 min of adsorption. [Reprinted with permission from Dong et al. (2011). Copyright 2011 by the American Chemical Society.]

when interacting with the surface (Dong et al., 2011). The HPNPs show faster adsorption kinetic and free energy gain upon particle adsorption due to the release of internal stress from the unfavorable internal packing geometry. Once adsorbed on the surface, HPNPs may experience greater structural relaxation and spreading than the CPNPs, which can lead to a lower quantity of adsorbed HPNPs. This is consistent with the high viscosity of liquid crystalline cubic phases compared to the hexagonal phase (Larsson, 1989).

The adhesion strength of the nanoparticle was found to be dependent on the lipid system (Dong et al., 2011). GMO-based CPNP showed weaker surface adhesion compared to the phytantriol (PHYT)-based CPNP even though the same stabilizer, Pluronic F-127, stabilizes both types of CPNPs. The difference in the adhesion strength was proposed to be dependent on the different interactions between the Pluronic stabilizer and the lipid crystalline structure. It is believed that the hydrophobic domain of Pluronic anchors to the crystalline particle, either by direct adsorption or by interpenetration into crystalline phases, thereby allowing the hydrophilic chains to extend into the surrounding bulk water to provide steric stabilization (Dong et al., 2011; Kaasgaard and Drummond, 2006). It has been proposed that strong interdigitation between Pluronic and GMO-based CPNP inhibits strong particle adsorption and results in weak particle surface adhesion. On the other hand, Pluronic loosely adheres on the surface of PHYT-based CPNP, which can be easily displaced to allow strong particle adhesion with the surface. The PHYT-based CPNP has been shown to withstand extensive agitation, which may be of interest for agriculture applications where mechanical resistance is desirable.

10.5 FORMATION OF SURFACES

Formation of liquid crystalline structures on surfaces can be obtained in many different ways. Depending on the desired application and result, different methods can be utilized, ranging from adsorption and deposition of liquid crystalline particles to the formation from chemical reaction.

10.5.1 Adsorption of LCNP

One of the most convenient methods to form a crystalline structure on a surface is through adsorption of LCNP. The use of stabilizer preserves the liquid crystalline structure of LCNP in a dispersed form, which in turn allows easy delivery and application. The property of the particle stabilizer, solution condition, and also surface chemistry can significantly affect the adsorption interaction. Studies have shown strong adsorption dependence on physical parameters such as particle size, ionic strength, and pH conditions (Svensson et al., 2008a,b; Vandoolaeghe et al., 2006, 2008, 2009b,c). GMO-based cubic LCNP stabilized with Pluronic F-127 was found to adsorb intact on hydrophilic silica in the presence of an electrolyte, except at high alkaline pH (Vandoolaeghe et al., 2009b). It was suggested that the adsorption is controlled by an interplay between two competing processes: the faster adsorption of excess free Pluronic stabilizer and the slower binding of the LCNP in the gaps. The adsorption of free block copolymers at the surface can hinder the adsorption of LCNP by steric repulsion and thus limit the adsorption of the particles. The interaction mechanism is therefore dictated by interplays between the particle stabilizer, the lipid matrix, and the substrate surface. Figure 10.4 shows the effect of surface chemistry and solvent condition on the adsorption of LCNP. The slightly negatively charged LCNP adsorbed significantly more on the cationic surface.

A recent study has utilized the layer-by-layer (LbL) adsorption technique to embed CPNPs in the capsule shell wall (Driever et al., 2011). Oppositely charged polyelectrolytes are assembled layer by layer onto a surface, with the layers forming due to electrostatic attraction. The study demonstrated CPNP incorporation within a multilayer assembly of oppositely charged polyelectrolytes on planar silica substrate as well as on silica microparticles. The SAXS measurements confirm the preservation of the cubic phase embedded within the polyelectrolyte matrix. Confocal microscopy of the microparticles shows the presence of fluorescent CPNP within the polymer shell of silica particles both before and after silica core removal. This study demonstrates that the LbL adsorption technique can be used to embed LCNP to form a surface layer of liquid crystalline structure.

10.5.2 Forming from Components

The deposition of LCNP can be achieved by depositing the components from a solvent. Such a process occurs, for example, when constructing a calibration curve for an ATR-FTIR experiment. For ATR-FTIR, a calibration curve of known LCNP mass on the surface is needed to accurately quantify the adsorbed mass. To construct such a calibration curve, the components of LCNP can be dissolved in chloroform and applied to the surface (Dong et al., 2011). The chloroform solvent is then evaporated to leave a known mass deposited on the crystal surface. The layer prepared in this way only contains the substances that make up the LCNP, which may retain internal crystalline structure once the layer is rehydrated.

Another possible mechanism of depositing LCNP is through the application of dry powder precursors on a surface. Dry powder precursors of CPNP have been produced using the spray–drying technique from starch-encapsulated dispersion as well as dextran-encapsulated emulsion (Spicer et al., 2002). These spray–dried precursors of CPNP have been shown to form the cubic crystalline structure upon hydration in water. This new formulation technique may allow the formation of a crystalline surface through deposition and then rehydration from powdered CPNP precursors on the surface. Such an application still requires an understanding of means to control the adhesion strength between the CPNP and the supporting surface.

The formation of a crystalline structure from the adsorption of lipid components has been demonstrated on mucosa membrane. Lipid components, such as GMO and glyceryl monolinoleate (GML), can form cubic phase liquid crystals when in contact with the aqueous mucosa (Nielsen et al., 1998). The crystalline phases are formed from the hydration of the lipid components such that a drug added to the precursor formulation is also incorporated in the cubic phase formed. Tensiometric measurements show that precursors of the cubic phase are more mucoadhesive than the cubic phase (Nielsen et al., 1998). This interesting finding suggests that formation of crystalline structure from lipid components can be a better approach for surfaces with high water content.

An interesting phenomena relating to the formation of a cubic liquid crystalline phase in a confinement between two surfaces is reported in a study using the interferometric surface force apparatus (SFA) and ellipsometry (Campos et al., 2002). The study utilized GMO, which has low solubility in aqueous solution, and, therefore, the solution was equilibrated with an excess of GMO cubic liquid crystal. GMO was found to adsorb from aqueous solution to a hydrophobized silica surface as a 2.5-nm monolayer. Surprisingly, the force versus distance curves between two hydrophobized mica surfaces in saturated solutions of GMO is characterized by a strong repulsive interaction beneath surface separation of 40 nm. If sufficient force is applied, the surfaces can be brought into (adhesive) contact. Here it should be noted that the surplus of GMO will float on the surface of the aqueous solution as a fully swollen cubic phase. This means that any disturbances of the system can cause a phase separation of the cubic phase. The confinement of a mixed solution/dispersion or a solution/gas close to saturation between two surfaces (in effect forming a capillary) can produce so-called capillary-induced phase separation (CIPS). CIPS has been demonstrated for a number of systems, which can drastically change the forces between the surfaces (Evans and Wennerström, 1999). This is exactly what can happen in the confinement between the surfaces, where a cubic phase is suggested to form between two surfaces in close contact in a solution saturated with GMO. The repulsive force observed by Campos et al. (2002) is likely to arise from the compression of the cubic phase GMO, which is known to be very stiff (Pitzalis et al., 2000).

10.5.3 Formation by Chemical Reactions (e.g., Lipolysis)

In the pioneering in vitro study of lipolysis of a droplet of triglycerides in an intestine-like environment, Patton and Carey observed, apart from the initially occurring crystalline phase, a viscous isotropic phase composed of monoglycerides and fatty acids, which is identical to the one formed in monoglyceride systems (Patton and Carey, 1979). In excess of bile salts, the lipolysis products, later defined as cubic phase, are rapidly solubilized in mixed micelles of lipids and bile salts. However, the bile acid amounts in vivo are not sufficient to solubilize all lipids after a meal rich in fats, which implies that the cubic liquid crystalline phases exist in vivo (Lindström et al.,

1981). Lipase and water must be free to diffuse through the phases formed by the lipolysis products, surrounding the diminishing fat droplet. Thus, the bicontinuity as well as the incorporation properties of the cubic monoglyceride phases are thought to be important features for the lipolysis process (Patton et al., 1985).

So far lipolysis has not been used for the surface modification of oil film. One reason might be that it can be hard to stop the lipolytic activity at the requested liquid crystalline structure. Recent interfacial studies of lipase activities on monolayers have provided some leads on how to control the lipase activity by modulating the lipid composition (Reis et al., 2009). Another obstacle is to prepare well-defined oil films. The recently developed spin–freeze–thaw technique has opened up the possibility of forming stable, uniform oil films with thicknesses of ~2 μm by spin coating the oil onto hydrophobically modified silicon substrates (Zarbakhsh et al., 2005). We are convinced that this technique has a so far unexploited potential for the future.

10.6   STRUCTURAL CHARACTERISTICS

10.6.1   Internal Structure of the Layer

The internal crystalline structure of the dispersed LCNP can be characterized with SAXS and cryo-TEM. These two techniques have been the most popular because they provide information on the internal crystalline structure and particle morphology (Boyd et al., 2009). In SAXS, the periodic crystalline structure gives rise to characteristic diffraction patterns. Indexing the peaks in the diffraction pattern to a specific space group allows identification of the crystalline phase. Cryo-TEM is a way to directly visualize the morphology and structure of the LCNP in the equilibrium state with minimal disruption to the morphology. The nanoparticles are frozen before imaging. The internal crystalline structure can be deduced from the cryo-TEM micrograph and can be further processed by Fourier transformation to reveal the crystalline phases. Cryo-TEM reveals the structure of individual particles, while SAXS measures on an assembly of particles with higher accuracy, which makes it less sensitive to individual variations. These complementary techniques have been widely utilized to investigate the internal structure of dispersed particles (Barauskas and Landh, 2003; Barauskas et al., 2005a,b, 2006a); however, they cannot easily provide the corresponding information on the internal structure of a crystalline layer on a surface.

Neutron reflectometry can provide some quantitative information on the internal structure of the layer at an interface. The wavelength of neutrons, ~5 Å, is similar to that of molecular length scales, providing enough resolution to detect crystalline structures at the interface (Nylander et al., 2008). Uniform thin films on surfaces can exhibit pronounced fringes in the NR curve, and repeating structures normal to the interface can generate Bragg diffraction peaks. Also, lateral inhomogeneity across the surface can appear as intensity signals in the off-specular scattering. Thus NR can help to reveal quantitative structural and compositional information on adsorption of LCNP.

Recently, NR measurements were carried out to investigate the structure of the interfacial layer upon adsorption of LCNP (Vandoolaeghe et al., 2008, 2009b,c). The reflectivity profiles show sharp peaks that are analogous to Bragg peaks in the diffraction from a crystal but in this case arise from one-dimensional order of repeating patterns. The intensity of the peak increases with time (Fig. 10.7a) revealing the increased presence of an organized repeating structure perpendicular to the plane of the surface, which arises from the ordered arrangement of the lipid molecules in the LCNP (Vandoolaeghe et al., 2009b). The NR curve fits nicely to the cubic phase model with the scattering length density profile varying as a sinusoidal function (Fig. 10.7b). The NR experiment demonstrates that LCNPs adsorb intact on the hydrophilic Si surfaces. The particles adsorb randomly at the interface with the internal cubic phase undisturbed and retain their internal crystalline structure at the interface for over 44 h (Vandoolaeghe et al., 2009b).

10.6.2 Phase Transitions at the Interface

The intact LCNPs can undergo phase transition when in contact with other lipids at an interface. As mentioned previously, the exchange of lipids can take place when GMO-based CPNP interacts with the supported lipid bilayer. The amount of lipid exchange can be quantified with NR by determining the composition of hydrogenous components of CPNP (GMO and F-127) and the deuterated bilayer (d-DOPC). The results show that the final composition of the lipid layer on a surface at the end of an exchange depends very much on the initial bilayer coverage (Vandoolaeghe et al., 2009c). At high bilayer coverage, the addition of CPNP leads to an extensive exchange of lipids. The final bilayer is composed of 72% of the CPNP components and the surface contains very little residual intact particles. At lower bilayer coverage, spreading of CPNP components fills in the defects of the bilayer. At the same time, a substantial adsorption of the intact particles can be seen by the presence of the Bragg diffraction peak (Vandoolaeghe et al., 2009c).

The lipid exchange can also change the composition of the adsorbed CPNP and induce a phase change. An NR study shows the incorporation of d-DOPC molecules from the bilayer into the GMO-based CPNP (Vandoolaeghe et al., 2009c). The incorporation of DOPC shifts the composition from the cubic phase toward the cubic/lamellar phase and possibly the lamellar phase based on the GMO/DOPC/D_2O phase diagram. Such a transition can lead to a contraction of the unit cell dimension of the cubic structure and shifts the Bragg diffraction peak toward higher momentum transfer, Q, values. The shift and broadening of the Bragg diffraction peak are observed over time, which are consistent with the progression of the lipid structure toward a new phase from the incorporation of d-DOPC in the GMO domains. This demonstrates that a phase transformation of LCNP can occur when an interfacial interaction alters the composition of the crystalline particles.

Phase transformation can also be triggered by an external stimulus such as a change in osmotic gradient. Studies have shown that GMO lipid membrane in a polymer scaffold is responsive (Aberg et al., 2008). When the osmotic gradient is above a critical transition point, a cubic to lamellar phase is induced at the surface of the membrane. This phase transformation drastically reduces the permeability and changes the diffusive transport behavior. Before the phase transformation, small compounds are rather permeable in the bicontinuous cubic phase. After the transformation, the lamellar phase reduces the diffusion transport (Sparr et al., 2009). The similar nature of the responding membrane is seen in skin, where gradients can alter the molecular structure in the membrane and thereby regulate the transport across a barrier (Sparr et al., 2009).

10.6.3 Effect of Including Guest Molecules

The internal phase of the dispersed LCNPs can be affected by the presence of a steric stabilizer, by the addition of guest molecules, and may also be changed upon interaction at the interface. A steric stabilizer is required for the stable dispersion of the bulk crystalline phases; however, the nature and quantity of the stabilizer may itself induce a change in the crystalline phases in the dispersed particles (Boyd et al., 2009). The effect of stabilizer is specific to the lipid–stabilizer combination. For example, the F-127 stabilizer can change the GMO-based particle from the diamond bicontinuous cubic phase (Pn3m) to the primitive bicontinuous cubic phase (Im3m) at high concentration but not with other lipids such as phytantriol or glyceryl monolinoleate (Boyd et al., 2009).

Inclusion of guest molecules such as drugs and other additives can also affect the internal crystalline structure of LCNP. Phase transformations have been observed in the bulk crystalline phases and also in the internal structure of the dispersed particles. For specific examples, see reviews by Drummond and Fong (1999), Shah et al. (2001), and Boyd et al. (2009) for bulk gel phase and dispersed particles, respectively. As a general rule, addition of hydrophobic molecules encourages transformation from reversed cubic to reversed hexagonal phases, while the inclusion of hydrophilic molecules induces curvature toward the lipid region and favors lamellar phases (Boyd et al., 2009; Shah et al., 2001). For example, addition of vitamin E acetate to phytantriol (Dong et al., 2006) and glyceryl trioleate to GMO (Mele et al., 2004) results in transformations of cubic phase to hexagonal phase. This is in contrast to the addition of charged membrane lipid, dipalmitoylphosphatidylserine (DPPS), to phytantriol, which converts the cubic phase into the lamellar phase and results in the increased presence of vesicles (Shen et al., 2010).

10.7 APPLICATIONS AND FUTURE OUTLOOK

Liquid crystalline nanoparticles have great potential in many different applications because of their structure and encapsulation ability. Two applications of interest are as delivery agents and as a surface functionalizers. To use an LCNP as a delivery agent, it is important to understand its physical property and interaction with the encapsulated molecules. Furthermore, it is crucial to understand the interfacial property between LCNP and the surface where the particles will interact. For drug delivery purposes, there is a need to develop new lipid compositions with low physiological toxicity. Systematic studies, as developed in some of the works cited, to monitor the interaction of the nanoparticles with model membranes can then be related to the synthesis of new formulations to assess their potential for delivery applications.

Liquid crystalline nanoparticles can also be used to prepare crystalline surface films with a well-defined nanostructure. These surface films can be formed through simple adsorption of the nanoparticles, direct deposition of the LCNP precursors, and also built up from the lipid components. The crystalline structure on a surface can be altered because phase transition can take place upon adsorption. The inclusion of guest molecules can also change the crystalline phases. Thus, the surface structure will need to be systematically monitored, as described previously, to ensure a well-defined crystalline structure is maintained.

ACKNOWLEDGMENTS

We are grateful to our colleagues Fredrik Tiberg, Justas Barauskas, Markus Johnsson, and Camilla Cervin for many stimulating discussions on lipid liquid crystalline nanoparticles and interfaces. Financial support was obtained from the Swedish Foundation for Strategic Research.

REFERENCES

Aberg, C., Pairin, C., Costa-Balogh, F. O., and Sparr, E. (2008). Responding double-porous lipid membrane: Lyotropic phases in a polymer scaffold. *Biochimica et Biophysica Acta—Biomembranes*, *1778*(2), 549–558.

Angelova, A., Angelov, B., Mutafchieva, R., Lesieur, S., and Couvreur, P. (2011). Self-assembled multicompartment liquid crystalline lipid carriers for protein, peptide, and nucleic acid drug delivery. *Accounts of Chemical Research*, *44*(2), 147–156.

Azzam, R. M. A., and Bashara, N. M. (1977). *Ellipsometry and Polarized Light*, North-Holland, Amsterdam.

Bangham, A. D., and Horne, R. W. (1964). Negative staining of phospholipids and their structural modification by surface-active agents as observed in the electron microscope. *Journal of Molecular Biology*, *8*, 9562–9565.

Barauskas, J., and Landh, T. (2003). Phase behavior of the phytantriol/water system. *Langmuir*, *19*(23), 9562–9565.

Barauskas, J., and Nylander, T. (2008). Lyotropic liquid crystals as delivery vehicles for food ingredients. In N. Garti (Ed.), *Delivery and Controlled Release of Bioactives in Foods and Nutraceuticals*. Woodhead Publishing Series in Food Science, Technology and Nutrition, Cambridge, England.

Barauskas, J., Johnsson, M., and Tiberg, F. (2005a). Self-assembled lipid superstructures: Beyond vesicles and liposomes. *Nano Letters*, *5*(8), 1615–1619.

Barauskas, J., Johnsson, M., Johnson, F., and Tiberg, F. (2005b). Cubic phase nanoparticles (Cubosome): Principles for controlling size, structure, and stability. *Langmuir*, *21*(6), 2569–2577.

Barauskas, J., Johnsson, M., Nylander, T., and Tiberg, F. (2006a). Hexagonal liquid-crystalline nanoparticles in aqueous mixtures of glyceryl monooleyl ether and pluronic F127. *Chemistry Letters*, *35*(8), 830–831.

Barauskas, J., Misiunas, A., Gunnarsson, T., Tiberg, F., and Johnsson, M. (2006b). "Sponge" nanoparticle dispersions in aqueous mixtures of diglycerol monooleate, glycerol dioleate, and polysorbate 80. *Langmuir*, *22*(14), 6328–6334.

Binks, B. P. (1999). *Modern Characterization Methods of Surfactant Systems*, Marcel Dekker, New York.

Boyd, B. J., Dong, Y. D., and Rades, T. (2009). Nonlamellar liquid crystalline nanostructured particles: Advances in materials and structure determination. *Journal of Liposome Research*, *19*(1), 12–28.

Campos, J., Eskilsson, K., Nylander, T., and Svendsen, A. (2002). On the interaction between adsorbed layers of monoolein and the lipase action on the formed layers. *Colloids Surfaces B: Biointerfaces*, *26*, 172–182.

Chang, D. P., Jankunec, M., Barauskas, J., Tiberg, F., and Nylander, T. (2012). Adsorption of lipid liquid crystalline nanoparticles on cationic, hydrophilic and hydrophobic surfaces. *ACS Applied Materials & Interfaces* (accepted).

Defeijter, J. A., Benjamins, J., and Veer, F. A. (1978). Ellipsometry as a tool to study adsorption behavior of synthetic and biopolymers at air-water-interface. *Biopolymers*, *17*(7), 1759–1772.

Dong, Y. D., Larson, I., Hanley, T., and Boyd, B. J. (2006). Bulk and dispersed aqueous phase behavior of phytantriol: Effect of vitamin E acetate and F127 polymer on liquid crystal nanostructure. *Langmuir*, *22*(23), 9512–9518.

Dong, Y. D., Larson, I., Bames, T. J., Prestidge, C. A., and Boyd, B. J. (2011). Adsorption of nonlamellar nanostructured liquid-crystalline particles to biorelevant surfaces for improved delivery of bioactive compounds. *ACS Applied Materials & Interfaces*, *3*(5), 1771–1780.

Driever, C. D., Mulet, X., Johnston, A. P. R., Waddington, L. J., Thissen, H., Caruso, F., and Drummond, C. J. (2011). Converging layer-by-layer polyelectrolyte microcapsule and cubic lyotropic liquid crystalline nanoparticle approaches for molecular encapsulation. *Soft Matter*, *7*(9), 4257–4266.

Drummond, C. J., and Fong, C. (1999). Surfactant self-assembly objects as novel drug delivery vehicles. *Current Opinion in Colloid & Interface Science*, *4*(6), 449–456.

Ericsson, B., Larsson, K., and Fontell, K. (1983). A cubic protein-monoolein-water phase. *Biochimica et Biophysica Acta*, *729*(1), 23–27.

Evans, D. F., and Wennerström, H. (1994). *The Colloidal Domain: Where Physics, Chemistry, Biology, and Technology Meet*. VCH, New York.

Evans, D. F., and Wennerström, H. (1999). *The Colloidal Domain: Where Physics, Chemistry, Biology, and Technology Meet*, 2nd ed., Wiley-VCH, New York.

Gustafsson, J., Ljusberg-Wahren, H., Almgren, M., and Larsson, K. (1996). Cubic lipid-water phase dispersed into submicron particles. *Langmuir*, *12*(20), 4611–4613.

Gustafsson, J., Ljusberg-Wahren, H., Almgren, M., and Larsson, K. (1997). Submicron particles of reversed lipid phases in water stabilized by a nonionic amphiphilic polymer. *Langmuir*, *13*(26), 6964–6971.

Hind, A. R., Bhargava, S. K., and McKinnon, A. (2001). At the solid/liquid interface: FTIR/ATR—the tool of choice. *Advances in Colloid and Interface Science*, *93*(1–3), 91–114.

Hook, F., Kasemo, B., Nylander, T., Fant, C., Sott, K., and Elwing, H. (2001). Variations in coupled water, viscoelastic properties, and film thickness of a Mefp-1 protein film during adsorption and cross-linking: A quartz crystal microbalance with dissipation monitoring, ellipsometry, and surface plasmon resonance study. *Analytical Chemistry*, *73*(24), 5796–5804.

Hughes, A. V., Howse, J. R., Dabkowska, A., Jones, R. A. L., Lawrence, M. J., and Roser, S. J. (2008). Floating lipid bilayers deposited on chemically grafted phosphatidylcholine surfaces. *Langmuir*, *24*, 1989–1999.

Hyde, S. T. (1996). Bicontinuous structures in lyotropic liquid crystals and crystalline hyperbolic surfaces. *Current Opinion in Solid State & Materials Science*, *1*(5), 653–662.

Hyde, S. T., Andersson, S., Ericsson, B., and Larsson, K. (1984). A cubic structure consisting of a lipid bilayer forming an infinite periodic minimum surface of the gyroid type in the glycerolmonooleate-water system. *Zeitschrift Für Kristallographie*, *168*(1–4), 213–219.

Israelachvili, J. N., Mitchell, D. J., and Ninham, B. W. (1976). Theory of self-assembly of hydrocarbon amphiphiles into micelles and bilayers. *Journal of Chemical Society Faraday Transactions II*, *72*, 1525–1568.

Jonsson, B. (1998). *Surfactants and Polymers in Aqueous Solution*. Wiley, Chichester.

Kaasgaard, T., and Drummond, C. J. (2006). Ordered 2-D and 3-D nanostructured amphiphile self-assembly materials stable in excess solvent. *Physical Chemistry Chemical Physics*, *8*(43), 4957–4975.

Landgren, M., and Jonsson, B. (1993). Determination of the optical-properties of Si/SiO₂ surfaces by means of ellipsometry, using different ambient media. *Journal of Physical Chemistry*, *97*(8), 1656–1660.

Landh, T. (1994). Phase-behavior in the system pine oil monoglycerides-poloxamer-407-water at 20-degrees-C. *Journal of Physical Chemistry*, *98*(34), 8453–8467.

Larsson, K. (1989). Cubic lipid-water phases—structures and biomembrane aspects. *Journal of Physical Chemistry*, *93*(21), 7304–7314.

Larsson, K. (1991). Emulsions of reversed micellar phases and aqueous dispersions of cubic phases of lipids—some food aspects. *ACS Symposium Series*, *448*, 44–50.

Larsson, K. (1999). Colloidal dispersions of ordered lipid-water phases. *Journal of Dispersion Science and Technology*, *20*(1–2), 27–34.

Larsson, K. (2000). Aqueous dispersions of cubic lipid-water phases. *Current Opinion in Colloid & Interface Science, 5*(1–2), 64–69.

Larsson, K. (2009). Lyotropic liquid crystals and their dispersions relevant in foods. *Current Opinion in Colloid & Interface Science, 14*(1), 16–20.

Larsson, K., and Tiberg, F. (2005). Periodic minimal surface structures in bicontinuous lipid–water phases and nanoparticles. *Current Opinion in Colloid & Interface Science, 9*, 365–369.

Larsson, K., Quinn, P., Sato, K., and Tiberg, F. (2006). *Lipids: Structure, Physical Properties and Functionality*, Oily Press, Bridgwater, England.

Lawrence, M. J. (1994). Surfactant systems—their use in drug-delivery. *Chemical Society Reviews, 23*(6), 417–424.

Lindström, M., Ljusberg-Wahren, H., Larsson, K., and Borgström, B. (1981). Aqueous lipid phases of relevance to intestinal fat digestion and absorption. *Lipids, 16*, 749–754.

Malmsten, M. (2007a). Phase transformations in self-assembly systems for drug delivery applications. *Journal of Dispersion Science and Technology, 28*, 63–72.

Malmsten, M. (2007b). Phase transformations in self-assembly systems for drug delivery applications. *Journal of Dispersion Science and Technology, 28*(1), 63–72.

Mele, S., Murgia, S., Caboi, F., and Monduzzi, A. (2004). Biocompatible lipidic formulations: Phase behavior and microstructure. *Langmuir, 20*(13), 5241–5246.

Mitchell, D. J., and Ninham, B. W. (1981). Micelles, vesicles and microemulsions. *Journal of the Chemical Society Faraday Transactions II, 77*, 601–629.

Mouritsen, O. G. (2005). *Life—As Matter of Fat. The Emerging Science of Lipidomics*. Springer, Heidelberg.

Neto, C., Aloisi, G., Baglioni, P., and Larsson, K. (1999). Imaging soft matter with the atomic force microscope: Cubosomes and hexosomes. *Journal of Physical Chemistry B, 103*(19), 3896–3899.

Nielsen, L. S., Schubert, L., and Hansen, J. (1998). Bioadhesive drug delivery systems. I. Characterisation of mucoadhesive properties of systems based on glyceryl mono-oleate and glyceryl monolinoleate. *European Journal of Pharmaceutical Sciences, 6*(3), 231–239.

Nylander, T., Campbell, R. A., Vandoolaeghe, P., Cardenas, M., Linse, P., and Rennie, A. R. (2008). Neutron reflectometry to investigate the delivery of lipids and DNA to interfaces. *Biointerphases, 3*(2), Fb64–Fb82.

Patton, J. S., and Carey, M. C. (1979). Watching fat digestion. *Science, 204*, 145–148.

Patton, J. S., Vetter, R. D., Hamosh, M., Borgström, B., Lindström, M., and Carey, M. C. (1985). The light microscopy of fat digestion. *Food Microstructure, 4*, 29–41.

PerkinElmer, I. (2005). Technical Notes: FT-IR Spectroscopy Attenuated Total Reflectance (ATR). PerkinElmer Life and Analytical Science, Shelton, CT.

Pitzalis, P., Monduzzi, M., Krog, N., Larsson, H., Ljusberg-Wahren, H., and Nylander, T. (2000). Characterization of the liquid–crystalline phases in the glycerol monooleate/diglycerol monooleate/water system. *Langmuir, 16*, 6358–6365.

Prasad, V., Semwogerere, D., and Weeks, E. R. (2007). Confocal microscopy of colloids. *Journal of Physics—Condensed Matter, 19*(11), 113102.

Razumas, V., Larsson, K., Miezis, Y., and Nylander, T. (1996a). A cubic monoolein cytochrome c water phase: X-ray diffraction, FT-IR, differential scanning calorimetric, and electrochemical studies. *Journal of Physical Chemistry, 100*(28), 11766–11774.

Razumas, V., Talaikyte, Z., Barauskas, J., Larsson, K., Miezis, Y., and Nylander, T. (1996b). Effects of distearoylphosphatidylglycerol and lysozyme on the structure of the monoolein-water cubic phase: X-ray diffraction and Raman scattering studies. *Chemistry and Physics of Lipids, 84*(2), 123–138.

Reimhult, E., Hook, F., and Kasemo, B. (2002). Vesicle adsorption on SiO_2 and TiO_2: Dependence on vesicle size. *Journal of Chemical Physics, 117*(16), 7401–7404.

Reimhult, E., Hook, F., and Kasemo, B. (2003). Intact vesicle adsorption and supported biomembrane formation from vesicles in solution: Influence of surface chemistry, vesicle size, temperature, and osmotic pressure. *Langmuir, 19*(5), 1681–1691.

Reis, P., Holmberg, K., Watzke, H., Leser, M. E., and Miller, R. (2009). Lipases at interfaces: A review. *Advances in Colloid and Interface Science, 147–148*, 237–250.

Richter, R. P., Berat, R., and Brisson, A. R. (2006). Formation of solid-supported lipid bilayers: An integrated view. *Langmuir, 22*, 3497–3505.

Rodahl, M., Hook, F., Fredriksson, C., Keller, C. A., Krozer, A., Brzezinski, P., Voinova, M., and Kasemo, B. (1997). Simultaneous frequency and dissipation factor QCM measurements of biomolecular adsorption and cell adhesion. *Faraday Discussions, 107*, 229–246.

Sackmann, E. (1996). Supported membranes: Scientific and practical applications. *Science, 271*, 43–48.

Sagalowicz, L., and Leser, M. E. (2010). Delivery systems for liquid food products. *Current Opinion in Colloid & Interface Science, 15*(1–2), 61–72.

Sagalowicz, L., Leser, M. E., Watzke, H. J., and Michel, M. (2006a). Monoglyceride self-assembly structures as delivery vehicles. *Trends in Food Science & Technology, 17*, 204–214.

Sagalowicz, L., Leser, M. E., Watzke, H. J., and Michel, M. (2006b). Monoglyceride self-assembly structures as delivery vehicles. *Trends in Food Science & Technology, 17*(5), 204–214.

Sagalowicz, L., Mezzenga, R., and Leser, M. E. (2006c). Investigating reversed liquid crystalline mesophases. *Current Opinion in Colloid & Interface Science, 11*(4), 224–229.

Sauerbrey, G. (1959). Verwendung Von Schwingquarzen Zur Wagung Dunner Schichten Und Zur Mikrowagung. *Zeitschrift fur Physik, 155*(2), 206–222.

Seddon, J. M. (1990). Structure of the inverted hexagonal (HII) phase, and non-lamellar phase transitions of lipids. *Biochimica et Biophysica Acta, 1031*, 1–69.

Shah, J. C., Sadhale, Y., and Chilukuri, D. M. (2001). Cubic phase gels as drug delivery systems. *Advanced Drug Delivery Reviews, 47*(2–3), 229–250.

Shen, H. H., Crowston, J. G., Huber, F., Saubern, S., McLean, K. M., and Hartley, P. G. (2010). The influence of dipalmitoyl phosphatidylserine on phase behaviour of and cellular response to lyotropic liquid crystalline dispersions. *Biomaterials, 31*(36), 9473–9481.

Sparr, E., Aberg, C., Nilsson, P., and Wennerstrom, H. (2009). Diffusional transport in responding lipid membranes. *Soft Matter, 5*(17), 3225–3233.

Spicer, P. T. (2005). Progress in liquid crystalline dispersions: Cubosomes. *Current Opinion in Colloid & Interface Science*, *10*(5–6), 274–279.

Spicer, P. T., Small, W. B., Lynch, M. L., and Burns, J. L. (2002). Dry powder precursors of cubic liquid crystalline nanoparticles (cubosomes). *Journal of Nanoparticle Research*, *4*(4), 297–311.

Svensson, O., Thuresson, K., and Arnebrant, T. (2008a). Interactions between chitosan-modified particles and mucin-coated surfaces. *Journal of Colloid & Interface Science*, *325*(2), 346–350.

Svensson, O., Thuresson, K., and Arnebrant, T. (2008b). Interactions between drug delivery particles and mucin in solution and at interfaces. *Langmuir*, *24*(6), 2573–2579.

Thomas, R. K. (2004). Neutron reflection from liquid interfaces. *Annual Review of Physical Chemistry*, *55*, 391–426.

Tiberg, F., Harwigsson, I., and Malmsten, M. (2000). Formation of model lipid bilayers at the silica-water interface by co-adsorption with non-ionic dodecyl maltoside surfactant. *European Biophysics Journal with Biophysics Letters*, *29*(3), 196–203.

Tien, H. T., and Ottova, A. L. (2001). The lipid bilayer concept and its experimental realization: From soap bubbles, kitchen sink, to bilayer lipid membranes. *Journal of Membrane Science*, *189*, 83–117.

Tompkins, H. G., and Irene, E. A. (2005). *Handbook of Ellipsometry*. Springer, Heidelberg, Germany.

Vandoolaeghe, P., Tiberg, F., and Nylander, T. (2006). Interfacial behavior of cubic liquid crystalline nanoparticles at hydrophilic and hydrophobic surfaces. *Langmuir*, *22*(22), 9169–9174.

Vandoolaeghe, P., Rennie, A. R., Campbell, R. A., Thomas, R. K., Hook, F., Fragneto, G., Tiberg, F., and Nylander, T. (2008). Adsorption of cubic liquid crystalline nanoparticles on model membranes. *Soft Matter*, *4*(11), 2267–2277.

Vandoolaeghe, P., Barauskas, J., Johnsson, M., Tiberg, F., and Nylander, T. (2009a). Interaction between lamellar (vesicles) and nonlamellar lipid liquid-crystalline nanoparticles as studied by time-resolved small-angle X-ray diffraction. *Langmuir*, *25*(7), 3999–4008.

Vandoolaeghe, P., Campbell, R. A., Rennie, A. R., and Nylander, T. (2009b). Adsorption of intact cubic liquid crystalline nanoparticles on hydrophilic surfaces: Lateral organization, interfacial stability, layer structure, and interaction mechanism. *Journal of Physical Chemistry C*, *113*(11), 4483–4494.

Vandoolaeghe, P., Rennie, A. R., Campbell, R. A., and Nylander, T. (2009c). Neutron reflectivity studies of the interaction of cubic-phase nanoparticles with phospholipid bilayers of different coverage. *Langmuir*, *25*(7), 4009–4020.

Wacklin, H. P. (2010). Neutron reflection from supported lipid membranes. *Current Opinion in Colloid & Interface Science*, *15*(6), 445–454.

Yang, D., Armitage, B., and Marder, S. R. (2004). Cubic liquid-crystalline nanoparticles. *Angewandte Chemie–International Edition*, *43*(34), 4402–4409.

Zarbakhsh, A., Querol, A., Bowers, J., Yaseen, M., Lu, J. R., and Webster, J. R. P. (2005). Neutron reflection from the liquid-liquid interface: Adsorption of hexadecylphosphorylcholine to the hexadecane-aqueous solution interface. *Langmuir 2005*, *21*, 11704–11709.

■■■■■ CHAPTER 11

Multicompartment Lipid Nanocarriers for Targeting of Cells Expressing Brain Receptors

CLAIRE GÉRAL, ANGELINA ANGELOVA, and SYLVIANE LESIEUR

CNRS, University of Paris Sud, Châtenay-Malabry, France

BORISLAV ANGELOV

Institute of Macromolecular Chemistry, Academy of Sciences of the Czech Republic, Prague, Czech Republic

VALÉRIE NICOLAS

Cell Imaging Platform, University of Paris Sud, Châtenay-Malabry, France

Abstract

Neurotrophic factors, such as the brain-derived neurotrophic factor (BDNF), are essential for the development and survival of human neurons. Their encapsulation in carrier lipid systems is anticipated to overcome the problems resulting from the pharmacokinetics of peptides in the cerebral circulation. Lipid nanocarriers containing an omega-3 polyunsaturated fatty acid [eicosapentaenoic acid] (EPA), showing neuroprotective effects, should trigger the BDNF activity. The purpose of this study is to design and characterize self-assembled lipid systems suitable for encapsulation and potentiation of neurotrophic peptide activity and study of multicompartment liquid crystalline formulations in vitro on a human neuronal cell line expressing BDNF receptors. Sterically stabilized nanodispersed lipid systems are prepared from a PEGylated (polyethylene glycol) liquid crystalline phase in excess water. Such lipid nanovectors, derived by self-assembly, are of ongoing interest thanks to their biocompatible compositions and the relatively low energy input required for their manufacture. The latter is an essential factor for the encapsulation of fragile and temperature-sensitive peptide molecules. Having induced the differentiation of a human neuroblastoma cell line SH-SY5Y, as a model of neurodegeneration, we examine in vitro the expression of the TrkB brain receptor of neurotrophins and the cytotoxicity of the designed multicompartment lipid nanocarriers to neuronal cells.

Self-Assembled Supramolecular Architectures: Lyotropic Liquid Crystals, First Edition.
Edited by Nissim Garti, Ponisseril Somasundaran, and Raffaele Mezzenga.
© 2012 John Wiley & Sons, Inc. Published 2012 by John Wiley & Sons, Inc.

319

11.1 INTRODUCTION

The development and maintenance of neuronal populations in the central nervous system (CNS) is controlled by the binding of neurotrophic factors to their membrane receptors in the brain (Malcangio and Lessmann, 2003; Nagahara et al., 2009; Nosheny et al., 2005; Pattarawarapan and Burgess, 2003; Wu and Pardridge, 1999). The failure of the involved ligand–receptor interactions is often associated to a deficiency of neurotrophin molecules and is responsible for several neurodegenerative diseases and certain psychiatric disorders (Desmet and Peeper, 2006; Fumagalli et al., 2006; Pillai and Mahadik, 2008; Sun and Wu, 2006). The therapeutic use of neurotrophins is unfortunately restrained by their biochemical instability and rapid degradation in the biological medium. Preclinical studies have indicated the low bioavailability of neurotrophins for the therapeutic targets (Tuszynski et al., 2005). In addition, uncontrolled administration of brain-derived neurotrophic factor (BDNF) might interfere with the mechanisms regulating the survival, differentiation, and maintenance of neuronal populations.

The vectorization of the neurotrophic peptides can augment their circulation times and thus improve their bioavailability by hampering the rapid peptide elimination after delivery. We focus on liquid crystalline lipid nanoparticle formulations consisting of monoolein, a PEGylated (polyethylene glycol) lipid, and an omega-3 polyunsaturated fatty acid (eicosapentaenoic acid, EPA) permitting to achieve nanovectors that not only can encapsulate and protect neurotrophins but themselves are active on neuronal cells via the incorporated omega-3 potentiating function. Monoolein is a monoglyceride lipid that allows the production of various types of biocompatible carriers for nondestructive peptide and protein administration (Caboi et al., 2001; Larsson, 1989). The use of a PEGylated lipid in the design of the nanovector compositions is expected to increase the steric stability and slow down the elimination of the carriers by the reticuloendothelial endoplasmic system. To model the neurodegeneration phenomenon with human neurons, the neuroblastoma cell line SH-SY5Y is differentiated by sequential treatment with retinoic acid and BDNF. It is established that the remedy of the human neuroblastoma cells with EPA-containg lipid nanoparticles induces a neuronal phenotype with neurite growth following intense TrkB receptor expression.

11.2 PEPTIDE TARGETING AND DELIVERY

11.2.1 Choice of Therapeutic Peptide with Neurotrophic Properties

Neurotrophins are peptide molecules that ensure the development and survival of neurons (Bemelmans et al., 1999; Burdick et al., 2006; Jaboin et al., 2002). They exert significant control over the life and death of the neuronal cells (Zuccato and Cattaneo, 2009). Among these modulators, the BDNF is a major regulator of the neuronal plasticity, survival, and differentiation (Fumagalli et al., 2006; Matsumoto et al., 1995; Vogelin et al., 2006). Studies have shown its involvement in the pathogenesis of neurodegenerative diseases such as the Huntington's disease, Alzheimer's disease, Parkinson's disease, as well as depression (Table 11.1). The peptide BDNF thus appears as a promising therapeutic candidate against neurodegenerative disorders. So far, the short plasma half-life and the low bioavailability of the molecule have limited its therapeutic development. However, the encapsulation in lipid nanocarriers should overcome the problems of its degradation after administration. The vectorization should permit to achieve a longer plasma half-life and better controlled biodistribution and release.

Crystallographic structural data (Fig. 11.1) have shown that neurotrophins are biomolecules that exist exclusively as dimers of about 27 kDa (Robinson et al., 1995). The neurotrophic peptides have a common structural motif of three disulfide bonds. The homologous sequence encompasses around 50% of all amino acid residues. Each monomer, consisting of about 120 amino acids, involves three pairs of antiparallel β sheets connected to four loops. The neurotrophins are active in the form of homodimers and operate by binding to two types of brain receptors (Pattarawarapan and Burges, 2003; Schramm et al., 2005). The common (lower affinity) receptor, the $p75^{NTR}$ protein, is a member of the family of tumor necrosis factor receptors.

The second type of receptor is the greater affinity receptor, tropomyosin-related kinase (Trk), belonging to the family of tyrosine kinase receptors. The

TABLE 11.1 Summary of Diseases Related to BDNF Deficiency in View of Disorders Involving TrkB Receptor Signaling, and Expected Roles of Therapeutic Molecules

Pathological States Suggested to Involve BDNF Mechanisms	Potential Roles for Therapeutic Molecular Ligands upon Binding to TrkB Brain Receptors
Alzheimer's disease	Agonist to compensate for BDNF deficiency, prevent neural degeneration, improve synaptic transmission, promote plasticity, promote neurogenesis
Huntington's disease	Agonist to compensate for BDNF deficiency and to promote striatal neuron function
Parkinson's disease	Agonist to upregulate D_3 receptor expression and to protect dopaminergic neurons
Rett syndrome	Agonist to compensate for BDNF deficiency
Motor neuron disease	Agonist to inhibit motor neuron death
Depression	Agonist to promote neurogenesis or upregulate neurotransmitter function
Ischemic stroke	Agonist to prevent neuronal death and/or to upregulate neuronal function and promote postinjury plasticity
HIV dementia	Agonist to protect neurons from HIV-1/gp120-induced death
Multiple sclerosis	Agonist to mimic immunomediated BDNF protective mechanisms in MS lesions
Spinal cord injury	Agonist to promote axonal regeneration, plasticity, and activation of TrkB without $p75^{NTR}$ liganding to overcome myelin inhibition
Nerve injury	Agonist to prevent postaxotomy neuron loss and/or promote regeneration
Hearing loss	Agonist to protect auditory neurons
Obesity, diabetes, and metabolic syndrome	Agonist to limit food intake and correct metabolic syndrome
Peripheral tissue ischemia	Agonist to promote angiogenesis
Epilepsy	Antagonist or partial agonist to inhibit hyperexcitable circuits in injured brain
Pain	Antagonist, partial agonist, or agonist to suppress pain transmission
Cancer	Agonist or partial agonist to inhibit TrkB-dependent tumor therapeutic resistance mechanisms

Source: Adapted from Longo (2010).

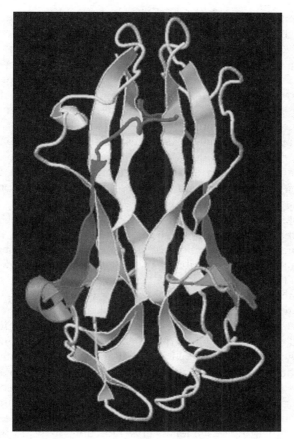

Figure 11.1 Crystallographic structure of a dimer formed by the neurotrophins BDNF and NT3 (PDB code: 1BND).

β sheets of the neurotrophins (Fig. 11.1) consist of highly conserved residues, which are important for maintaining their tertiary structure. The amino acid sequences of the N- and C-terminal regions and of the variable loops play a functional role in binding and activation of the brain receptor TrkB. The peptide BDNF (Fig. 11.1) displays a net positive charge and therefore can be efficiently entrapped in anionic lipid carriers.

11.2.2 Current Problems in Peptide Delivery with Lipid Nanocarriers—Nanosystems Obtained from Polar Lipids Hydrated in Aqueous Medium

Bioactive peptides, issued by advanced biotechnologies or biochemical synthesis, are of increasing interest for therapeutic applications. However, the use of peptides and proteins in medicine is hampered by their minimal permeability and passage through biological barriers, short half-life in the systemic circulation, and low bioavailability resulting from degradation due to either proteolysis or hydrolysis. With sterically stabilized lipid nanovector systems, the challenges for protein delivery, imposed by enzymatic degradation, early elimination from the circulation, capturing by the reticuloendothelial system, accumulation at nontargeted organs or tissues, and risk of immune response could be progressively surmounted.

Lipid nanoparticulate carrier systems of therapeutic biomolecules can provide a reservoir for peptides with a high degree of protection against enzymatic degradation and other destructive factors. The supramolecular architecture of the lipid carriers constraints the direct contact of the peptide molecules with the circulation medium and can be designed and built up so as to withstand the destabilizing influence of the biological environment. Self-assembly offers additional advantages through the association of several kinds of active and matrix components in one liquid crystalline nanocarrier object. The release properties and the biocompatibility of the self-assembled lipid nanocarriers with cells and tissues can be specifically optimized. All these advantages make the self-assembled nanodispersed lipid systems very interesting vectors for the administration of peptide and protein molecules.

Lipids of various types have been studied for the vectorization of peptides with different routes of administration (Angelova, 2005a,b, 2008; Jorgensen et al., 2006; Nylander et al., 1996; Rizwan et al., 2009; Shah et al., 2001). In the presence of an aqueous medium, amphiphilic lipids self-organize so as to minimize the exposure of their nonpolar chains to the surrounding water phase, whereas their polar head groups remain in contact with the solution (Drummond and Fong, 1999). Hydrated lipids may adopt different structural organizations, for instance, lamellar, inverted cubic, inverted hexagonal, and sponge phases (Fig. 11.2).

The arrangement of amphiphilic lipids in aggregates is governed by intermolecular hydrophobic and electrostatic interactions, temperature, degree of hydration, and the geometry of the molecules. The supramolecular packing is determined through the balance between the attractive interactions of the hydrophobic residues, on one hand, and the repulsive interactions between the polar head groups, on the other hand. The "critical packing parameter" is defined as $p = v/al$, where v and l is the volume and length of the apolar part of the amphiphile, respectively, and a corresponds to the area occupied by the lipid molecule at the hydrophobic–hydrophilic interface (Israelachvili et al., 1976). The steric packing constrains give indications for predicting the aggregation state of the lipids in aqueous medium. Depending on the critical packing parameter, one may distinguish globular micelles ($p < \frac{1}{3}$), cylindrical geometries such as elongated micelles or worms in a hexagonal phase ($\frac{1}{3} < p < \frac{1}{2}$), quasi-plane bilayers in vesicles or multilamellar phases ($p \sim 1$), inverted hexagonal ($p > 1$) and bicontinuous cubic membranes ($p > 1$), or sponge phase

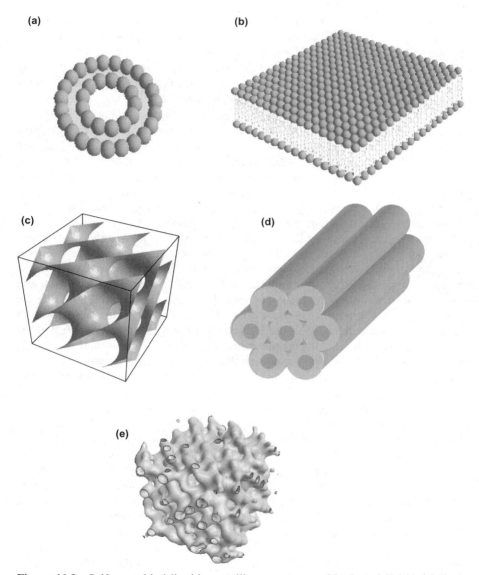

Figure 11.2 Self-assembled liquid crystalline structures of hydrated lipids: (a) liposome, (b) lamellar phase, (c) bicontinuous cubic phase, (d) inverted hexagonal phase, and (e) sponge phase.

structures ($p > 1$). The hydrated lipids display structural properties that are intermediate between those of disordered liquid states and well-ordered crystalline solid states. They are referred to as liquid crystalline phases or mesophases (Fig. 11.2). By their particular architecture and capacity of incorporating guest substances, liquid crystalline phases of lipids represent a source of nanostructures for vectorization of peptides and proteins.

11.2.2.1 *Vesicles and Liposomes*
Among the known lipid carriers, vesicles and liposomes are the most widely employed in drug delivery (Gabizon et al., 1994; Kawasaki et al., 2002; Klibanov et al., 1990; Martina et al., 2008) and have characteristics suitable for peptide and protein encapsulation. Such nanoparticles can be obtained from hydrated phospholipids or lipid mixtures forming lamellar phases (Fig. 11.2b). They are biocompatible, biologically inert, and display little toxicity and antigenic reactions. Liposomes are closed bilayer structures, composed of one or more phospholipid membranes, enclosing an aqueous volume (Fig. 11.2a). Lipophilic substances can be incorporated into the lipid bilayer, whereas hydrophilic compounds can be entrapped in the aqueous core. The aqueous compartments are used for the encapsulation and protection of soluble peptides or proteins (Walde and Ichikawa, 2001). There are numerous techniques for vesicle and liposome preparation, but some of them are well-suited for manipulations compatible with the maintaining of the integrity of the protein or peptide molecules (Table 11.2). A major obstacle for the development of liposomal formulations has been imposed by aggregation phenomena and a risk of degradation during storage due to physicochemical instability (Torchilin, 2005).

Several studies have been conducted on encapsulation of peptides or proteins in liposomes (Gregoriadis et al., 1999; Martins et al., 2007). Liposomes are often produced by hydration of a lipid film followed by energy input, for instance, sonication. Upon mechanical agitation, the hydrated lipid sheets disconnect and self-assemble to form vesicles. The bilayer structure is maintained by the hydrophobic effect that inhibits the interaction between the lipid hydrocarbon chains and water. The morphology of the vesicles is influenced by the chemical composition and the charge of the lipids, which in some cases can be controlled by pH (Kawasaki et al., 2002; Oberdisse et al., 1996). The diameter of a lipid vesicle can vary from 20 nm to several hundred micrometers (Walde and Ichikawa, 2001). The release of active biomolecules, nanoparticle (NP) stability, and the in vivo biodistribution are determined by the vesicles surface charge, stealth properties, size, and membrane fluidity (Torchilin, 2005). The membrane permeability can be modulated by variations in the lipid bilayer composition, phospholipids nature, inclusion of additives such as cholesterol or dioleoylphosphatidylethanolamine, DOPE, as examples (Gregoriadis et al., 1999). PEGylated derivatives can be incorporated in the lipid bilayer in order to avoid a premature capture of the liposomes by the reticuloendothelial system (Gabizon et al., 1994; Klibanov et al., 1990). The development of sterically stabilized NPs (stealth liposomes) can be further

TABLE 11.2   Examples of Liposome Preparation Compatible with Peptide and Protein Encapsulation

Peptide/Protein Embedded in Lipid Carriers	Liposome Fabrication Method	Biomolecule-to-Vector Association Efficiency (%)
Adamantyltripeptides	Dry lipid hydration	—
Antiovalbumin antibodies	Dry lipid hydration	—
Basic fibroblast growth factor	Freeze–thawing extrusion	75–80
Bovine serum albumin (BSA)	Reversal evaporation	25–71
	Double emulsification	43–71
	Freeze–thawing	20–45
Calcitonin	Dry lipid hydration	20
Enkephalin	Double emulsification	50–85
Epidermal growth factor receptor	Freeze–thawing extrusion	20–30
Hemoglobin	Dry lipid hydration extrusion	37–62
Horseradish peroxidase	Extrusion	2–5
Human γ-globulin	Dehydration–rehydration	30–31
Insulin	Reverse phase evaporation, freeze–thawing	30–82
Leishmania antigen	Freeze–thawing extrusion	—
Leridistim	Double emulsification	≥70
Leuprolide	Dry lipid hydration, reverse phase evaporation	37–76
		33–72
Nerve growth factor	Reverse phase evaporation	24–34
Octreotide	Double emulsification	50–85
Progenipoietin	Double emulsification	80–90
Superoxide dismutase	Dry lipid hydration	1–13
	Dehydration–rehydration	2–3
	Pro-liposome	39–65
TAT peptide	Extrusion	—

Source: Adapted from Martins et al. (2007).

extended to targeting strategies employing a conjugation with specific monoclonal antibodies (Martin and Papahadjopoulos, 1982).

11.2.2.2 Nanostructured Lipid Carriers Obtained from Nonlamellar Liquid Crystalline Phases   Mariani et al. (1988) and Luzzati (1997) have stressed the rich polymorphism of lipids, which can self-assemble into either lamellar or nonlamellar phases including different inverse bicontinuous cubic phases (Q^{II}), an inverse hexagonal (H_{II}) phase, a sponge (L_3) phase, reverse micellar cubic phases, and so forth. By analogy with the liposomal dispersions, prepared from a lamellar phase in excess water, these liquid crystalline phases can be fragmented, with the help of dispersing agents under agitation, to form nanosized colloidal-type amphiphilic aggregates. The internal structures of the obtained lipid nanoparticles may adopt various kinds of supramolecular organizations (Larsson, 2000; Larsson and Tieberg, 2005; Seddon and Templer, 1995), which can accommodate peptides either in the aqueous channel compartments, in the lipid bilayers, or at the lipid–water interfaces. In cubic phases (Fig. 11.2c), the created three-dimensional lipid membrane surface area essentially exceeds that of a liposome interface. Hence, the capacity for incorporation of guest molecules in nonlamellar, multicompartment-type lipid carriers is considerably enhanced (Angelova et al., 2003; Barauskas et al., 2005; Caboi et al., 2001; Shah et al., 2001).

In the majority of studies on nonlamellar liquid crystalline nanoparticle preparation, monoolein (MO) has been used as a biodegradable, biocompatible, and nontoxic lipid, which is listed in the Food and Drug Administration (FDA) regulation as an inactive component. MO has a low solubility in water, but it easily swells in contact with an aqueous phase and, therefore, exhibits a lyotropic behavior (Hyde et al., 1984; Larsson, 1989; Luzzati, 1997). Its phase diagram is well characterized (Qiu and Caffrey, 2000). The long-chain unsaturated monoglyceride forms liquid crystalline phases upon hydration, in particular inverse bicontinuous cubic phases of the Ia3d (Q^{230}) and Pn3m (Q^{224}) types (Angelov et al., 2006, 2007, 2009; Mariani et al., 1988; Qiu and Caffrey, 2000). In mixed systems with soluble polymeric surfactant, another cubic liquid crystalline phase of the Im3m (Q^{229}) symmetry has been described (Gustafsson et al., 1997; Larsson, 2000; Nakano et al., 2001).

Cubosomes Cubosomes are multicompartment lipid particles with the internal structure of an inverted bicontinuous cubic liquid crystalline type (Angelov et al., 2006; Gustafsson et al., 1996; Nakano et al., 2001; Siekmann et al., 2002; Spicer and Hayden, 2001). The cubic lipid membrane is periodically organized and divides intertwined networks of aqueous channels (Fig. 11.3). The structural properties of the cubosomes characterize them as multicompartment vehicles for the delivery of peptides, proteins, and other molecules (Angelova et al., 2003, 2005a,b, 2008, 2011a; Caboi et al., 2001; Nylander et al., 1996, Rizwan et al., 2009). The cubosome structure offers the possibility to entrap hydrophilic substances in the water channels and to control the biomolecular release by a slow diffusion process (Anderson and Wennerström, 1990; Boyd, 2003; Fong et al., 2009). Modification of the charges in the liquid crystalline phase can influence the kinetics of release (Lynch et al., 2003). Toward parenteral administration, the colloidal dispersions of cubosome particles can be stabilized by the inclusion of amphiphilic polymers (Gustafsson et al., 1997; Murgia et al., 2010; Nakano et al., 2001; Spicer and Hayden, 2001). Investigations with rats on the interaction of plasma components with cubosomes, composed of monoolein and Pluronic F-127, have shown a prolonged circulation (Leesajakul et al., 2004).

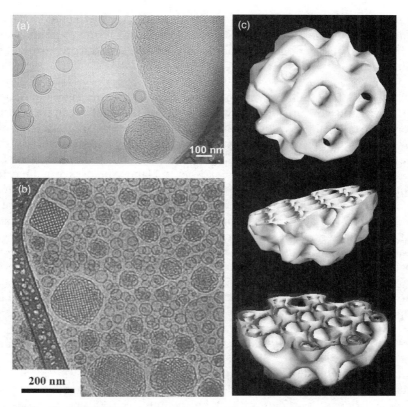

Figure 11.3 Cryo-TEM images of cubic-phase liquid crystalline particles, cubosomes, in (a) MO/OA (70/30 mol/mol) dispersion [adapted from Ferreira et al. (2006)] and in (b) MO/Poloxamer 407 (92/8 wt/wt) dispersion [adapted from Spicer and Hayden, (2001)]. The cubosomes appear to coexist with (a) particles with transient structures or with (b) vesicles. (c) Cubosome particle with a diamond-type internal cubic structure. The cross sections of the particle show a network of aqueous nanochannels [adapted from Angelova et al. (2005b)].

A number of molecules have been incorporated in nanostructured cubic-phase carriers (Angelova et al., 2011a) and tested for intravenous, nasal, and intragastric administration (e.g., insulin, somatostatin, and cyclosporine) (Drummond and Fong, 1999). The liquid crystalline phase has been found to enhance the stability of encapsulated active molecules such as enzymes or coenzymes (Larsson, 2000; Nylander et al. 1996). However, the solubilized guest molecules can modify the internal cubic lattice structure of the cubosomes. A new category of particles, "proteocubosomes," obtained by entrapment of proteins in lipid cubic lattice networks, has been established (Angelova et al., 2005b, 2011a). The proteocubosomes appeared to be built up by the assembly of porous nanodroplets ("nanocubosomes") with diameters between 30 and 60 nm. Large protein molecules have been suggested to be confined at the interfaces between the nanocubosomes inside the multicompartment proteocubosome particles.

Hexosomes Depending on temperature and water composition, monoglyceride lipids and phytantriol, in mixtures with coamphiphiles, can form also an inverted hexagonal phase (Angelov et al., 2011b; Dong et al., 2006; Fong et al., 2009; Qiu and Caffrey, 2000; Rizwan et al., 2009). By adding surfactants and supplying mechanical energy, this liquid crystalline phase can be dispersed in excess aqueous medium to form hexosomes. Nanostructures, obtained from an inverted hexagonal phase (Fig. 11.2d), have been reported as vehicles for the delivery of proteins and peptides. In particular, the encapsulation of lysozyme has received attention as a model protein (Mishraki et al., 2010; Zabara et al., 2011). The preparation of aqueous colloidal dispersions from a reverse hexagonal phase has been evidenced by electron microscopy (Ferreira et al., 2006; Gustafsson et al., 1997). The hexosome nanoparticles, formed by self-assembly of monoolein and oleic acid (Ferreira et al., 2006), have found application as nanodispersions for local delivery of the peptide cyclosporine A (Lopes et al., 2006).

Sponge Phase and Spongosomes The sponge (L_3) phase has been regarded as a bicontinuous bilayer phase (Fig. 11.2e) that is a disordered version of bicontinuous cubic lipid phases (Bender et al., 2008; Merclin et al., 2004; Wadsten et al., 2006). Landh and Larsson (1993) have patented the use of colloidal L_3-phase particles with stabilized interfaces. Sponge phases have been widely investigated for their swelling properties and drug encapsulation (Alfons and Engstrom, 1998; Angelov et al., 2009, 2011b; Merclin et al., 2004). To our knowledge, no study has described yet the peptide or protein release from colloidal L_3-phase nanoparticles (spongosomes). Such particles could coexist with cubosomes, exemplifying the possibility of developing self-assembled colloidal systems with multiphase release characteristics (Barauskas et al., 2006). The manufacturing of thermodynamically stable vesicles from a sponge phase also presents interest for future pharmaceutical applications (Imura et al., 2005).

11.2.2.3 Other Lipid Systems for Peptide and Protein Vectorization

Solid lipid nanoparticles (Almeida and Sauto, 2007; Martins et al., 2007; Mehner and Mäder, 2001) have been studied as carriers for proteins and peptides such as insulin or somatostatin. Nanoemulsions, microemulsions, and emulsions have been used as vehicles for ovokinine, which is a vasodilator peptide (Fujita et al., 1995; Jorgensen et al., 2006; Martins et al., 2007). Other vectors, called "layersomes," which are vesicles surface coated by polymer multilayers (Ciobanu et al., 2007), as well as solid triglyceride nanoparticles, have served for the encapsulation of calcitonin (Garcia-Fuentes et al., 2005).

The formulations of solid lipid nanoparticles have generally been used for the administration of peptides by the oral route. The advantages of emulsified microemulsions (EMEs) (Yaghmur and Glatter, 2009) for peptide delivery remain to be explored.

11.2.3   Human Cellular Model for Expression of Brain Receptors

Cell cultures are considered as a good model to study the activity of therapeutic molecules prior to preclinical trials. A human cell line expressing the tyrosine kinase receptor TrkB was selected as a model for in vitro screening of the ligand binding to brain receptors. The SH-SY5Y cell line has been previously employed in studies of neuroprotectants, antidepressants, and anti-Parkinson's, as well as for the expression of the amyloid precursor protein, which is the main constituent of the plaques associated with Alzheimer's disorders (Donnici et al., 2008; Presgraves et al., 2004; Ruiz-Leon and Pascual, 2003).

Despite the fact that they are not identical to mature neurons in the human brain, the differentiated SH-SY5Y cells are quite suitable to study the interactions of therapeutic formulations with neurotrophic pathways (see Fig. 11.15). Their advantage over primary cell cultures is that they provide an almost infinite amount of human cells with the same biochemical characteristics and do not require the use of antimitotics as with other cellular models. Owing to their dependence on trophic BDNF after treatment, the SH-SY5Y cells, presenting TrkB receptor populations, permit to study the mechanisms of neuroprotection (Kaplan et al., 1993; Serres and Carney, 2006). Therefore, this model system appears valuable for exploring the therapeutic potential of neuroprotectants in neurodegenerative disorders. A protocol to quantify neuritogenesis and perform automated image capture with these cells has been set up by the method of fluorescent immunocytochemistry (Simpson et al., 2001). It may allow screening of small-molecule mimetics of neurotrophic growth factors.

For SH-SY5Y cells, a protocol of differentiation, based on a sequential treatment with retinoic acid and BDNF, has been proposed (Cernianu et al., 2008; Encinas et al., 2000; Nishida et al., 2008). It yields a homogeneous population of differentiated neuronal cells expressing the TrkB receptor with synchronized cell cycles. The treated neuroblastoma cells are strictly dependent on BDNF for survival. When the short- and long-term treatment with BDNF is interrupted, the cells enter in a process of programmed cell death with apoptosis patterns. This treatment allows the growth of neurites, forming extensive networks. The stability of the cell culture is ensured for at least 3 weeks in the presence of BDNF (Encinas et al., 2000). Cernianu et al. (2008) have reported latent cells after treatment with retinoic acid, which has caused an arrest of the cell proliferation and maintenance of the viability even after 10 days of remedy. In addition, the treatment with an omega-3 polyunsaturated fatty acid (EPA) for 3 days was found to increase the cellular viability and the expression of neurotrophin (TrkB) receptors on the SH-SY5Y cells, differentiated by retinoic acid and BDNF (Kou et al., 2008). The role of the omega-3 fatty acid in the presence of BDNF has been explained for its effects on the synaptic plasticity and cognition (Kaplan et al., 1993; Matsumoto et al., 1995). Polyunsaturated fatty acids have shown efficiency in inhibiting neuronal apoptosis (Kim et al., 2000).

In the neuronal phenotype, differences have been discovered between two subclones of the SH-SY5Y cells induced by differentiation: (i) the SH-SY5Y-E cells, which differentiate upon alternative treatment by retinoic acid and BDNF, and (ii) SH-SY5Y-A cells, which differentiate easier via the formation of a neuronal network after 5 days of treatment with retinoic acid. An increased neuronal phenotype has been found when a sequential treatment with retinoic acid and BDNF has been conducted over 3 days (Nishida et al., 2008).

It should be emphasized that the treatment of the human neuroblastoma cell line SH-SY5Y by retinoic acid allows its differentiation by slowing down or stopping the cellular proliferation mechanism. As a result, the characteristic properties of the human neuronal cells, when they are affected by neurodegeneration, are reproduced in the chosen cellular model. BDNF has been evidenced to stimulate the neuronal growth of differentiated human neuroblastoma cells. Thus, the SH-SY5Y cell line, treated with retinoic acid, represents an appropriate cellular model to evaluate in vitro the activity of neurotrophic factors.

11.3 EXPERIMENTAL SECTION

11.3.1 Preparation of Lipid Nanoparticle Dispersions

Toward the objective to conceive anionic, multicompartment liquid crystalline type of nanocarriers, constituted by the hydrated monoglyceride MO as the main cubic-phase forming lipid, we employed the self-assembly method. Oleic acid (OA) and EPA, which is an omega-3 polyunsaturated fatty acid (20:5), were added to MO, in various proportions, in order to functionalize the cubic-phase nanovector. A PEGylated phospholipid was also included in the self-assembled lipid mixture in an attempt to generate sterically stabilized multicompartment nanoparticles.

Monoolein (1-oleyl-*rac*-glycerol), oleic acid (OA), *cis*-5,8,11,14,17 eicosapentaenoic acid (20:5, EPA), dimyristoylphosphatidylcholine (DMPC), and retinoic acid were purchased from Sigma-Aldrich (St. Quentin, France). The 1,2-dioleyl-*sn*-glycero-3-phosphoethanolamine-N-[methoxy (polyethylene glycol)-2000] (DOPE-PEG$_{2000}$) was a product of Avanti Polar Lipids. Recombinant human BDNF and a primary monoclonal antihuman TrkB antibody (MAB397) were purchased from R&D Systems (Lille, France). The affinity-purified fluorescent secondary antibody Alexa Fluor 633 goat antimouse IgG$_{2b}$ (A21146) (2 mg/ml) was purchased from Invitrogen (Cergy Pontoise, France). Mono- and dibasic sodium phosphate salts were dissolved in MilliQ water (Millipore Co., Molsheim, France) to prepare the aqueous buffer phase, which was sterilized.

Chloroform solutions of MO, OA, EPA, and DOPE-PEG$_{2000}$ were prepared at desired concentrations by weighing the lipid powders or oils; the addition of solvent (chloroform) was done using a Hamilton syringe. After mixing, the solvent was evaporated from the MO/DOPE-PEG$_{2000}$/OA/EPA systems under a gentle stream of nitrogen gas to form fine and homogeneous lipid films, which were lyophilized overnight and stored at 4°C. The hydration of the mixed lipid layers was performed by adding a phosphate buffer. Nanoparticulate dispersions were obtained by vortexing and ultrasonic irradiation in an ice bath for less than 20 min (Branson 2510 ultrasonic bath, "set sonics" mode, power 100 W). Filtration through a 0.22-μm filter (Minisart High Flow, Sartorius, Palaiseau, France) was then carried out under a laminar flow hood in order to sterilize the samples, which were stored in glass vials at room temperature before the physicochemical and cellular tests.

11.3.2 Characterization of Lipid Nanocarriers

11.3.2.1 Quasi-Elastic Light Scattering (QELS) and Optical Density Measurements
The hydrodynamic diameter of the generated lipid nanoobjects was determined using a Nanosizer apparatus (Nano-ZS90, Malvern, Orsay, France) at a scattering angle of 90° and at 25°C. The average hydrodynamic diameter, d_h, was calculated considering the mean translational diffusion coefficient, D, of the particles in accordance with the Stokes–Einstein law for spherical particles in the absence of interactions: $d_h = k_B T/3\eta\pi D$, where k_B is the Boltzmann constant, T is temperature, and η is the viscosity of the aqueous medium.

The size measurements were performed after dilution of 150 μl of a lipid NP sample in 1 ml of phosphate buffer or 0.2 ml of sample in 2 ml of cell culture medium. The results were analyzed using the MALVERN Zetasizer software (version 6.11). The particle distributions were expressed via the "volume" analysis mode by averaging three measurements with the same sample. Optical density (OD) measurements were conducted on a double-beam ultraviolet (UV)/visible spectrophotometer PerkinElmer Lambda 2 (PerkinElmer, Courtaboeuf, France) at 25°C.

11.3.2.2 Small-Angle X-ray Scattering (SAXS)
The small-angle X-ray scattering measurements with the MO/DOPE-PEG$_{2000}$/EPA (83/2/15 mol %) and MO/DOPE-PEG$_{2000}$ (98/2 mol %) samples were performed at the ID22 beam line of the Diamond Light Source (Didcot, UK). The investigated lipid dispersions were liquid solutions of nanoparticles that were filled in borosilicate glass capillaries (diameter 1.5 mm, 10-μm wall thickness). The samples were mounted in a holder in air and at room temperature (21°C), without a special vacuum chamber. The detector at the I22 station was a gas wire RAPID 2D detector (www.diamond.ac.uk/Home/Beamlines/I22/tech/detectors.html),

the area of which was divided into 512×512 pixels. The wavelength of the incident X-ray beam was 1.55 Å during the experiments and the beam size at sample was 320×70 μm². The sample to SAXS detector distance was 1314 mm. The accessible q range was from 0.009 to 0.417 Å⁻¹. The X-ray flux (~10^{12} photons/s) was strong enough to allow direct sample exposures. During the measurements, a metal attenuator was installed in order to prevent the liquid crystalline nanoparticle samples from X-ray damages. The exposure time was about 1 s. Silver behenate was used for the q-range calibration and glassy carbon for the intensity normalization. After the data acquisition, the obtained two-dimensional (2D) images were integrated into one-dimensional (1D) scattering curves by means of the Fit2D software (www.esrf.eu/computing/scientific/FIT2D/). The backgrounds coming from the glass capillary and the solvent were measured and subtracted using conventional procedures.

11.3.3 Cell Culture Assays

11.3.3.1 Treatment and Differentiation of Human Neuroblastoma SH-SY5Y Cells
The neuroblastoma cell line SH-SY5Y was kindly provided by Dr. Jean-Marc Muller of the Institute of Physiology and Cell Biology (CNRS UMR 6187, Poitiers, France). The cells were grown in culture medium DMEM (Dulbecco's modified Eagle's medium), provided by Lonza (Belgium), and supplemented with 1% pyruvate, 1% streptomycin (PennStrep), and 10% fetal calf serum, after decomplementation by heat (all percentages are expressed in vol/vol). The cell cultures were maintained at 37°C in a saturated humid atmosphere containing 95% air and 5% CO_2. Cell divisions were made when the cells arrived at confluence, using trypsin for detachment. The doubling time of the investigated neuroblastoma cells was variable (between 2 and 14 days), and therefore the confluence of the cells had to be regularly monitored.

For the differentiation of the neuroblastomas, the cells were seeded, at an initial density of 10^4 cells/cm², in coated sterile plates. When they were estimated to reach 80% confluence, retinoic acid was added to the culture medium, at a concentration of 10 μM, by dilution from a stock ethanol solution with concentration of 10^{-3} M. After 5 days of incubation in the presence of retinoic acid, the cells were washed three times with DMEM. Then, they were incubated for 3 days, in the absence or presence of BDNF (2 ng/ml), with added monoolein-based lipid NP systems (containing EPA or OA), which were prediluted in DMEM at different lipid concentrations.

11.3.3.2 Confocal Fluorescence Microscopy Imaging of TrkB Receptor Expression in Differentiated Neuroblastoma SH-SY5Y Cells
The cells were seeded on glass slides (0.17 mm thick and 12 mm in a diameter) in 24-well plates sterilized in autoclave. When the cells were estimated to be at 80% confluence, a treatment with retinoic acid was effected for 5 days, and then the cells were fixed a few hours before the confocal microscopy imaging. The fixation was performed by removal of the medium, addition of 4% paraformaldehyde for 10 min, followed by saturation of the latter with 50 mM

NH$_4$Cl for 10 min. Three rinses were finally made with phosphate-buffered saline (PBS) (provided by Lonza, Belgium). The nonspecific sites were blocked using 3% BSA solution (Sigma Aldrich, France) for 30 min. Then, a primary antibody anti-TrkB was added (at a concentration of 10 μg/ml) to the fixed cells for 1 h in a humid chamber at 37°C. After three successive washes with a solution of 1% BSA, the cells were incubated with a secondary antibody (diluted at 1/500), in the dark for 45 min, and then washed three times with PBS. The recovered strips were mounted on microscope slides using a drop of Vectashield mounting liquid that preserves the fluorescence of the secondary antibody. The samples were sealed with varnish to prevent premature drying of the preparation and stored at 4°C in the dark.

The imaging was performed on the day of the sample preparation using a confocal microscope ZEISS (model LSM 510, Zeiss, Germany). The experiments were set with a fluorescence excitation wavelength of 633 nm. Images were recorded with fluorescence emission collected from 650 nm. Differential interference contrast (DIC), or Nomarski type, images were obtained in parallel. An oil objective 63×/1.4 was employed.

11.3.3.3 Cellular Viability Assessed by MTT Tests The cells viability was determined based on the activity of the mitochondrial dehydrogenase enzyme that cleaves the tetrazolium salt 3-[4,5-dimethylthiazol]-2,5-diphenyltetrazolium (MTT) into a formazan product. The cells were seeded in 96-well plates and grown to 80% confluence in culture medium. For differentiation, they were incubated for 5 days after adding, in each well, a volume of 200 μl retinoic acid solution (10 μM) prepared in culture medium. Two washes were then accomplished with serum-free DMEM before replacing it with DMEM (without serum) containing increasing concentrations of lipid formulations and/or BDNF (2 ng/ml). Every experimental condition was carried out on 8 wells that constituted one column of the 96-well plate. After 72 h of incubation, 20 μl of MTT (5 mg/ml), prepared in PBS, were added to each well, with the exception for the control wells. Two hours later, the medium was aspirated. The cells and the formazan crystals were solubilized by adding 200 μl of dimethyl sulfoxide (DMSO). The plate was stirred for a few minutes and the optical density was determined using a plate reader operated at a wavelength 570 nm. The results are presented in percentages of cell viability compared to controls.

11.4 PHYSICOCHEMICAL AND BIOLOGICAL INVESTIGATIONS OF NANOPARTICULATE MIXED LIPID SYSTEMS

11.4.1 Nanoparticles Prepared from MO/DOPE-PEG$_{2000}$/OA/EPA Self-Assembled Lipid Mixtures

The first objective of the study was to determine the molar ratios between the investigated lipids favoring the formation of multicompartment nanoparticles

with liquid crystalline organizations, suitable for encapsulation of neuroprotective peptides. The fragmentation and dispersion of the bicontinuous cubic
phase of MO to form nanometer-sized objects was realized using the PEGylated
amphiphile DOPE-PEG$_{2000}$. The latter can self-associate in surfactant micelles
at concentrations above its critical micellar concentration (cmc ~ 2×10^{-5} M).
The molar fraction of DOPE-PEG$_{2000}$ in the lipid mixtures was set at 2 mol %
with respect to monoolein. A higher PEGylation percentage could essentially
influence the mean packing parameter of the lipid assembly and may favor
the formation of small vesicles rather than particles from a dispersed cubic
phase. Electrostatic stabilization of the multicompartment NPs was achieved
by varying the molar fractions of the charged lipid molecules. It was taken into
account that the fatty acids, OA and EPA, are negatively charged at pH 7.4.
Such co-lipids are anticipated to contribute to the formation of cubosomes
(Nakano et al., 2001).

11.4.1.1   *Characterization of Nanoparticulate Systems with Increasing Molar Fractions of the Charged Co-Lipid Oleic Acid (OA) in the Mixtures MO/DOPE-PEG$_{2000}$/OA*   Nanoparticulate MO/DOPE-PEG$_{2000}$/OA samples
with increasing molar fractions of oleic acid (0, 5, 10, 15, 20, and 30 mol %
OA) were studied by QELS (at dilution 150-µl sample/1-ml buffer) as a function of the time elapsed after preparation. The results for the particles size
distribution in freshly prepared samples (J0) as well as after 2 days (J2) of
particle storage at room temperature showed a coexistence of two major
populations of nanoobjects (Fig. 11.4). The corresponding mean NP diameters
were about 40 and 100 nm at low molar OA molar fractions and can be
ascribed to the presence of vesicle and multicompartment NPs in the aqueous
dispersions. The NP dispersions showed an overall stability with the time
elapsed after preparation. However, a tendency to form larger aggregates
(most probably cubosomes) can be observed as a function of the co-lipid molar
fraction (Fig. 11.4).

Other cryo-TEM (transmission electron microscopy) studies (Ferreira
et al., 2006; Spicer and Hayden, 2001) have demonstrated that cubosome particles, with internal structure reflecting the organization of the original cubic
phase of monoolein, can coexist with smaller (vesicle or transient) particles
even in the presence of 30 mol % OA (see Fig. 11.3). Formation of coexisting
vesicles and dispersed particles from the monoolein cubic phase has been
induced also by the co-surfactant poloxamer 407 (Gustafsson et al., 1997;
Spicer and Hayden, 2001).

11.4.1.2   *Stability of Dispersed NP Systems upon Dilution*   The stabil-
ity of the MO/DOPE-PEG$_{2000}$ (98/2 mol/mol) NP sample was studied by
QELS, on the seventh day after their preparation, upon dilutions of 1/10 and
1/1000 in buffer medium (Fig. 11.5). The volume distribution profile of the NPs
was kept almost unchanged upon dilution. The obtained data confirmed that
the lipid organization in this sample is not a micellar one. In fact, the sample

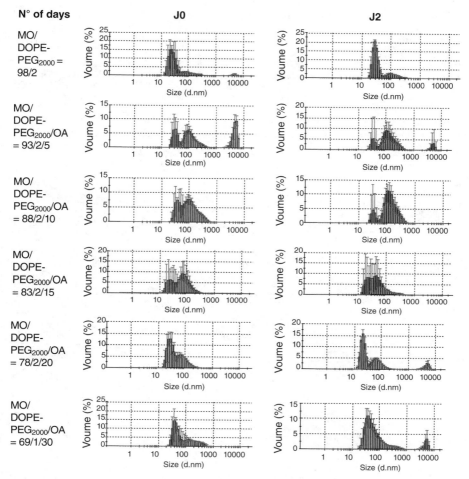

Figure 11.4 Stability over time of NP size distributions in lipid dispersions MO/
DOPE-PEG$_{2000}$/OA containing increasing molar fractions of oleic acid (0, 5, 10, 15, 20,
30 mol % OA). J0 denotes the day of the NP preparation, whereas J2 denotes 2 days
of NP storage at room temperature.

dilution beyond 1/30 should provoke the dissociation of eventually presenting
micelles in the system as the cmc of DOPE-PEG$_{2000}$ is about 2×10^{-5} M.
Micelles could not be measured by QELS upon dilution to lower DOPE-
PEG$_{2000}$ concentrations. Based on the determined characteristic sizes, the
results in Figure 11.5 should be attributed to the formation of PEGylated
multicompartment lipid particles. The aggregation of the MO/DOPE-PEG$_{2000}$
(98/2 mol/mol) NPs into larger liquid crystalline objects upon dilution is typical
for cubosome and spongosome organizations.

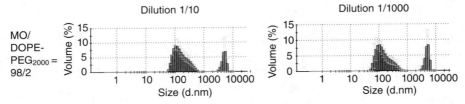

Figure 11.5 Stability of NP size (expressed in vol %) distribution in the lipid dispersion MO/DOPE-PEG$_{2000}$ (98/2 mol/mol) determined by QELS upon dilutions of 1/10 and 1/1000 in buffer.

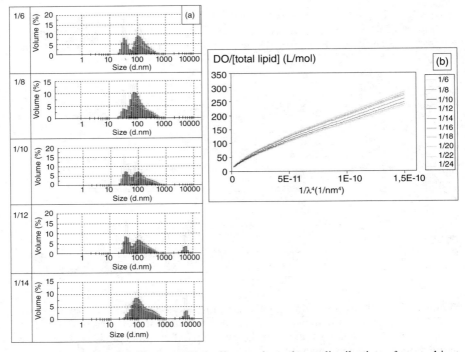

Figure 11.6 Effect of dilution with buffer on the volume distribution of nanoobject sizes in the MO/DOPE-PEG$_{2000}$/OA (88/2/10 mol %) dispersed sample established by (a) QELS and (b) by optical density (OD) measurements.

For the example of charged multicompartment lipid nanocarriers, QELS and OD measurements were performed with the MO/DOPE-PEG$_{2000}$/OA (88/2/10 mol %) sample on the fourth day after its preparation in order to study its NP size distribution stability over time and upon dilution (Fig. 11.6). The two types of NP populations remain in coexistence in the aqueous medium upon dilution. Aggregates of particles begin to grow at dilutions beyond 1/10. Indeed, the dilution can modify the distribution of DOPE-PEG$_{2000}$ between the lipid membrane and the aqueous buffer, leading to a certain destabilization of lipid NPs with regard to steric shielding. The diluted lipid objects show a preference of merging into multicompartment particles with larger sizes and forming a more stable (bulk) phase, which for the monoglyceride lipid MO is the bicontinuous cubic phase.

Figure 11.6a shows that the proportion of the small NPs (vesicles or cubosomal intermediates) with respect to the large particles (multicompartment cubosomes or spongosomes) changes with the dilution, but the two types of NP populations are maintained, upon dilution, through the electrostatic stabilization provided by the charged lipid OA. This conclusion is confirmed by the analysis of the OD curves determined at different concentrations (Fig. 11.6b). The scattering of light by lipid objects can be considered, in a first approximation, as proportional to the factor V^2/λ^4, where V is the volume of the particles and λ is the optical wavelength. The changes in the OD, normalized by the total lipid concentration, showed very similar profiles as a function of the dilution (Fig. 11.6b). This confirms the overall conservation of the lipid particle organizations despite the dilution. A closer examination of the curves in Figure 11.6b reveals an increase in the {OD/[total lipid]} factor when the sample concentration diminishes, which indicates a slight tendency of the lipid NP systems to reorganize into populations with a larger average particle size upon dilution. A fusion of small (vesicle) particles into larger multicompartment particles (cubosomes) may explain the observed aggregation upon dilution. This hypothesis is consistent with the cryo-TEM observations reported in the literature (Ferreira et al., 2006).

11.4.1.3 *Incorporation of a Bioactive Co-Lipid (EPA) in the PEGylated Liquid Crystalline Nanosystems*

The charged component (oleic acid) in the self-assembled NPs of MO/DOPE-PEG$_{2000}$/OA was progressively substituted by the bioactive, polyunsaturated fatty acid, EPA. This omega-3 fatty acid (20:5) appears to exhibit a high degree of conformational flexibility, which can affect the organization and curvature of the mixed lipid membranes. The total molar fraction of the charged lipids (i.e., the sum of OA and EPA) was fixed at 15 mol % with respect to the MO in the mixtures. It was found that this molar fraction results in mean NP sizes, which are optimal (~120 nm) for passage through membranes. Keeping constant the molar fraction (15 mol %) of the charged molecules in the lipid carriers, the NP size distribution was investigated as a function of the EPA content in the MO/DOPE-PEG$_{2000}$/OA/EPA systems. Figure 11.7 summarizes the QELS results about the time stability of the NP size distributions in MO/DOPE-PEG$_{2000}$/OA/EPA mixtures containing increasing amounts of a polyunsaturated fatty acid (EPA). The lipid nanodispersions prepared with higher molar fractions of EPA, with respect to OA, showed a more narrow size distribution after 4 days of storage (Fig. 11.7). Correspondingly, the NP system MO/DOPE-PEG$_{2000}$/EPA (83/2/15 mol %) showed a higher stability over time as compared to the system containing 15 mol % of OA only.

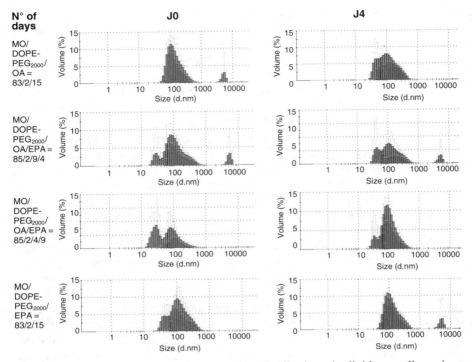

Figure 11.7 Time stability of the NP size distributions in lipid nanodispersions obtained from MO/DOPE-PEG$_{2000}$/OA/EPA mixtures determined by QELS on the day of the NP preparation (J0) and after 4 days (J4) of NP storage at room temperature.

11.4.1.4   Stability of the Dispersed MO/DOPE-PEG$_{2000}$/OA and MO/DOPE-PEG$_{2000}$/EPA Systems upon Dilution in DMEM Culture Medium   Particle size distribution stability in MO/DOPE-PEG$_{2000}$/OA (83/2/15 mol %) and MO/DOPE-PEG$_{2000}$/EPA (83/2/15 mol %) multicompartment nanocarrier systems was investigated upon dilution in serum-free culture medium (DMEM) (Fig. 11.8). The formulation containing OA showed excellent stability in the DMEM culture medium irrespective of the dilution. The EPA-containing formulations showed the reappearance of small nanoobjects upon dilution in DMEM. The population of larger, multicompartment lipid objects, supposed to be cubosomes, remained unchanged upon dilution in cell culture medium.

11.4.2   SAXS of Sterically Stabilized Nanodispersed Lipid Systems

Small-angle X-ray scattering curves of PEGylated liquid crystalline NP functionalized by 15 mol % of omega-3 polyunsaturated fatty acid (EPA)

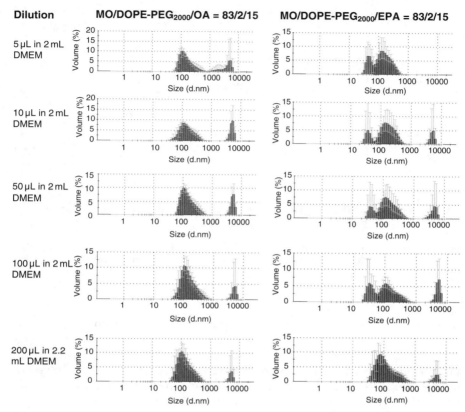

Figure 11.8 Stability of NP size (expressed in vol %) distributions in lipid dispersions MO/DOPE-PEG$_{2000}$/OA (83/2/15 mol %) and MO/DOPE-PEG$_{2000}$/EPA (83/2/15 mol %) determined by QELS upon sample dilutions in DMEM culture medium.

are presented in Figure 11.9. The scattering curves of the aqueous dispersions of MO/DOPE-PEG$_{2000}$/EPA (83/2/15 mol %) and MO/DOPE-PEG$_{2000}$ (98/2 mol %) self-assembled NP systems show similar structural features and trends. It can be concluded that the incorporation of 15 mol % of a bioactive component EPA does not dramatically influence the internal organization of the nanocarriers. The polyunsaturated fatty acid appears to mix homogeneously with the main monoglyceride component (MO) in the generated multicompartment lipid particles.

11.4.3 In Vitro Assays with Human Neuroblastoma SH-SY5Y Cell Line

11.4.3.1 *TrkB Receptor Expression in Differentiated Human Neuroblastoma Cells* The expression of the neurotrophin receptor TrkB,

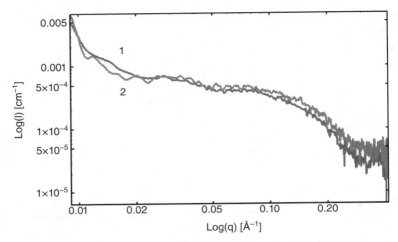

Figure 11.9 Small-angle X-ray scattering (SAXS) curves of diluted lipid nanocarrier systems: MO/DOPE-PEG$_{2000}$ (98/2 mol %) (curve 1) and MO/DOPE-PEG$_{2000}$/EPA (83/2/15 mol %) (curve 2).

following the SH-SY5Y cell differentiation by treatment with retinoic acid, was evidenced by confocal fluorescence microscopy (for the employed fluorescently labeled antibody, $\lambda_{ex} = 633$ nm and $\lambda_{em} > 650$ nm). The images were taken from SH-SY5Y cells treated and fixed on glass slides in order to visualize the expression of TrkB receptors by detecting the fluorescence of a secondary antibody that recognizes the anti-TrkB antibody, which was bound to the neurotrophin receptor TrkB. Controls were acquired in order to ensure the nonfluorescent character of the cells themselves and to check that the employed primary antibody was specific to the TrkB receptor. The untreated SH-SY5Y cells had no autofluorescence at the studied excitation wavelength (Fig. 11.10a). No fluorescence was detected when the cells were treated by retinoic acid and when the secondary antibody was incubated with the cells in the absence of a primary antibody (anti-TrkB). These results confirmed the antibodies specificity and the nonfluorescence of the cells treated with retinoic acid in the absence of a primary antibody (Fig. 11.10b). The unusual morphology of the fixed cells was explained by the drastic shock employed for their fixation on microscopic slides.

The expression of the TrkB receptors was established by superimposing the fluorescence images of the SH-SY5Y cells with those recorded in a Nomarski mode. The fluorescently labeled secondary antibody was proven to be specific to the primary anti-TrkB antibody as it did not bind to the cells in the lack of a primary antibody (that recognizes the expressed neurotrophin receptor). The confocal fluorescence images of cells, treated for 4 days with retinoic acid (10 µM) are shown in Figure 11.11. The cells exhibited fluorescence due to the Alexa Fluor 633 anti-IgG antibody, which binds to the antihuman TrkB mouse

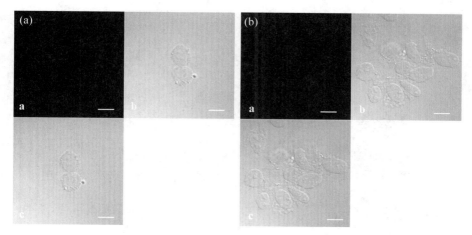

Figure 11.10 Confocal microscopy images (oil objective 63×/1.4) of cells fixed on glass slides in control experiments: (a) fluorescence image (zoom 2×), (b) Nomarski (DIC) image, and (c) superposition of the images (a) and (b). Images of untreated cells (A), and of cells treated with retinoic acid (10 μM) during 5 days and incubated with a secondary antibody in the absence of a primary antibody (B). Image size: 73.25 × 73.25 μm². The scale bar corresponds to 10 μm.

IgG (primary antibody). An overlay of the fluorescence and the Nomarski images of the cells was achieved. Thus, the SH-SY5Y cells were demonstrated to express the TrkB receptor of the neurotrophin BDNF upon treatment with retinoic acid (10 μM) for 4 days. The expression of the BDNF receptor could be demonstrated also by other methods such as the Western blot analysis of ribonucleic acid (RNA) of the protein (Kaplan et al., 1993, Encinas et al., 2000, Kou et al., 2008; Matsumoto et al., 1995).

11.4.3.2 Cellular Viability upon Treatment with Neurotrophic Peptide BDNF and Multicompartment Liquid Crystalline Lipid Nanocarriers, as Evidenced by MTT Tests

The viability of differentiated SH-SY5Y cells after 5 days of treatment with retinoic acid (10 μM) was estimated to be $44 \pm 6\%$ compared to untreated cells. The obtained result reveals an essential decrease in the cell proliferation in the absence of neuroprotective treatment. Therefore, the created model with differentiated neuroblastoma cells is substantially relevant to the study of neuroprotective strategies in view of potential therapies of neurodegenerative diseases.

MTT tests were accomplished with differentiated SH-SY5Y cells at decreasing concentrations of MO/DOPE-PEG$_{2000}$/OA (83/2/15 mol %), MO/DOPE-PEG$_{2000}$/EPA (83/2/15 mol %), and DMPC/DOPE-PEG$_{2000}$/EPA (83/2/15 mol %) lipid nanocarrier samples of analogous compositions. The initial number of cells for the MTT tests was 20,000 cells/well/200 μl. Figure

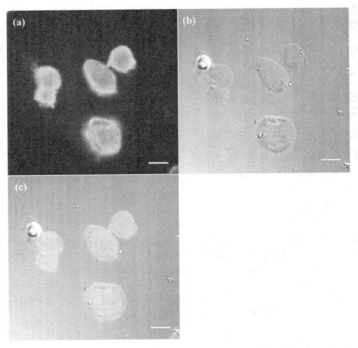

Figure 11.11 Confocal fluorescence microscopy images (oil objective 63×/1.4) of SH-SY5Y cells incubated with retinoic acid (10 μM) at 37°C under 5% CO_2 for 4 days and then fixed on glass slides: (a) Fluorescence image (zoom × 2), (b) Nomarski image, and (c) the superposition of the images (a) and (b). Image size: 73.25 × 73.25 μm². The scale bar corresponds to 10 μm.

11.12 shows the results for the relative viability of the cells treated with the peptide BDNF alone or with decreasing concentrations of the MO/DOPE-PEG_{2000}/EPA (83/2/15 mol %) nanodispersion in the presence or absence of BDNF. The control experiments corresponded to differentiated cells, which were not subjected to neuroprotective treatment. The viability of cells treated with BDNF alone was $108 \pm 6\%$ ($C_{lipid} = 0$) (Fig. 11.12). It evidenced that the neurotrophic factor BDNF essentially increases the survival of differentiated neuroblastoma cells, which exhibited initial cellular viability around 40%. Thus, BDNF was proven to act as a neuroprotectant that maintains cellular survival and promotes the recovery of the SH-SY5Y cells.

A maximal cellular viability of 123% was found for treatment with BDNF added to MO/DOPE-PEG_{2000}/EPA (83/2/15 mol %) nanocarrier formulations with a total lipid concentration of 4.21×10^{-7} M (i.e. 2×10^{-6} nmol lipid/cell). In the absence of BDNF, the cellular viability decreased to 90% for the same formulation at equal concentration of the neurotrophin potentiating agent

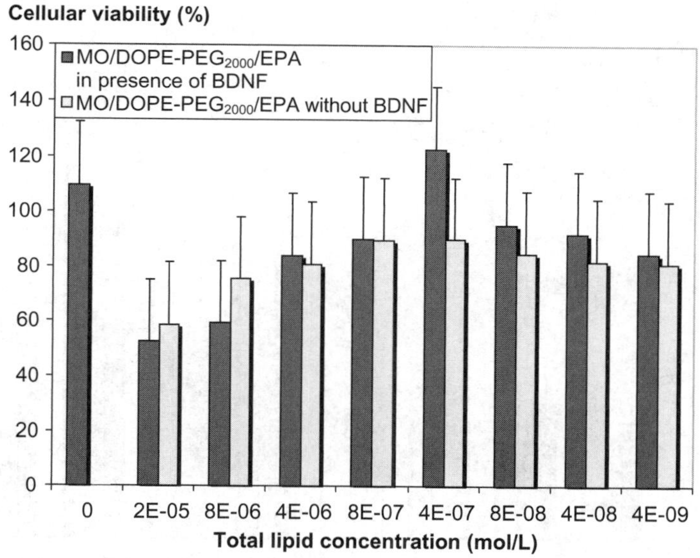

Figure 11.12 Viability of differentiated SH-SY5Y cells treated with BDNF alone or by MO/DOPE-PEG$_{2000}$/EPA (83/2/15 mol %) NPs in the absence or in the presence of BDNF (2 ng/ml or 4×10^{-8} ng/cell). The MTT test was conducted at 37°C. The histograms are presented in a sequence of decreasing total lipid concentrations and expressed relative to the results obtained with the differentiated cells without neuroprotective treatment.

(EPA). Hence, the bioactive component EPA was evidenced to potentiate the effects of BDNF under the investigated conditions.

The relative cellular viability, determined in the presence of BDNF for the formulation MO/DOPE-PEG$_{2000}$/EPA (83/2/15 mol %) is compared in Figure 11.13 to the viability of cells treated with MO/DOPE-PEG$_{2000}$/OA (83/2/15 mol %) and DMPC/DOPE-PEG$_{2000}$/EPA (83/2/15 mol %) formulations. In control experiments, the differentiated SH-SY5Y cells were treated with BDNF only. It was established that cells treated with the MO/DOPE-PEG$_{2000}$/OA (83/2/15 mol %) formulation present a lower cell viability compared to those treated with MO/DOPE-PEG$_{2000}$/EPA (83/2/15 mol %). This confirms that the activity of BDNF is potentiated by the EPA and not by other lipids of the nanocarriers. The maximum cellular viability of 150% was achieved with the DMPC/DOPE-PEG$_{2000}$/EPA (83/2/15 mol %) NPs formulation at a concentration of 4.21×10^{-8} M. The phospholipid DMPC appears to participate in a less significant interaction with the cell membranes and thus displays a lower cytotoxicity compared to monoolein at the same lipid concentrations. However, the peptide encapsulation efficiency in phospholipid nanostructures could also be lower.

Cellular viability (%)

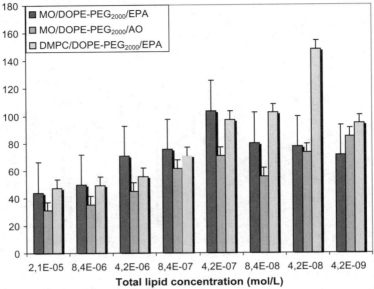

Figure 11.13 Viability of differentiated SH-SY5Y cells treated with MO/DOPE-PEG$_{2000}$/EPA (83/2/15 mol %), MO/DOPE-PEG$_{2000}$/OA (83/2/15 mol %) and DMPC/DOPE-PEG$_{2000}$/EPA (83/2/15 mol %) in the presence of BDNF (2 ng/ml or 4×10^{-8} ng/cell) assessed by MTT test at 37°C. The histograms are presented in a sequence of decreasing total lipid concentrations and expressed relative to the results obtained for treatment of differentiated cells with BDNF alone.

The cell viability, established for the MO/DOPE-PEG$_{2000}$/EPA (83/2/15 mol %) and DMPC/DOPE-PEG$_{2000}$/EPA (83/2/15 mol %) nanodispersions, was compared to that determined for the ethanol solution of EPA (Fig. 11.14). The EPA solution considerably enhances the effects of BDNF and nearly doubles the overall cellular viability of the differentiated cells, except for 10 µM concentration, which leads to about 50% decrease in the BDNF-mediated cellular viability. These results are in correlation with those for differentiated cells treated with EPA and 1 ng/ml BDNF (Kou et al., 2008). The comparison suggests that the neuroprotection phenomenon is dependent on the concentration of BDNF and that the cellular viability, enhanced by EPA, should be greater in the presence of 2 ng/ml BDNF. The two lipid systems, containing 15 mol % EPA, do not appear toxic at low concentrations of EPA up to 6×10^{-7} M. Beyond this concentration, the NP formulations induce 50% cell toxicity in the investigated model.

The morphology of alive, unfixed SH-SY5Y cells was studied by phase contrast microscopy. Images were acquired for different treatments as indicated in Figure 11.15. A change in the morphology (more elongated cells) and

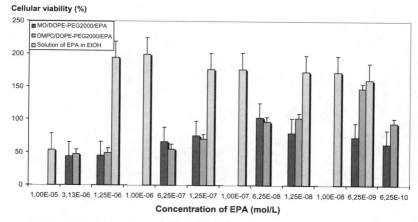

Figure 11.14 Viability of differentiated SH-SY5Y cells treated with MO/DOPE-PEG$_{2000}$/EPA (83/2/15 mol %), DMPC/DOPE-PEG$_{2000}$/EPA (83/2/15 mol %) and an ethanol solution of EPA in the presence of BDNF (2 ng/ml or 4×10^{-8} ng/cell) evaluated by MTT test at 37°C. The histograms are presented in a sequence of decreasing concentrations of EPA and expressed relative to the results obtained for treatment of differentiated cells with BDNF only.

formation of neurites was established after treatment with retinoic acid at a concentration of 10 μM during 1 and 5 days (Fig. 11.15b and 11.15c). The cells, treated with BDNF, seem to have more rounded shapes, approaching those of undifferentiated SH-SY5Y cells (Fig. 11.15d). Very long and branched neurites were observed when the neuroblastoma cells were treated with functionalized lipid nanocarrier systems as well as with BDNF alone. The latter finding is consistent with the literature reports (Matsumoto et al., 1995). The treatments with lipid nanocarriers at a concentration of 4×10^{-7} M do not seem to affect the cells morphology (Figs. 11.15e, 11.15f, and 11.15g), which confirms the nontoxic character of the lipid formulations as demonstrated by the performed MTT tests.

11.5 CONCLUSION

The investigated mixed lipid liquid crystalline NP systems were found to involve two populations of nanoobjects that were interpreted as coexisting small vesicles and multicompartment particles (cubosomes). Structural transitions of the cubosomes were favored at increasing molar fractions of the negatively charged fatty acid components. The formulations MO/DOPE-PEG$_{2000}$/OA/EPA appeared to be more stable on storage and on dilution when they were functionalized by 15 mol % of EPA. In vitro tests under confocal fluorescent microscope showed the differentiation of the neuroblastoma

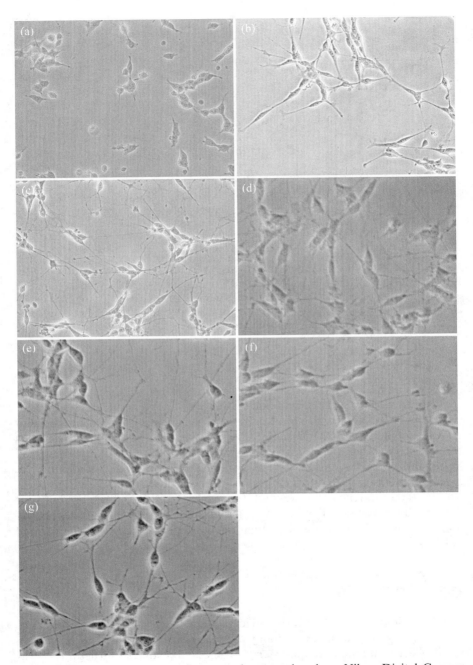

Figure 11.15 Phase contrast microscopy images taken by a Nikon Digital Camera DXM1200C with an objective of 40×/0.60. The SH-SY5Y cells are alive in their culture medium (a), differentiated by treatment with retinoic acid (10 μM) during 1 day (b) or during 5 days (c), and then treated, after 5 days of differentiation, by 2 ng/ml BDNF alone (d) or with BDNF and MO/DOPE-PEG$_{2000}$/EPA (83/2/15 mol %) (e), MO/DOPE-PEG$_{2000}$/OA (83/2/15 mol %) (f), or DMPC/DOPE-PEG$_{2000}$/EPA (83/2/15 mol %) (g), nanodispersions with a total lipid concentration of 4×10^{-7} M.

SH-SY5Y cells by retinoic acid and the expression of the BDNF receptor (TrkB). After the induction of a reduced cellular viability by the treatment with retinoic acid, the phenomenon of neuroprotection by BDNF and the effect of potentiation of this neurotrophic factor by EPA and liquid crystalline lipid nanocarrier formulations (containing MO/DOPE-PEG$_{2000}$/EPA or DMPC/DOPE-PEG$_{2000}$/EPA) were evidenced. The human neuroblastoma SH-SY5Y cell line, treated with retinoic acid, was proven as a good model to study the efficiency and the toxicity of neuroprotective agents associated with functionalized lipid NP formulations. The polyunsaturated fatty acid EPA showed a real effect of potentiation of the neurotrophic factor BDNF alone and in a combination with liquid crystalline monoolein-based nanoassembles.

ACKNOWLEDGMENT

Dr. Jean-Marc Muller (CNRS UMR 6187) is gratefully acknowledged for providing the neuroblastoma cell line SH-SY5Y. The experimental facilities at the synchrotron Diamond Light Source (UK) and the Cell-Imaging platform of IFR 141 (IPSIT), "Institute of Therapeutic Innovation: From Fundamental to Medicines," are acknowledged. This contribution is dedicated to the memory of Dr. Michel Ollivon.

REFERENCES

Alfons, K., and Engstrom, G. (1998). Drug compatibility with the sponge phases formed in monoolein, water, and propylene glycol or poly(ethylene glycol). *Journal of Pharmaceutical Sciences, 87*, 1527–1530.

Almeida, A. J., and Sauto, E. (2007). Solid lipid nanoparticles as a drug delivery system for peptides and proteins. *Advanced Drug Delivery Reviews, 59*, 478–490.

Anderson, D. M., and Wennerström H. (1990). Self diffusion in bicontinuous cubic phases, L3 phases and microemulsions. *Journal of Physical Chemistry, 94*, 8683–8694.

Angelov, B., Angelova, A., Papahadjopoulos-Sternberg, B., Lesieur, S., Sadoc, J.-F., Ollivon, M., and Couvreur, P. (2006). Detailed structure of diamond-type lipid cubic nanoparticles. *Journal of the American Chemical Society, 128*, 5813–5817.

Angelov, B., Angelova, A., Garamus, V. M., Le Bas, G., Lesieur, S., Ollivon, M., Funari, S. S., Willumeit, R., and Couvreur, P. (2007). Small-angle neutron and X-ray scattering from amphiphilic stimuli-responsive diamond type bicontinuous cubic phase. *Journal of the American Chemical Society, 129*, 13474–13479.

Angelov, B., Angelova, A., Vainio, U., Garamus, V. M., Lesieur, S., Willumeit, R., and Couvreur P. (2009). Long living intermediates during a lamellar to a diamond-cubic lipid phase transition: A SAXS investigation. *Langmuir, 25*, 3734–3742.

Angelov, B., Angelova, A., Mutafchieva, R., Lesieur, S., Vainio, U., Garamus, V. M., Jensen, G. V., and Pedersen, J. S. (2011b). SAXS investigation of a cubic to a sponge

(L3) phase transition in self-assembled lipid nanocarriers. *Physical Chemistry Chemical Physics*, *13*, 3073–3081.

Angelova, A., Ollivon, M., Campitelli., A., and Bourgaux, C. (2003). Lipid cubic phases as stable nanochannel network structures for protein biochip development: X-ray diffraction study. *Langmuir*, *19*, 6928–6935.

Angelova, A., Angelov, B., Papahadjopoulos-Sternberg, B., Bourgaux, C., and Couvreur, P. (2005a). Protein driven patterning of self-assembled cubosomic nanostructures: Long oriented nanoridges. *Journal of Physical Chemistry B*, *109*, 3089–3093.

Angelova, A., Angelov, B., Papahadjopoulos-Sternberg, B., Ollivon, M., and Bourgaux, C. (2005b). Proteocubosomes: Nanoporous vehicles with tertiary organized fluid interfaces. *Langmuir*, *21*, 4138–4143.

Angelova, A., Angelov, B., Lesieur, S., Mutafchieva, R., Ollivon, M., Bourgaux, C., Willumeit, R., and Couvreur, P. (2008). Dynamic control of nanofluidic channels in protein drug delivery vehicles. *Journal of Drug Delivery Science & Technology*, *18*, 41–45.

Angelova, A., Angelov, B., Mutafchieva, R., Lesieur, S., and Couvreur, P. (2011a). Self-assembled multicompartment liquid crystalline lipid carriers for protein, peptide, and nucleic acid drug delivery. *Accounts of Chemical Research*, *44*, 147–156.

Barauskas, J., Johnsson, M., Johnson, F., and Tiberg, F. (2005). Cubic phase nanoparticles (Cubosome): Principles for controlling size, structure, and stability. *Langmuir*, *21*, 2569–2577.

Barauskas, J., Misiunas, A., Gunnarsson, T., Tiberg, F., and Johnsson, M. (2006). "Sponge" nanoparticle dispersions in aqueous mixtures of diglycerol monooleate, glycerol dioleate and polysorbate 80. *Langmuir*, *22*, 6328–6334.

Bemelmans, A. P., Horellou, P., Pradier, L., Brunet, I., Colin, P., and Mallet, J. (1999). Brain-derived neurotrophic factor-mediated protection of striatal neurons in an excitotoxic rat model of Huntington's disease, as demonstrated by adenoviral gene transfer. *Human Gene Therpy*, *10*, 2987–2997.

Bender, J., Jarvoll, P., Nydén, M., and Engström, S. (2008). Structure and dynamics of a sponge phase in the methyl δ-aminolevulinate/monoolein/water/propylene glycol system. *Journal of Colloid and Interface Science*, *317*, 577–584.

Boyd, B. J. (2003). Characterisation of drug release from cubosomes using the pressure ultrafiltration method. *International Journal of Pharmacy*, *260*, 239–247.

Burdick, J. A., Ward, M., Liang, E., Young, M. J., and Langer, R. (2006). Stimulation of neurite outgrowth by neurotrophins delivered from degradable hydrogels. *Biomaterials*, *27*, 452–459.

Caboi, F., Amico, G. S., Pitzalis, P., Monduzzi, M., Nylander, T., and Larsson, K. (2001). Addition of hydrophilic and lipophilic compounds of biological relevance to the monoolein/water system. I. Phase behavior. *Chemistry and Physics of Lipids*, *109*, 47–62.

Cernaianu, G., Brandmaier, P., Scholz, G., Ackermann, O. P., Alt, R., Rothe, K., Cross, M., Witzigmann, H., and Trobs, R. B. (2008). All-trans retinoic acid arrests neuroblastoma cells in a dormant state. Subsequent nerve growth factor/brain-derived neurotrophic factor treatment adds modest benefit. *Journal Pediatric Surgery*, *43*, 1284–1294.

Ciobanu, M. B., Heurtault, P., Schultz, C., Ruhlmann, C., Muller, D., and Frisch, B. (2007). Layersome: Development and optimization of stable liposomes as drug delivery system. *International Journal of Pharmaceutics*, *344*, 154–157.

Desmet, C. J., and Peeper, D. S. (2006). The neurotrophic receptor TrkB: A drug target in anti-cancer therapy? *Cell Molecular Life Science*, *63*, 755–759.

Dong, Y. D., Larson, I., Hanley, Y., and Boyd, B. J. (2006). Bulk and dispersed aqueous phase behavior of phytantriol: Effect of vitamin E acetate and F127 polymer on liquid crystal nanostructure. *Langmuir*, *22*, 9512–9518.

Donnici, L., Tiraboschi, E., Tardito, D., Musazzi, L., Racagni, G., and Popoli, M. (2008). Time-dependent biphasic modulation of human BDNF by antidepressants in neuroblastoma cells. *BMC Neuroscience*, *9*, 61–67.

Drummond, C. J., and Fong, C. (1999). Surfactant self-assembly objects as novel drug delivery vehicles. *Current Opinion Colloid Interface Science*, *4*, 449–456.

Encinas, M., Iglesias, M., Liu, Y. H., Wang, H. Y., Muhaisen, A., Cena, V., Gallego, C., and Comella, J. X. (2000). Sequential treatment of SH-SY5Y cells with retinoic acid and brain-derived neurotrophic factor gives rise to fully differentiated, neurotrophic factor-dependent, human neuron-like cells. *Journal of Neurochemistry*, *75*, 991–1003.

Ferreira, D. A., Bentley, M., Karlsson, G., and Edwards, K. (2006). Cryo-TEM investigation of phase behaviour and aggregate structure in dilute dispersions of monoolein and oleic acid. *International Journal of Pharmacy*, *310*, 203–212.

Fong, W. K., Hanley, T., and Boyd, B. J. (2009). Stimuli responsive liquid crystals provide "on-demand" drug delivery in vitro and in vivo. *Journal of Controlled Release*, *135*, 218–226.

Fujita, H., Sasaki, R., and Yoshikawa, M. (1995). Potentiation of the antihypertensive activity of orally administered ovokinin, a vasorelaxing peptide derived from ovalbumin, by emulsification in egg phosphatidylcholine. *Bioscience Biotechnology Biochemistry*, *59*, 2344–2345.

Fumagalli, F., Racagni, G., and Riva, M. A. (2006). The expanding role of BDNF: A therapeutic target for Alzheimer's disease? *Pharmacogenomics Journal*, *6*, 8–15.

Gabizon, A., Catane, R., Uziely, B., Kaufman, B., Safra, T., Cohen, R., Martin, F., Huang, A., and Barenholz, Y. (1994). Prolonged circulation time and enhanced accumulation in malignant exudates of doxorubicin encapsulated in polyethylene-glycol coated liposomes. *Cancer Research*, *54*, 987–992.

Garcia-Fuentes, M., Torres, D., and Alonso, M. J. (2005). New surface-modified lipid nanoparticles as delivery vehicles for salmon calcitonin. *International Journal of Pharmaceutics*, *296*, 122–132.

Gregoriadis, G., McCormack, B., Obrenovic, M., Saffie, R., Zadi, B., and Perrie, Y. (1999). Vaccine entrapment in liposomes. *Methods*, *19*, 156–162.

Gustafsson, J., Ljusberg-Wharen, H., Almgrem, M., and Larsson, K. (1996). Cubic lipid-water phase dispersed into submicron particles. *Langmuir*, *12*, 4611–4613.

Gustafsson, J., Ljusberg-Wahren, H., Almgren, M., and Larsson, K. (1997). Submicron particles of reversed lipid phases in water stabilized by a nonionic amphiphilic polymer. *Langmuir*, *13*, 6964–6971.

Hyde, S. T., Andersson, S., Ericsson, B., and Larsson, K. (1984). A cubic structure consisting of a lipid bilayer forming an infinite periodic minimal surface of the gyroid type in the glycerol monooleate water system. *Zeitschrift für Kristallographie, 168,* 213–219.

Imura, T., Yanagishita, H., Ohira, J., Sakai, H., Abe, M., and Kitamoto, D. (2005). Thermodynamically stable vesicle formation from glycolipid biosurfactant sponge phase. *Colloid & Surfaces B: Biointerfaces, 43,* 115–121.

Israelachvili, J. N., Mitchell, D. J., and Ninham, B. W. J. (1976). Theory of self-assembly of hydrocarbon amphiphiles into micelles and bilayers. *Chemical Society: Faraday Transactions II 72,* 1525–1568.

Jaboin, J., Kim, C. J., Kaplan, D. R., and Thiele, C. J. (2002). Brain-derived neurotrophic factor activation of TrkB protects neuroblastoma cells from chemotherapy-induced apoptosis via phosphatidylinositol 3′-kinase pathway. *Cancer Research, 62,* 6756–6763.

Jorgensen, L., Moeller, E. H., van de Weert, M., Nielsen, H. M., and Frokjaer, S. (2006). Preparing and evaluating delivery systems for proteins. *European Journal of Pharmaceutical Sciences, 29,* 174–182.

Kaplan, D. R., Matsumoto, K., Lucarelli, E., and Thiele, C. J. (1993). Induction of TrkB by retinoic acid mediates biologic responsiveness to BDNF and differentiation of human neuroblastoma-cells. *Neuron, 11,* 321–331.

Kawasaki, H., Souda, M., Tanaka, S., Nemoto, N., Karlsson, G., Almgren, M., and Maeda, H. (2002). Reversible vesicle formation by changing pH. *Journal of Physical Chemistry B, 106,* 1524–1527.

Kim, H.-Y., Akbar, M., Lau, A., and Lisa, E. (2000). Inhibition of neuronal apoptosis by docosahexaenoic acid (22:6n-3). *Journal of Biological Chemistry, 275,* 35215–35223.

Klibanov, A. L., Maruyama, K., Torchilin, V. P., and Huang, L. (1990). Amphipathic polyethyleneglycols effectively prolong the circulation time of liposomes. *FEBS Letters, 268,* 235–237.

Kou, W., Luchtman, D., and Song, C. (2008). Eicosapentaenoic acid (EPA) increases cell viability and expression of neurotrophin receptors in retinoic acid and brain-derived neurotrophic factor differentiated SH-SY5Y cells. *European Journal of Nutrition, 47,* 104–113.

Landh, T., and Larsson, K., inventors (1993). Particles, method of preparing said particles and uses thereof. Canadian Patent WO93/06921, April 15, 1993, GS Development AB, SE.

Larsson, K. (1989). Cubic lipid-water phases—Structural and biomembrane aspects. *Journal of Physical Chemistry, 93,* 7304–7314.

Larsson, K. (2000). Aqueous dispersion of cubic lipid-water phases. *Current Opinion of Colloid & Interface Science, 5,* 64–69.

Larsson, K., and Tiberg, F. (2005). Periodic minimal surface structures in bicontinuous lipid–water phases and nanoparticles. *Current Opinion of Colloid & Interface Science, 9,* 365–369.

Leesajakul, W., Nakano, M., Taniguchi, A., and Handa, T. (2004). Interaction of cubosomes with plasma components resulting in the destabilization of cubosomes in plasma. *Colloids & Surfaces B: Biointerfaces, 34,* 253–258.

Longo, F. M. (2010). Pharmaceutical formulations comprising neurotrophin mimetics. U.S. patent application 20100267727.

Lopes, L. B., Ferreira, D., Daniel de Paula, M., Tereza, J., Garcia, J. A., Thomazini, Fantini, M. C. A., and Bentley, M. V. L. B. (2006). Reverse hexagonal phase nano-dispersion of monoolein and oleic acid for topical delivery of peptides: in vitro and in vivo skin penetration of cyclosporin A. *Pharmaceutical Research*, *23*, 1332–1342.

Luzzati, V. (1997). Biological significance of lipid polymorphism: The cubic phases. *Current Opinion in Structural Biology*, *7*, 661–668.

Lynch, M. L., Ofori-Boateng, A., Hippe, A., Kochvar, K., and Spicer, P. T. (2003). Enhanced loading of water-soluble actives into bicontinuous cubic phase liquid crystals using cationic surfactants. *Journal of Colloid & Interface Science*, *260*, 404–413.

Malcangio, M., and Lessmann, V. (2003). A common thread for pain and memory synapses? Brain-derived neurotrophic factor and trkB receptors. *Trends Pharmacological Science*, *24*, 116–121.

Mariani, P., Luzzati, V., and Delacroix, H. (1988). Cubic phases of lipid-containing systems. Structure analysis and biological implications. *Journal of Molecular Biology*, *204*, 165–189.

Martin, F. J. and Papahadjopoulos, D. (1982). Irreversible coupling of immunoglobulin fragments to preformed vesicles—An improved method for liposome targeting. *Journal of Biological Chemistry*, *257*, 286–288.

Martina, M. S., Wilhelm, C., and Lesieur, S. (2008). The effect of magnetic targeting on the uptake of magnetic-fluid-loaded liposomes by human prostatic adenocarcinoma cells. *Biomaterials*, *29*, 4137–4145.

Martins, S., Sarmento, B., Ferreira, D. C., and Souto, E. B. (2007). Lipid-based colloidal carriers for peptide and protein delivery—Liposomes versus lipid nanoparticles. *International Journal of Nanomedicine*, *2*(4), 595–607.

Matsumoto, K., Wada, R. K., Yamashiro, J. M., Kaplan, D. R., and Thiele, C. J. (1995). Expression of brain-derived neurotrophic factor and p145(TrkB) affects survival, differentiation, and invasiveness of human neuroblastoma-cells. *Cancer Research*, *55*, 1798–1806.

Mehner, W., and Mäder, K. (2001). Solid lipid nanoparticles: Production, characterization and applications. *Advanced Drug Delivery Reviews*, *47*, 165–196.

Merclin, N., Bender, J., Sparr, E., Guy, R. H., Ehrsson, H., and Engstrom, S. (2004). Transdermal delivery from a lipid sponge phase—iontophoretic and passive transport in vitro of 5-aminolevulinic acid and its methyl ester. *Journal of Controlled Release*, *100*, 191–198.

Mishraki, T., Libster, D., Aserin, A., and Garti, N. (2010). Lysozyme entrapped within reverse hexagonal mesophases: Physical properties and structural behavior. *Colloids and Surfaces B: Biointerfaces*, *75*, 47–56.

Murgia, S., Falchi, A. M., Mano, M., Lampis, S., Angius, R., Carnerup, A. M., Schmidt, J., Diaz, G., Giacca, M., Talmon, Y., and Monduzzi, M. (2010). Nanoparticles from lipid-based liquid crystals: Emulsifier influence on morphology and cytotoxicity. *Journal of Physical Chemistry B*, *114*, 3518–3525.

Nagahara, A. H., Merrill, D. A., Coppola, G., Tsukada, S., Schroeder, B. E., Shaked, G. M., Wang, L., Blesch, A., Kim, A., Conner, J. M., Rockenstein, E., Chao, M. V., Koo, E. H., Geschwind, D., Masliah. E., Chiba, A. A., and Tuszynski, M. H. (2009). Neuroprotective effects of brain-derived neurotrophic factor in rodent and primate models of Alzheimer's disease. *Nature Medicine*, *15*, 331–337.

Nakano, M., Sugita, A., Matsuoka, H., and Handa, T. (2001). Small angle x-ray scattering and ^{13}C NMR investigation on the internal structure of "cubosomes." *Langmuir, 17*, 3917–3922.

Nishida, Y., Adati, N., Ozawa, R., Maeda, A., Sakaki, Y., and Takeda, T. (2008). Identification and classification of genes regulated by phosphatidylinositol 3-kinase- and TRKB-mediated signalling pathways during neuronal differentiation in two subtypes of the human neuroblastoma cell line SH-SY5Y. *BMC Research Notes, 1*, 95–106.

Nosheny, R. L., Mocchetti, I., and Bachis, A. (2005). Brain-derived neurotrophic factor as a prototype neuroprotective factor against HIV-1-associated neuronal degeneration. *Neurotoxic Research, 8*, 187–198.

Nylander, T., Mattisson, C., Razumas, V., Meizis, Y., and Hakansson, B. (1996). A study of entrapped enzyme stability and substrate diffusion in a monoglyceride-based cubic liquid crystalline phase. *Colloids & Surfaces B, 114*, 311–320.

Oberdisse, J., Couve, C., Appell, J., Berret, J. F., Ligoure, C., and Porte, G. (1996). Vesicles and onions from charged surfactant bilayers: A neutron scattering study. *Langmuir, 12*, 1212–1218.

Pattarawarapan, M., and Burgess, K. (2003). Molecular basis of neurotrophin—Receptor interactions. *Journal of Medical Chemistery, 46*, 5277–5291.

Pillai, A., and Mahadik, S. P. (2008). Increased truncated TrkB receptor expression and decreased BDNF/TrkB signaling in the frontal cortex of reeler mouse model of schizophrenia. *Schizophrenia Research, 100*, 325–333.

Presgraves, S. P., Borwege, S., Millan, M. J., and Joyce, J. N. (2004). Involvement of dopamine D-2/D-3 receptors and BDNF in the neuroprotective effects of S32504 and pramipexole against 1-methyl-4-phenylpyridinium in terminally differentiated SH-SY5Y cells. *Experimental Neurology, 190*, 157–170.

Qiu, H., and Caffrey, M. (2000). The phase diagram of the monoolein/water system: Metastability and equilibrium aspects. *Biomaterials, 21*, 223–234.

Rizwan, S. B., Hanley, T., Boyd, B. J., Rades, T., and Hook, S. (2009). Liquid crystalline systems of phytantriol and glyceryl monooleate containing a hydrophilic protein: Characterisation, swelling and release kinetics. *Journal of Pharmacological Science, 98*, 4191–4204.

Robinson, R. C., Radziejewski, C., Stuart, D. I., and Jones, E. Y. (1995). Structure of the brain-derived neurotrophic factor/neurotrophin 3 heterodimer. *Biochemistry, 34*, 4139–4146.

Ruiz-Leon, Y., and Pascual, A. (2003). Induction of tyrosine kinase receptor B by retinoic acid allows brain-derived neurotrophic factor-induced amyloid precursor protein gene expression in human SH-SY5Y neuroblastoma cells. *Neuroscience, 120*, 1019–1026.

Schramm, A., Schulte, J. H., Astrahantseff, K., Apostolov, O., van Limpt, V., Sieverts, H., Kuhfittig-Kulle, S., Pfeiffer, P., Versteeg, R., and Eggert, A. (2005). Biological effects of TrkA and TrkB receptor signaling in neuroblastoma. *Cancer Letters, 228*, 143–153.

Seddon, J. M., and Templer, R. H. (1995). In R. Lipowsky and E. Sackmann (Eds.), *Handbook of Biological Physics*, Elsevier Science, London, pp. 97–153.

Serres, F., and Carney, S. L. (2006). Nicotine regulates SH-SY5Y neuroblastoma cell proliferation through the release of brain-derived neurotrophic factor. *Brain Research, 1101*, 36–42.

Shah, J. C., Sadhale, Y., and Chilukuri, D. M. (2001). Cubic phase gels as drug delivery systems. *Advanced Drug Delivery Reviews, 47*, 229–250.

Siekmann, B., Bunjes, H., Koch, M. H. J., and Westesen, K. (2002). Preparation and structural investigations of colloidal dispersions prepared from cubic monoglyceride-water phases. *International Journal of Pharmacy, 244*, 33–43.

Simpson, P. B., Bacha, J. L., Palfreyman, E. L., Woollacott, A. J., McKernan, R. M., and Kerby, J. (2001). Retinoic acid-evoked differentiation of neuroblastoma cells predominates over growth factor stimulation: An automated image capture and quantitation approach to neuritogenesis. *Analitical Biochemistry, 298*, 163–169.

Spicer, P. T., and Hayden, K. L. (2001). Novel process for producing cubic liquid cristalline nanoparticles (cubosomes). *Langmuir, 17*, 5748–5756.

Sun, Y. E., and Wu, H. (2006). The ups and downs of BDNF in Rett syndrome. *Neuron, 49*, 321–323.

Torchilin, V. P. (2005). Recent advances with liposomes as pharmaceutical carriers. *Nature Reviews in Drug Discovery, 4*, 145–160.

Tuszynski, M. H., Thal, L., Pay, M., Salmon, D. P., Sang, U. H., Bakay, R., Patel, P., Blesch, A., Vahlsing, H. L., Ho, G., Tong, G., Potkin, S. G., Fallon, J., Hansen, L., Mufson, E. J., Kordower, J. H., Gall, C., and Conner, J. (2005). A phase 1 clinical trial of nerve growth factor gene therapy for Alzheimer disease. *Nature Medicine, 11*, 551–555.

Vogelin, E., Baker, J. M., Gates, J., Dixit, V., Constantinescu, M. A., and Jones, N. F. (2006). Effects of local continuous release of brain derived neurotrophic factor (BDNF) on peripheral nerve regeneration in a rat model. *Experimental Neurology, 199*, 348–353.

Wadsten, P., Wöhri, A. B., Snijder, A., Katona, G., Gardiner, A. T., Cogdell, R. J., Neutze, R., and Engström S. (2006). Lipidic sponge phase crystallization of membrane proteins. *Journal of Molecular Biology, 364*, 44–53.

Walde, P., and Ichikawa, S. (2001). Enzymes inside lipid vesicles: Preparation, reactivity and applications. *Biomolecular Engineering, 18*, 143–177.

Wu, D. F., and Pardridge, W. M. (1999). Neuroprotection with noninvasive neurotrophin delivery to the brain. *Proceedings of the National Academy of Sciences USA, 96*, 254–259.

Yaghmur, A., and Glatter, O. (2009). Characterization and potential applications of nanostructured aqueous dispersions. *Advances in Colloid Interface Science, 147*(48), 333–342.

Zabara, A., Amar-Yuli, I., and Mezzenga, R. (2011). Tuning in-meso-crystallized lysozyme polymorphism by lyotropic liquid crystal symmetry. *Langmuir, 27*, 6418–6425.

Zuccato, C., and Cattaneo, E. (2009). Brain-derived neurotrophic factor in neurodegenerative diseases. *Nature Reviews Neurology, 5*, 311–322.

Self-Assembled Supramolecular Architectures: Lyotropic Liquid Crystals, First Edition.
Edited by Nissim Garti, Ponisseril Somasundaran, and Raffaele Mezzenga.
© 2012 John Wiley & Sons, Inc. Published 2012 by John Wiley & Sons, Inc.